New Perspectives in Turbulence

Lawrence Sirovich
Editor

New Perspectives in Turbulence

With 131 illustrations

Springer-Verlag
New York Berlin Heidelberg London
Paris Tokyo Hong Kong Barcelona

Lawrence Sirovich
Center for Fluid Mechanics, Division of Applied Mathematics, Brown University,
Providence, RI 02912

Mathematics Subject Classification Codes: 76Fxx, 76D05, 76E25, 60Gxx, 82A50

Library of Congress Cataloging-in-Publication Data

New perspectives in turbulence / Lawrence Sirovich, editor.
  p. cm.
  Based on papers from a meeting held at Salve Regina College,
Newport, R.I., June 12–15, 1989.
Includes bibliographical references.
  1. Turbulence–Congresses. I. Sirovich. L. 1933– .
QA913.N49 1991
532'.0527–dc20   91-2648

Photocomposed copy prepared from author files using TEX, LATEX, and $\mathcal{A}_{\mathcal{M}}\mathcal{S}$TEX.
Printed and bound by Edwards Brothers, Inc., Ann Arbor, Michigan.
Printed in the United States of America.

9 8 7 6 5 4 3 2 1   Printed on acid-free paper.

ISBN 0-387-97559-4 Springer-Verlag New York Berlin Heidelberg
ISBN 3-540-97559-4 Springer-Verlag Berlin Heidelberg New York

Preface

This collection of articles has its origin in a meeting which took place June 12–15, 1989, on the grounds of Salve Regina College in Newport, Rhode Island. The meeting was blessed by beautiful, balmy weather and an idyllic setting. The sessions themselves took place in Ochre Court, one of the elegant and stately old *summer cottages* for which Newport is acclaimed. Lectures were presented in the grand ballroom overlooking the famous Cliff Walk and Block Island Sound.

Counter to general belief, the pleasant surroundings did not appear to encourage truancy or in any other way diminish the quality of the meeting. On the contrary, for the four days of the meeting there was a high level of excitement and optimism about the *new perspectives in turbulence*, a tone that carried over to lively dinner and evening discussions. The participants represented a broad range of backgrounds, extending from pure mathematics to experimental engineering. A dialogue began with the first speakers which cut across the boundaries and gave to the meeting a mood of unity which persisted.

Alas, the written word cannot bring to life the *real time* quality of the meeting. Nevertheless, the reader of this book will learn that the articles that follow are all of a high caliber. (It should be mentioned that George Zaslovsky, who was unable to attend the meeting, also contributed a manuscript.) Articles were interchanged among the authors and an effort made to have the articles reflect some of the interaction that took place.

A strenuous attempt was made to get manuscripts from all the speakers. However, as comparison with the speaker list shows some contributions are missing. Rather than delay the appearance of this volume any longer, it was decided to bring closure to the project at this point.

Thanks are due to D.Y. Hsieh, for not only serving as Organizing Chairman, but also for his splendid and insightful report on the tragic events which had taken place in China only days before the meeting, and from which he had just returned. Very special thanks are also due to Julia Nesbitt, who was greatly responsible for having made this a very special meeting, and to Madeline Brewster, who made the preparation of this book possible.

L.S., Saltaire, May 1990

Contents

Contributors

D.J. Carruthers Department of Applied Mathematics and Theoretical Physics, University of Cambridge, Silver Street, Cambridge CB3 9EW, United Kingdom

Ashvin B. Chabra Mason Laboratory, Box 2159, Yale University, New Haven, Connecticut 06520

Peter Constantin Department of Mathematics, University of Chicago, Chicago, Illinois 60637

G. Erlebacher Institute for Computer Applications in Science and Engineering, NASA Langley Research Center, Hampton, Virginia 23665

J.C.H. Fung Department of Applied Mathematics and Theoretical Physics, University of Cambridge, Silver Street, Cambridge CB3 9EW, United Kingdom

J.P. Gollub Physics Department, Haverford College, Haverford, Pennsylvania 19041 and Physics Department, The University of Pennsylvania, Philadelphia, Pennsylvania 19104

Hyder S. Husain Department of Mechanical Engineering, University of Houston, Houston, Texas 77204-4792

Fazle Hussain Department of Mechanical Engineering, University of Houston, Houston, Texas 77204-4792

M.Y. Hussaini Institute for Computer Applications in Science and Engineering, NASA Langley Research Center, Hampton, Virginia 23665

J.C.R. Hunt Department of Applied Mathematics and Theoretical Physics, University of Cambridge, Silver Street, Cambridge CB3 9EW, United Kingdom

Eric Jackson Applied & Computational Mathematics, Princeton University, Princeton, New Jersey 08554

Leo P. Kadanoff The Research Institutes, University of Chicago, Chicago, Illinois 60637

Robert H. Kraichnan 303 Portillo Drive, Los Alamos, New Mexico 87544

John L. Lumley Sibley School of Mechanical and Aerospace Engineering, Cornell University, Ithaca, New York 14853

Mogens V. Melander Department of Mathematics, Southern Methodist University, Dallas, Texas

Steven A. Orszag Applied & Computational Mathematics, Princeton University, Princeton, New Jersey 08554

R. Ramshankar Carrier Corporation

R.Z. Sagdeev Space Research Institute, Profsoyuznaya 84/32, Moscow 117810, USSR

S. Sarkar Institute for Computer Applications in Science and Engineering, NASA Langley Research Center, Hampton, Virginia 23665

Zhen-Su She Applied & Computational Mathematics, Princeton University, Princeton, New Jersey 08554

Lawrence Sirovich Center for Fluid Mechanics, Division of Applied Mathematics, Brown University, Providence, Rhode Island 02912

J.T. Stuart Division of Applied Mathematics, Brown University, Providence, Rhode Island 02912 and Mathematics Department, Imperial College, London SW7 2BZ, United Kingdom

G.M. Zaslavsky Space Research Institute, Profsoyuznaya 84/32, Moscow 117810, USSR

Presentations

Stochastic Modeling of Isotropic Turbulence
Robert H. Kraichnan
Kraichnan Associates

Rapid Distortion Theory as a Means of Exploring the Structure of Turbulence
Julian C.R. Hunt, D.J. Carruthers and J.C.H. Fung
University of Cambridge

Order and Disorder in Turbulent Flows
John L. Lumley
Cornell University

The Turbulent Fluid as a Dynamical System
David Ruelle
Institut des Hautes Etudes Scientifiques

Empirical Eigenfunctions and Low Dimensional Systems
Lawrence Sirovich
Brown University

Stochastic Structures in Turbulence
Ronald J. Adrian
University of Illinois

Spatiotemporal Chaos in Interfacial Waves
Jerry P. Gollub and R. Ramshankar
Haverford College

The Complementary Roles of Experiments and Simulation in Coherent Structure Studies
Mogens V. Melander, Hyder S. Husain and Fazle Hussain
University of Houston

Renormalization Group and Probability Distributions in Turbulence
Victor Yakhot
Princeton University

Remarks on the Navier-Stokes Equations
Peter Constantin
University of Chicago

Novel Experiments Aimed at Controlling Turbulent Shear Flow
I. Wygnanski
University of Arizona

Scaling and Structures in the Hard Turbulence Region of Rayleigh–Bénard Convection
Leo P. Kadanoff
University of Chicago

Probabilistic Multifractals and Negative Dimensions
Ashvin B. Chhabra and K.R. Sreenivasan
Yale University

Simulation of Transition to Turbulence
Leonhard Kleiser
Institute for Theoretical Fluid Mechanics

Simulation of Transition to Turbulence
M. Yousuff Hussaini
NASA Langley Research Center

Statistical Aspects of Vortex Dynamics in Turbulence
Zhen-Su She, Eric Jackson and Steven A. Orszag
Princeton University

The Lagrangian Picture of Fluid Motion and its Implication for Flow Structures
J.T. Stuart
Imperial College

Appearance of Transition in Boundary Layers
Thorwald Herbert
Ohio State University

Broadband Instabilities in Bounded Elliptical Vortex Flow: Early Theories and Experiments
Willem V.R. Malkus
Massachusetts Institute of Technology

Participants

Abernathy, Frederich
Adams, Eric
Adrian, Ronald
Alving, Amy
Aubry, Nadine
Balachandar, S.
Ball, Kenneth
Bisshopp, Fred
Brasseur, James
Castaing, Bernard
Chasnov, Jeffrey
Chauve, M.P.
Chen, Haibo
Chen, Hudong
Chen, Shiyi
Childress, Stephen
Cimbala, John
Cole, Julian
Constantin, Peter
Cross, Michael
Deane, Anil
Everson, Richard
Fernholz, H.H.
Fineberg, Jay
Fong, Jefferson
Gollub, Jerry
Handler, Robert
Hartke, Greg
Herbert, Thorwald
Hermanson, James
Hohenberg, Pierre
Holt, Steve

Hsieh, Din-Yu
Hsu, Guan-Shong
Hunt, Julian
Hussain, Fazle
Hussaini, Yousuff
Jackson, Eric
Kadanoff, Leo
Karlsson, Sture
Kiad, Shigeo
Kirby, Michael
Kleiser, Leonhard
Knight, Bruce
Kollmann, Wolfgang
Korman, Murray
Kraichnan, Robert
Kumar, Ajay
Lasheras, Juan
Lin, C.C.
Liu, Joseph
Lumley, John
Malkus, Willem
Maxey, Martin
Mayer-Kress, G.
McMichael, James
Meiburg, Eckart
Meng, James
Mohring, Willi
Morkovin, Mark
Nachman, Arje
Nadolink, Richard
Nelkin, Mark
Newton, Paul

Ng, Bart
Nikjooy, Mohammad
Orszag, Steve
Panchapakesan, N.R.
Park, Hung Mok
Platt, Nathan
Polifke, Wolfgang
Raghn, Surya
Reisenthel, Patrick
Ruelle, David
Sanghi, Sanjeev
She, Zheu-Su
Shi, Xungang
Sirovich, Lawrence
Sreenivasan, Katepalli
Stuart, J. Trevor
Su, Chau-Hsing
Tangborn, Andrew
Tarman, Hakan
Titi, Edriss
Veerman, Peter
Voth, Eric
Waleffe, Fabian
Warhaft, Zellman
Wark, Candace
Woodruff, Steve
Wygnanski, I.
Wyngaard, John
Yakhot, Victor
Yang, Zhongmin
Yoon, Kyung Hwan
Zhou, Heng

Chapter 1

Stochastic Modeling of Isotropic Turbulence

Robert H. Kraichnan

This paper reviews some stochastic models for Navier-Stokes turbulence and related problems. Discussion is confined to exactly soluble dynamical models that have intrinsic consistency properties and are constructed systematically from the equations of motion. Such models provide a unified framework for the examination of a number of approximation methods developed during the past few decades. The stochastic models that have been most studied predict the evolution of low-order moments. Their achievements include the description of error growth (sensitivity to perturbation), the deduction of generalized (non-local) eddy damping from the equations of motion, the representation of energy cascade and vorticity intensification, and discrimination between the behaviors of Eulerian and Lagrangian time correlations. Fairly good quantitative predictions of wavenumber spectra have been obtained at all Reynolds numbers. A current challenge is the intermittency of small scales at moderate as well as high Reynolds numbers. New models, based on nonlinear mappings of stochastic fields, offer closures for probability distributions and are a promising tool for attacking the intermittency problem.

1. Introduction

This paper is a limited examination of the past, and possible future, of one branch of the deductive theory of isotropic turbulence. By "deductive theory" I mean the body of analysis that proceeds in some recognizably systematic mathematical fashion from the Navier-Stokes (NS) equation to predictions of turbulence phenomena of physical interest. No attempt will made here to give a comprehensive, evenhanded survey of the many lines of research[2,50,51,56,57,65] that have arisen since the pioneering work by G. I. Taylor.[68] The treatment will center around dynamical stochastic modeling: the systematic construction of physically relevant, but exactly soluble, dynamical systems that automatically embody certain consistency properties. Apart from their intrinsic value, such models provide a framework for the critical discussion of several other approaches. The strengths and weaknesses of the stochastic models help point to possible new directions for future work.

One of the most fascinating characteristics of turbulence is the interplay of order and randomness: well-formed structures, like ropes of intense vorticity, exist within and interact with an ocean of random-appearing excitation. The structures themselves exhibit random plasticity of shape. Randomness in turbulence is associated with extreme sensitivity to perturbations. At the same time, suitably chosen statistics are robustly stable against perturbation.

One of the principal puzzles is that such a complex set of phenomena yields as much as it does to ideas of great simplicity and crudity, like eddy viscosity and step-wise cascade. Stochastic models, and related analysis, have made very significant, but yet sharply limited, contributions to the qualitative and quantitative understanding of these questions. Stochastic models studied in the past have led to the elucidation and calculation of low-order moments. Their achievements include:

(a) Derivation of eddy viscosity, and its generalization, from the NS equation.

(b) Description of the cascade of energy from low to high wavenumbers.

(c) Quantitative predictions of energy spectra and two-time covariances at all Reynolds numbers.

(d) Predictions of the growth of initial, statistically described perturbations of turbulence.

(e) Distinction between the statistical effects of sweeping of flow structures past fixed points and the distortion of flow structures by straining.

(f) Capture of some characteristic departures of turbulence statistics from Gaussian, including skewness of the velocity derivatives and depression of overall nonlinearity.

Shortcomings and failures of stochastic models and related approaches that yield low-order moments include:

(g) There are little or no meaningful internal estimates of errors.

(h) Essential higher statistical structure, especially the intermittency of small scales, is not captured.

(i) The typical space-time structures are not portrayed.

The stochastic model that leads to the direct-interaction approximation plays a central role. This model has simply described properties and many closure theories can profitably be discussed by comparison with it.

It is likely that effective treatment of intermittency and other higher statistical structure by stochastic models, or by other means, requires an analytical environment based on probability distribution functions (PDF's) in physical space, rather than on moments. The final Sec. of this paper describes recent work on new stochastic models that yield closures for PDF's, rather than moments. These models have the power to analyze intermittency phenomena, but work on them is still at a very early stage. A low-level application to Burgers' equation has given verified predictions of an exponential tail to the PDF of velocity gradient and a power-law tail to the PDF of fluid density.

2. Closure by Dynamical Stochastic Models

The incompressible NS equation may be written[56]

$$\left(\frac{\partial}{\partial t} - \nu\nabla^2\right)\mathbf{u}(\mathbf{x}, t) = -\lambda\mathbf{u}(\mathbf{x}, t)\cdot\nabla\mathbf{u}(\mathbf{x}, t) - \nabla p(\mathbf{x}, t), \qquad (2.1)$$

$$\nabla\cdot\mathbf{u} = 0, \qquad (2.2)$$

where $\mathbf{u}$ is the velocity field, ν is kinematic viscosity, λ is an ordering parameter (equal to unity), p is pressure, and the uniform fluid density is set equal to unity. It is well-known that the advection term $\mathbf{u}\cdot\nabla\mathbf{u}$ dynamically couples velocity-field moments of different orders, with the result that the entire initial statistical distribution affects the evolution of any given moment over a finite time. This is commonly called the closure problem of turbulence theory.[2,56] The linear viscous term also presents a closure problem, but one that shows up only in PDF formulations and is invisible at the level of moment analysis: the viscous term couples single-point PDF's to three-point PDF's.[52]

At the level of moment analysis, attempts have been made to apply standard statistical approximations to (2.1) and to treat the nonlinear term by various perturbation methods.[2,41,44,50,51,56,57,64,65] Both kinds of treatment are essentially uncontrolled at moderate and large Reynolds numbers because the nonlinear term, which causes departure of moments from Gaussian values, is not small and the perturbation series are not convergent. The problems that arise are qualitative rather than quantitative. Thus the assumption of zero fourth-order cumulants of the velocity PDF can lead to negative energy spectra[57] and various plausible-appearing primitive and renormalized perturbation treatments can lead to equations that blow up.[30]

The guiding idea of closure by dynamical stochastic modeling is to pose a soluble dynamic problem that is internally consistent because of built-in invariance, conservation and realizability properties.[30] The resulting statistics may be inaccurate or wrong, but the models cannot blow

up or have grossly unphysical properties like negative energy spectra. The
stochastic models to be reviewed in this Sec. are workable because they
create an effective small parameter that vanishes in a limit, no matter how
high the Reynolds number. This permits exact solution of the models by
perturbation methods despite the fact that such methods are not rationally
applicable to the original NS dynamics. Later discussion will include new
kinds of stochastic models based on non-perturbative, nonlinear mappings
of Gaussian stochastic processes into processes that may be very far from
Gaussian.

A basic stochastic model for (2.1) can be constructed by first invoking
a collection of M identical, but uncoupled, flow systems all residing in the
same physical space. The equations of motion are

$$\left(\frac{\partial}{\partial t} - \nu \nabla^2\right) \mathbf{u}^n(\mathbf{x}, t) = -\lambda \mathbf{u}^n(\mathbf{x}, t) \cdot \nabla \mathbf{u}^n(\mathbf{x}, t) - \nabla p^n(\mathbf{x}, t), \qquad (2.3)$$

$$\nabla \cdot \mathbf{u}^n = 0, \qquad (2.4)$$

where $n = 1, 2, ..., M$ labels the different systems. The statistics of the
individual flow systems are Gaussian, independent and identical at $t = 0$.
The key steps now are to replace (2.3) by model dynamical equations that
couple the M systems in a stochastic fashion and then to take the limit
$M \to \infty$. The model equations are:[17]

$$\left(\frac{\partial}{\partial t} - \nu \nabla^2\right) \mathbf{u}^n(\mathbf{x}, t) =$$

$$-\lambda M^{-1} \sum_{rs} \phi_{nrs} \mathbf{u}^r(\mathbf{x}, t) \cdot \nabla \mathbf{u}^s(\mathbf{x}, t) - \nabla p^n(\mathbf{x}, t), \qquad (2.5)$$

with (2.4) unchanged. Here ϕ_{nrs} is a constant coefficient assigned the value
$+1$ or -1 at random, subject only to invariance under any permutation of
its three indices. The factor M^{-1} serves to make the variances of the total
advection terms in (2.3) and (2.4) equal at $t = 0$. The ϕ_{nrs} vary only with
the indices nrs. When a statistical ensemble of realizations of the entire
collection of systems is formed, ϕ_{nrs} has precisely the same value in every
realization in that ensemble.

Eq. (2.5) has been called the random-coupling (RC) model. It can
be formulated in a number of different variations. In particular, there are
some formal advantages in dealing with collective variables over the collection of systems.[30] The form (2.5) is the simplest for present purposes.
It does not invoke collective variables. The effective small parameter is
$1/M$. If $M \to \infty$, then (a) each individual interaction of three systems,
represented by the terms in (2.5) involving a given ϕ_{nrs} and all of its index-
permutations, can be treated as an infinitesimal perturbation on the total

nonlinear interaction; (b) it can be asserted that the statistical dependence among any finite subset of flow systems induced by the total interaction is infinitesimal. The result is a closed set of equations that involve two fundamental statistical quantities: The velocity covariance in any of the flow systems and the mean response of the velocity amplitude in any flow system to an infinitesimal force added to the right side of (2.5). These quantities are independent of the flow-system labeling index n. The exactness of the perturbation analysis is a consequence solely of the limit $M \to \infty$ and does not depend on the size of the Reynolds number.

The final closed equations for velocity covariance and mean infinitesimal response functions have been called the direct-interaction approximation (DIA),[29] because the departures from independent Gaussian statistics arise from the direct interaction of small subsets (triads) of flow systems acting within the sea of interaction among all systems. As formulated above, the RC model and the DIA are not restricted to either isotropy or homogeneity. However, the final, closed equations are most easily written for homogeneous, isotropic statistics within a large cyclic box.[29,31]

Some important properties of the final DIA equations can be inferred from (2.5) without further analysis. First, (2.5) evolves actual mode amplitudes so that the resulting velocity covariance automatically satisfies all realizability conditions. Second, the symmetry imposed on the ϕ_{nrs} maintains conservation of kinetic energy and other quadratic constants of motion under the nonlinear interaction. Third, the symmetry of the ϕ_{nrs} further assures that the absolute equilibrium canonical ensembles associated with mode truncation of the Euler equation carry over to the model.[31] A fourth essential property is plausible just from the fact that (2.5) describes a large, nonlinearly coupled system, but it must be verified by analysis. This is that (2.5), like (2.1), exhibits sensitivity to small perturbation when the Reynolds number is large.[35] A fifth property is associated with a major deficiency of the DIA. This is the break-up of coherence effects in the self-advection of the velocity field. In (2.5), the advection of $\mathbf{u}^n(\mathbf{x}, t)$, the velocity of system n, via the substantial derivative is replaced by a nonlinear term that involves all the other systems r and s. The result cannot be interpreted as advection of $\mathbf{u}^n(\mathbf{x}, t)$ by a model velocity field.[23,32,44] These properties and their implications will be discussed in more detail in Sec. 3. For this purpose, the final DIA equations for isotropic turbulence[31] will now be written explicitly.

Take cyclic boundary conditions on a box of side $L \to \infty$ and introduce Fourier amplitudes by

$$u_i^n(\mathbf{x}, t) = \sum_{\mathbf{k}} u_i^n(\mathbf{k}; t) e^{i\mathbf{k} \cdot \mathbf{x}}. \tag{2.6}$$

The n-independent modal intensity scalar $U(k; t, t')$ may be defined by

$$\tfrac{1}{2}P_{ij}(\mathbf{k})U(k; t, t') = (L/2\pi)^3 \langle u_i(\mathbf{k}; t)u_j^*(\mathbf{k}; t') \rangle \quad (L \to \infty), \qquad (2.7)$$

where $\langle\rangle$ denotes ensemble average and $P_{ij}(\mathbf{k}) = \delta_{ij} - k_i k_j/k^2$. U is related to the energy spectrum in three dimensions by

$$E(k, t) = 2\pi k^2 U(k; t, t). \qquad (2.8)$$

Eq. (2.5) rewritten in the Fourier representation becomes

$$\left(\frac{\partial}{\partial t} + \nu k^2 \right) u_i^n(\mathbf{k}; t) =$$

$$-i\lambda \sum_{rs} \phi_{nrs} P_{ijm}(\mathbf{k}) \sum_{\mathbf{p}} u_j^r(\mathbf{p}; t)u_m^s(\mathbf{k} - \mathbf{p}; t), \qquad (2.9)$$

where

$$P_{ijm}(\mathbf{k}) = \tfrac{1}{2}[k_m P_{ij}(\mathbf{k}) + k_j P_{im}(\mathbf{k})].$$

Multiplication of (2.9) by $u_i^{n*}(k, t')$ followed by averaging gives

$$\left(\frac{\partial}{\partial t} + \nu k^2 \right) U(k; t, t') = S(k; t, t'), \qquad (2.10)$$

where $S(k; t, t')$ denotes the sum over triple moments thereby obtained on the right side. Further manipulations yield an equation for $G(k; t, t')$, the function that gives the mean response of (2.9) to infinitesimal perturbation:

$$\left(\frac{\partial}{\partial t} + \nu k^2 \right) G(k; t, t') + \int_{t'}^{t} \eta(k; t, s)G(k; s, t')ds = 0$$

$$[t \geq t', \quad G(k; t', t') = 1], \qquad (2.11)$$

where $\eta(k; t, s)$ is an average involving an off-diagonal infinitesimal response matrix. The DIA gives the following expressions for $S(k; t, t')$ and $\eta(k; t, s)$; they form a closed system with (2.10) and (2.11):

$$S(k; t, t') =$$

$$\pi k\lambda^2 \iint_{\triangle} pqdpdq \left[\int_0^{t'} a(k, p, q)G(k; t', s)U(p; t, s)U(q; t, s)ds \right.$$

$$\left. - \int_0^{t} b(k, p, q)U(k; t', s)G(p; t, s)U(q; t, s)ds \right], \qquad (2.12)$$

$$\eta(k; t, s) = \pi k\lambda^2 \iint_{\triangle} pqdpdq \, b(k, p, q)G(p; t, s)U(q; t, s), \qquad (2.13)$$

where $\iint_\triangle$ denotes integration over all wavenumbers p, q such that k, p, q can form a triangle, x, y, z are the cosines of the internal angles opposite sides k, p, q, and

$$a(k, p, q) = \tfrac{1}{2}(1 - xyz - 2y^2z^2), \ b(k, p, q) = (p/k)(xy + z^3). \qquad (2.14)$$

The DIA integro-differential equations (2.10)–(2.14) can be associated with a Langevin model equation,[44]

$$\left(\frac{\partial}{\partial t} + \nu k^2\right) u_i(\mathbf{k}; t) + \int_0^t \eta(k; t, s) u_i(\mathbf{k}; s) ds = b_i(\mathbf{k}; t), \qquad (2.15)$$

where

$$(L/2\pi)^3 \langle b_i(\mathbf{k}, t) b_i^*(\mathbf{k}, t') \rangle \equiv B(k; t, t')$$

$$= \pi k \lambda^2 \iint_\triangle pq\, dp\, dq\, a(k, p, q) U(p; t, t') U(q; t, t'). \qquad (2.16)$$

Here $b_i(k; t)$ is a random, zero-mean forcing term. $G(k; t, t')$ as given by (2.11) is the response function for (2.15). Note that (2.15) is a linear dynamical equation in any single realization because $\eta(k; t, s)$ and $b_i(\mathbf{k}; t)$ are determined by ensemble averages and are infinitesimally affected by the value of $\mathbf{u}(\mathbf{k}; s)$ in any one realization. Eqs. (2.10) and (2.11) follow from (2.15) upon simple manipulations based on the fact that the contribution of the random force $b_i(k, t)$ to $\partial U(k; t, t')/\partial t$ is

$$\int_0^{t'} G(k; t', s) B(k; s, t) ds.$$

Eqs. (2.10) and (2.11) may be integrated forward in time from specified initial values of $U(k; 0, 0)$. The energy balance equation is

$$\left(\frac{\partial}{\partial t} + 2\nu k^2\right) E(k, t) = T(k, t), \qquad (2.17)$$

where

$$T(k, t) = 4\pi k^2 S(k; t, t) \qquad (2.18)$$

measures the rate of transfer of energy into wavenumber k by nonlinear terms.

The random force may be interpreted as

$$b_i(\mathbf{k}; t) = -i\lambda P_{ijm}(\mathbf{k}) \sum_p \xi_j(\mathbf{p}; t) \xi_m(\mathbf{k} - \mathbf{p}; t), \qquad (2.19)$$

where $\xi_i(\mathbf{k}; t)$ is an isotropic, homogeneous Gaussian velocity field that satisfies

$$\langle \xi_i(\mathbf{k}; t) \xi_j^*(\mathbf{k}; t') \rangle = \langle u_i(\mathbf{k}; t) u_j^*(\mathbf{k}; t') \rangle. \qquad (2.20)$$

Thus $b_i(\mathbf{k};t)$ has precisely the form of the entire nonlinear term in the wavevector form of the original NS equation. However, in distinction to the exact $\mathbf{u}(\mathbf{x},t)$, the field $\boldsymbol{\xi}(\mathbf{x},t)$ remains precisely Gaussian under the dynamics. The $\eta(k;t,s)$ term in (2.15) then is required to maintain energy conservation.

3. The Direct-Interaction Approximation

3.1. *Time Covariances and Energy Transfer*

The basic structure of energy transfer and time covariances in the DIA can be seen from (2.10)–(2.14) and, most clearly, from (2.15). In the latter, the nonlinear interaction is represented by the η and b terms. The b term is a Gaussian force. If the η term were absent, this random force would, at all times, pump mean energy into each wavenumber mode. The η term is a dynamical damping term that acts against the b term to maintain the overall energy conservation in the mean by the total nonlinear interaction. These η and b terms together determine not only the energy transfer but also the dependence of the covariance $U(k;t,t')$ on its time arguments

Let the *rms* value of any vector component of the turbulent velocity be v_0 and let a characteristic energy-range wavenumber be k_0. In the exact dynamics of decaying, isotropic turbulence at moderate and large Reynolds numbers, the difference-time behavior of $U(k;t,t')$ for $k \geq k_0$ almost certainly is dominated by dephasing associated with random sweeping of structures of size $1/k$ by the large-scale velocity, of order v_0.[75,77] The expected characteristic decorrelation time of $U(k;t,t')$ is then $O(1/v_0 k)$. The mean infinitesimal response function $G(k;t,t')$ also expresses sweeping decorrelation and, in addition, it exhibits viscous damping. The order of magnitude of the characteristic decay time is then the lesser of $1/v_0 k$ and $1/\nu k^2$. The solutions of the DIA equations reproduce this behavior[31] and thereby give what is almost certainly a qualitatively faithful approximation to difference-time behavior at all turbulent Reynolds numbers. This comes about analytically because, for $k \geq k_0$, (2.16) is dominated by contributions in which p or q is $\approx k_0$, and this implies a corresponding time dependence for $b_i(k;t)$.

The energy transfer implied by the DIA is a more complicated matter. The general qualitative ideas of vortex stretching and associated energy cascade at high Reynolds number suggest that there should be an inertial range with energy transfer that is local in wavenumber. The DIA expression for energy transfer, given by (2.18) and (2.12), does give an inertial range with local transfer. However, the transfer is represented in a peculiar way, and the inertial range exponent[29] is $-3/2$ instead of $-5/3$ as called for by qualitative reasoning like that of Kolmogorov and Oboukov.[2]

What happens is that, for k in the inertial range, both the positive a term and the negative b term in $S(k;t,t)$, as given by (2.12), are individually dominated by contributions with p or q in the energy range. These energy-range contributions to the two terms strongly cancel, however, if $k \gg k_0$, with the result that the principal net contributions to $S(k;t,t)$ do come from p/k and q/k both O(1). The strong, canceling contributions involving energy range wavenumbers are a reflection of the slow modulation (within, say, an octave wavenumber band centered on k) that is associated with the advection of small-scale structures by the spatially non-uniform energy-range velocity.

The DIA inertial-range spectrum has the specific form

$$E(k) = C'(\epsilon v_0)^{1/2} k^{-3/2} \approx C' \epsilon^{2/3} (k/k_0)^{1/6} k^{-5/3}, \tag{3.1}$$

where C' is a dimensionless, order-unity constant and ϵ is the mean rate of energy dissipation by viscosity, per unit mass. In constrast, the spectrum form implied by the Kolmogorov-Oboukov ideas of local cascade is[2]

$$E(k) = C \epsilon^{2/3} k^{-5/3}, \tag{3.2}$$

where C is another dimensionless constant. The energy-range velocity v_0 appears in (3.1) because the effective time correlations in (2.12) all are determined by the Eulerian two-time functions of form $U(k;t,t')$ and $G(k;t,t')$. As already noted, DIA correctly assigns to these functions characteristic inertial-range difference times $t - t'$ of order $1/v_0 k$. However, such characteristic times are not reasonably connected with energy transfer in the inertial range. Instead, energy transfer should be associated with characteristic times for vortex stretching, which depend on the rate of strain in the fluid, not the *rms* velocity v_0. This important matter will be returned to in Sec. 3.3.

The dissipation-range spectrum predicted by DIA exhibits an essentially exponential falloff. The characteristic dissipation wavenumber is $k_d = \mathcal{R}_0^{2/3} k_0$, where $\mathcal{R}_0 = v_0 k_0^{-1}/\nu$, in contrast to the Kolmogorov dissipation wavenumber $k_s = \mathcal{R}_0^{3/4} k_0 = (\epsilon/\nu^3)^{1/4}$. For $k \gg k_d$ the DIA energy spectrum at all values of $\mathcal{R}$ has the form[29]

$$E(k) \propto k^3 \exp(-ck/k_d), \tag{3.3}$$

where c is a dimensionless order-unity constant.

Numerical solutions of the DIA equations at moderate values of the Taylor microscale Reynolds number $\mathcal{R}_\lambda$ give satisfactory quantitative as well as qualitative agreement with a number of statistics formed from full computer simulations of isotropic turbulence that decays from Gaussian initial conditions.[17,58] In particular, the DIA predictions for evolution of $\epsilon(t)$, $E(k,t)$ and for the difference-time dependence of $U(k;t,t')$ are good when $\mathcal{R}_\lambda \leq 35$, initially, and the initial spectrum $E(k,0)$ is compact. See Figs. 1 and 2. The DIA predictions for the evolution of third- and fourth-order cumulants at moderate $\mathcal{R}_\lambda$ will be discussed in Sec. 3.5.

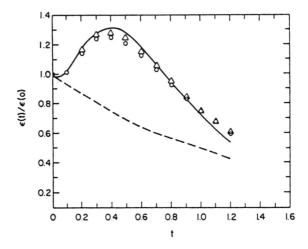

Fig. 1. Energy dissipation rate $\epsilon(t)$ *vs* time t in decaying isotropic turbulence for two simulation runs (data points) and for DIA (solid line). The dashed line shows the decay in the absence of nonlinear terms. The initial spectrum had the form $E(k,0) \propto k^4 \exp[-2(k/k_{peak})^2]$ with initial $\mathcal{R}_\lambda \approx 40$. From Herring, Riley, Patterson & Kraichnan.[18]

The DIA equations for $\nu = 0$ exhibit relaxation of initial spectra to the three-dimensional (3D) absolute equilibrium form $E(k) \propto k^2$, when the system is truncated at some cut-off wavenumber.[31] This behavior is associated with a Liouville theorem that holds for the Euler equation and survives the model construction (2.5). It is confirmed by direct computer simulations.[17]

3.2. Eddy Viscosity

The eddy-viscosity concept arises by analogy between the actions of small spatial scales of turbulence and those of thermal fluctuations.[2,43,50] There are obvious limitations to the analogy because turbulence exhibits a continuous range of scales so that no clean separation can be made between those scales of motion that act like an eddy viscosity and those that are acted upon. In contrast, there usually is a clean scale separation between the

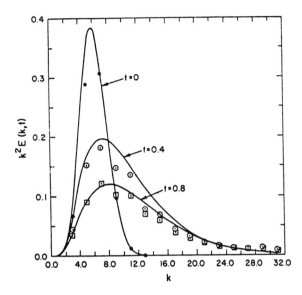

Fig. 2. Evolution of the dissipation spectrum $k^2 E(k, t)$ for one of the runs shown in Fig. 1 and for DIA (solid lines). From Herring, Riley, Patterson & Kraichnan.[18]

significantly excited modes of turbulent motion and the modes responsible for most of the molecular viscosity.

The energy transfer out of a given wavenumber k into wavenumbers $p, q \gg k$ appears in the DIA equations as contribution to the $b(k, p, q)$ term in (2.12) and as a contribution to the η term in (2.15).[29,31,51] The latter term looks something like the ν term, but it exhibits crucial qualitative differences: $\eta(k; t, s)$ does have a simple k^2 dependence and there is an integration over past time. Both of these differences are reflections of the fact that, in general, there is no scale separation among interacting modes. However, the part of the η term that arises from p and/or $q > K$ does look precisely like an augmentation of viscous damping if $k \ll K$, so that there is a clean separation of space and time scales. With the aid of (2.14), the asymptotic ($k \ll K$) contribution can be reduced to the form[33] $\nu(k|K)k^2 u_i(\mathbf{k}, t)$ where

$$\nu(k|K) = \frac{2\pi}{15} \int_K^\infty \left(7 \int_0^t G(p; t, s)U(p; t, s)ds - p\Delta(p, t) \right) p^2 dp, \quad (3.4)$$

$$\Delta(p,t) = \left[\frac{\partial}{\partial p} \int_0^t [G(p;t,s)U(q;t,s) - G(q;t,s)U(p;t,s)]ds \right]_{p=q}.$$

(3.5)

The integration over the past history of $u_i(\mathbf{k},t)$ has now disappeared because the fall-off time of this contribution to $\eta(k;t,s)$ as a function of $t-s$ is small compared to characteristic times of variation of $u_i(\mathbf{k},t)$. If the Reynolds number large, these DIA expressions have the flaw, noted in Sec. 3.1, that the characteristic times for convective dephasing control the size of the time integrals.

More generally, the integration over past time that appears in (2.15) may have important physical implications. If DIA is used to express the effects of subgrid scales in a large-eddy simulation, the nonlocalness in time means that the subgrid scales exert a reactive (visco-elastic) as well as a resistive effect on the explicit scales. The reactive effects may be important in determining stability properties of the explicit scales.

Another implication of the DIA stochastic model, particularly evident from the Langevin form (2.15), is that large wavenumbers exert some level of effective random forcing on lower wavenumbers. This forcing may have significant effects when large-spatial-scale structures are unstable.

Exact (but complicated) expressions can be obtained for the action of homogeneous turbulence on an infinitesimal mean velocity gradient.[43] In the limit where this gradient varies very slowly in space, the action is precisely that of an augmented viscosity. The exact expression for the effective eddy viscosity is a third-order moment, the ensemble average of an expression that is quadratic in velocity amplitudes and linear in the unaveraged response function of the turbulent velocity field to infinitesimal perturbation.

3.3. *Violation of Random Galilean Invariance by DIA*

As noted above, the Eulerian convective-dephasing time $1/v_0 k$ appears in the DIA energy transfer expression $T(k,t)$. This is contrary to the Kolmogorov concept of a cascade in wavenumber that is fully determined by the local statistical state in the wavenumber space and should not depend on v_0. Apart from theoretical considerations, the DIA formula for $T(k,t)$ gives an inertial-range energy transfer that is clearly too small in comparison with experimental results at high $\mathcal{R}_\lambda$.

Correctness of the Kolmogorov localness hypothesis cannot be asserted, but the DIA transfer expression is demonstrably incorrect: it violates the Galilean invariance property of the NS equation. If $\mathbf{u}(\mathbf{x},t)$ is a solution of (2.1) in infinite space or in a cyclic box, then $\mathbf{u}(\mathbf{x} - t\mathbf{V}(t), t) + \mathbf{V}(t)$ is also a solution, where $\mathbf{V}(t)$ is a spatially uniform velocity. Suppose now that $\mathbf{u}(\mathbf{x}, t)$ is a realization of a statistically homogeneous ensemble and that $\mathbf{V}(t)$ is Gaussianly distributed over the ensemble and is statistically independent of $\mathbf{u}(\mathbf{x}, t = 0)$. The addition of $\mathbf{V}(t)$ may be called a random Galilean transformation (RGT) of $\mathbf{u}(\mathbf{x}, t)$. The various moments of $\mathbf{u}(\mathbf{x}, t)$ either are invariant under a RGT, or else have precisely determined transformation properties.[32]

Clearly the RGT cannot affect energy transfer among wavevector modes in the exact dynamics because the random translating velocity exerts no straining. However this is not the case in the RC model (2.5). A different random translation affects each term in the sum over rs in (2.5) and the result is a destructive interference with the build up of the correlations associated with energy transfer. This shows up clearly in the final formula (2.12) for $S(k, t, t)$. If $\langle |\mathbf{V}|^2 \rangle$ is large compared to the mean-square turbulent velocity v_0^2, the correlation time of $U(k; t, t')$ becomes $O(1/\langle |\mathbf{V}|^2 \rangle^{1/2} k)$ under the RGT, and there is a corresponding depression of energy transfer.

Violation of random Galilean invariance (RGI) by the DIA is inevitable because the latter involves only Eulerian two-time statistics: The functions $G(k; t, t')$ and $U(k; t, t')$ do not carry enough information to distinguish between decorrelation due to simple convective dephasing and decorrelation due to the intrinsic distortion and change of velocity-field structures that induces energy transfer. An indicated need is to introduce Lagrangian statistics that measure properties along particle trajectories. Such statistics are invariant under RGT.

A general formal basis for the construction of Lagrangian velocity statistics can be constructed as follows.[23,32] Define $\mathbf{u}(\mathbf{x}, t|s)$ as the velocity measured at time s on the fluid-element trajectory that passes through $\mathbf{x}$ at time t. Then $\mathbf{u}(\mathbf{x}, t) = \mathbf{u}(\mathbf{x}, t|t)$. The classically defined Lagrangian velocity $\mathbf{v}(\mathbf{a}, t)$ is the velocity at time t of the fluid element whose initial position was $\mathbf{a}$. Thus $\mathbf{v}(\mathbf{a}, t) = \mathbf{u}(\mathbf{a}, 0|t)$. The equation of motion for the generalized velocity field $\mathbf{u}(\mathbf{x}, t|s)$ is

$$\left(\frac{\partial}{\partial t} + \lambda \mathbf{u}(\mathbf{x}, t) \cdot \nabla \right) \mathbf{u}(\mathbf{x}, t|s) = 0. \qquad (3.6)$$

Eq. (3.6) states simply that the velocity measured at time s on a given fluid element trajectory is independent of the particular spacetime point $(\mathbf{x}, t)$ on the trajectory that is used for labeling. The ordering parameter $\lambda = 1$ is included in (3.6) to facilitate later discussion of Lagrangian-based closures.

The DIA equations can be modified so that Eulerian time correlations are replaced by Lagrangian time correlations built from $\mathbf{u}(\mathbf{x}, t|s)$ (Refs. 14, 19, 23–25, 32, 39, 45). The equations are then properly invariant to RGT and yield an inertial range that obeys Kolmogorov scaling. The modified DIA will be discussed in Sec. 4.

3.4 *Sensitivity and Error Growth*

Let $\mathbf{u}^1(\mathbf{x})$ and $\mathbf{u}^2(\mathbf{x})$ represent two solutions of the NS equation that have identical isotropic, homogeneous statistics and whose initial values $\mathbf{u}^1(\mathbf{x}, 0)$ and $\mathbf{u}^2(\mathbf{x}, 0)$ are imperfectly correlated. The difference field $\delta\mathbf{u}(\mathbf{x}, t) \equiv \mathbf{u}^1(\mathbf{x}, t) - \mathbf{u}^2(\mathbf{x}, t)$ can be called the error velocity field. It can be used to construct statistical measures of the growth of error (discrepancy between the two evolved fields) with time. A simple error measure is the spectral covariance $\Delta(k; t, t')$ defined by

$$\tfrac{1}{2}P_{ij}(\mathbf{k})\Delta(k; t, t') = (L/2\pi)^3 \langle \delta u_i(\mathbf{k}; t)\delta u_j^*(\mathbf{k}; t') \rangle \quad (L \to \infty). \tag{3.7}$$

A related measure is the covariance scalar $W(k; t, t')$, defined by

$$\tfrac{1}{2}P_{ij}(\mathbf{k})W(k; t, t') = (L/2\pi)^3 \langle v_i^1(\mathbf{k}; t)\delta u_j^{2*}(\mathbf{k}; t') \rangle \quad (L \to \infty), \tag{3.8}$$

that describes the correlation between fields 1 and 2. It is important to note that the evolution of finite errors, rather than infinitesimal, is of principal physical interest. Analysis of Lyapunov exponents is little help in exploring the growth of finite errors. The RC model and associated DIA equations generalize straightforwardly to the simultaneous analysis of fields 1 and 2. They provide approximations to $\Delta(k; t, t')$ and $W(k; t, t')$ when errors are finite as well as when they are infinitesimal.[35,48,49]

The general character of the DIA for error evolution can be seen by writing the model Langevin equation (2.5) for each of the two fields 1 and 2 and using the representation (2.19) for $b_i(k; t)$. It is easily verified that $\eta(k; t, s)$ is identical for the two fields, while

$$b_i^n(\mathbf{k}; t) = -i\lambda P_{ijm}(\mathbf{k}) \sum_p \xi_j^n(\mathbf{p}; t)\xi_m^n(\mathbf{k} - \mathbf{p}; t) \quad (n = 1, 2), \tag{3.9}$$

Now suppose that there is initial imperfect correlation of the form

$$W(k; 0, 0) = \alpha U(k; 0, 0) \quad (0 < \alpha < 1). \tag{3.10}$$

Then it follows from (3.9) and the Gaussian statistics of $\boldsymbol{\xi}(\mathbf{x}, 0)$ that

$$\langle b_i^1(\mathbf{k}; 0)b_j^{n*}(\mathbf{k}; 0) \rangle = \alpha^2 \langle b_i^n(\mathbf{k}; 0)b_j^{n*}(\mathbf{k}; 0) \rangle \quad (n = 1 \text{ or } 2). \tag{3.11}$$

Since $\alpha^2 < \alpha$, this means that the initial forcings in the Langevin equations for fields 1 and 2 are less correlated than the initial fields $\mathbf{u}^1(\mathbf{k}; 0)$ and $\mathbf{u}^2(\mathbf{k}; 0)$ themselves. Consequently, fields 1 and 2 become less correlated under the equations of motion, and $\Delta(k; t, t)/U(k; t, t)$ increases with time.

The model equation gives conservation of energy by the nonlinear interactions, but it does not give conservation of error energy. In general, the latter tends to grow.

Numerical solutions of the DIA equations for error evolution have been compared with full computer simulations of decaying isotropic turbulence at moderate Reynolds number.[18] The DIA results are in qualitative and reasonably accurate quantitative agreement with the simulation results. An interesting feature of the calculations is that there is error growth that eventually dominates the entire spectrum even when the support of the initial error spectrum is confined to wavenumbers strongly damped by viscosity. See Fig. 3. The quantitative errors of the DIA usually amount to underestimation of error growth. Detailed examination of the approximation reveals a theoretical basis for this tendency.

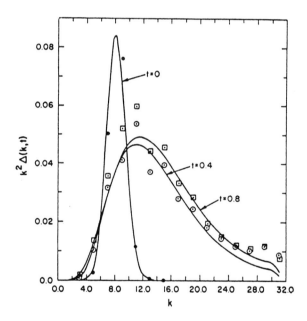

Fig. 3. Evolution of weighted error energy spectrum $k^2 \Delta(k,t)$ according to DIA (solid lines) and for the pair of simulations shown in Fig. 1. Dots are for $t = 0$, circles for $t = 0.4$ and squares for $t = 0.8$.

3.5 Departures from Gaussian Statistics

The most obvious non-Gaussian feature of isotropic turbulence from Gaussian form is energy transfer: The mean energy transfer function $T(k,t)$

is a third-order moment,[2,50,56] which vanishes in a Gaussian distribution. Other non-Gaussian features include spatial and temporal intermittency of viscous dissipation and other small-scale structures (Refs. 1, 3, 5, 12, 13, 20–22, 26–28, 38, 47, 54, 55, 67, 69, 70, 71, 74); a tendency for velocity field and vorticity field in x space to be parallel or antiparallel (enhanced helicity variance);[7] and depression of the mean-square of the total nonlinear term in x space (including pressure) below the value implied by the energy spectrum $E(k,t)$ under Gaussian statistics.[7,46,78] The mean-square of the total nonlinear term is a fourth-order moment.

An ubiquitous, but often overlooked, non-Gaussian phenomenon is associated with the advection of small-spatial-scale structures by the large-scale velocity field.[32] In a Gaussian velocity field, the velocity in one wavenumber band has no effect at all on that in another band. There is no advection. Advection effects on the time-covariance function of high-k modes show up, in lowest order, as two-time third-order moments.

The agreement of DIA results for $E(k,t)$ and $T(k,t)$ with simulation results at moderate $\mathcal{R}_\lambda$ implies that triple moments are predicted fairly well. An overall measure of straining, energy transfer, and vorticity enhancement is the skewness of the longitudinal velocity derivative $S(t) = -\langle(\partial u_1/\partial x_1)^3\rangle/\langle(\partial u_1/\partial x_1)^2\rangle^{3/2}$. The DIA prediction for this quantity[17,58] is qualitatively similar to simulation results and underestimates $S(t)$ by about 15% at t when the turbulence is well evolved from an initial $\mathcal{R}_\lambda \approx 35$. See Fig. 4. It is not an accident that DIA underestimates $S(t)$. As $\mathcal{R}_\lambda$ increases, the underestimate becomes more severe. It is associated with the failure of DIA to distinguish sweeping and distortion effects.

Recently, the DIA for decaying isotropic turbulence at moderate $\mathcal{R}_\lambda$ has been found to give an excellent prediction of the depression of the total nonlinear term below Gaussian values.[7,46,78] Both simulations and DIA give normalized variances of the nonlinear term of about 50% of Gaussian value at times when the turbulence is well evolved. See Fig. 5. In contrast, DIA gives zero for the fourth-order velocity-field cumulants that describe the departure, from Gaussian values, of pressure variance $\langle p^2\rangle$ and helicity variance $\langle(\mathbf{u}\cdot\boldsymbol{\omega})^2\rangle$, where $\boldsymbol{\omega}$ is vorticity.[7]

The DIA totally fails to capture the intermittency of dissipation in x space.[46] It gives zero for the relevant fourth-order cumulants of the velocity field. This is something of a puzzle. Why should this approximation give good predictions for certain fourth-order cumulants, those that determine the variance of the total nonlinear term, and completely ignore the cumulants associated with intermittency? Part of the answer probably is that the mean-square of the total nonlinear term is a statistic of immediate

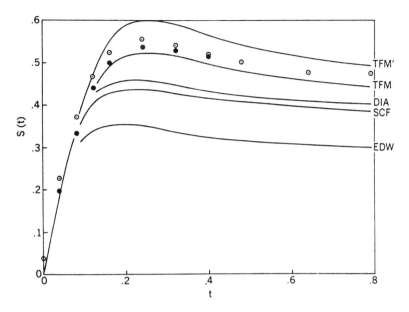

Fig. 4. Evolution for skewness $S(t)$ for two simulation runs (data points) with the spectrum form of Fig. 1 and initial $\mathcal{R}_\lambda \approx 34.5$. The solid curves show results for DIA, for Edwards' (EDW) closure (4.10), for two variants of the TFM closure (4.12) to (4.15) and for Herring's self-consistent field (SCF) closure. From Herring and Kraichnan.[17]

dynamical significance, for the RC model (2.5) as well as for the NS equation itself. The suppression of overall nonlinearity appears to be a generic statistical-mechanical effect controlled by the magnitudes of modal coupling coefficients and by where, in the mode space, the energy is initially concentrated. The quantities $\langle (\mathbf{u} \cdot \boldsymbol{\omega})^2 \rangle$, $\langle |\boldsymbol{\omega}|^4 \rangle$, $\langle \epsilon'^2 \rangle$, where ϵ' is the unaveraged dissipation at a point in x space, do not appear to have immediate dynamical significance in the DIA equations.

4. Related Approximations

A variety of other theoretical approaches can be related to the RC model and the associated DIA integro-differential equations for $U(k;t,t')$ and $G(k;t,t')$. Some of these approaches involve modified stochastic models that either remedy deficiencies in the physics of the RC model or are simpler to solve.[11,16,36,37,50,57] Other approaches, like renormalized perturbation theory (RPT)[44,51] and renormalization group (RG) treatments,[9,40,42,72,73] relate to DIA because the latter embodies the analytical structure of second-order perturbation theory. Finally, the DIA equations may be modified to

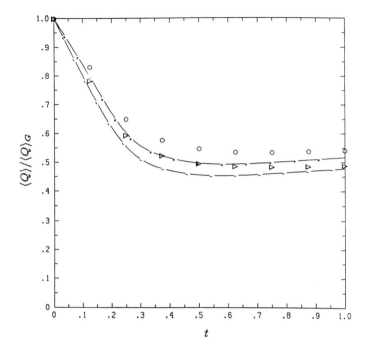

Fig. 5. Depression of nonlinearity below Gaussian values for isotropic turbulence decay with the spectrum of Fig. 1 and initial $\mathcal{R}_\lambda \approx 35.0$. The circles and line with plusses show simulation and DIA results for W/W_G, where $W = \langle |\mathbf{u}\Delta\nabla\mathbf{u} - \nabla p|^2 \rangle$ and W_G is the value of W for a Gaussian field with the same instantaneous spectrum as the N-S field. The triangles and line with minuses show $\langle |\partial\mathbf{u}/\partial t|^2 \rangle / \langle |\partial\mathbf{u}/\partial t|^2 \rangle_G$. From Chen, Herring, Kerr & Kraichnan.[7]

a Lagrangian formulation that resolves the problem of separating sweeping and distortion effects on the statistics. This Sec. starts with a discussion of the so-called Markovianized models derived from DIA, continues with the relation of DIA to RPT and RG, and concludes with a summary of the Lagrangian variants of DIA. Sec. 5 is devoted to decimation models, an alternate stochastic modeling approach that relates directly to (2.15), the Langevin model of DIA, and embeds DIA in a sequence of systematically improved approximations.[41,44]

4.1 *Markovianized Models*

The basic stochastic model (2.9) may be replaced by one that looks more complicated but is much easier to solve. This is the "Markovianized" model. It offers a more flexible physics, but at the cost of an introduced arbitrariness. The basic Markovian model equation is[11,36]

$$\left(\frac{\partial}{\partial t} + \nu k^2\right) u_i^n(\mathbf{k}; t) =$$

$$-i\lambda w(t) \sum_{rs} \phi_{nrs} P_{ijm}(\mathbf{k}) \sum_{\mathbf{p}} [\theta_{kpq}(t)]^{1/2} u_j^r(\mathbf{p}; t) u_m^s(\mathbf{k} - \mathbf{p}; t). \tag{4.1}$$

Here $w(t)$ is a Gaussian white-noise process that satisfies

$$\langle w(t)w(t')\rangle = \delta(t - t'). \tag{4.2}$$

The quantity $\theta_{kpq}(t)$ is a virtual memory time. It must be positive and symmetric in its three indices but otherwise is arbitrary. The $w(t)$ factor in (4.1) reduces the dynamical memory time to zero. The result is that response functions drop out of the final model equations for moments and closed equations for the single-time quantities $U(k; t, t)$ can be written in which the $\theta_{kpq}(t)$ appear as externally determined parameters. The final equations that replace (2.10)–(2.13) are

$$\left(\frac{\partial}{\partial t} + \nu k^2\right) U(k; t, t') + \eta(k; t)U(k; t, t') = 0 \quad (t \geq t'), \tag{4.3}$$

$$\left(\frac{\partial}{\partial t} + \nu k^2\right) G(k; t, t') + \eta(k; t)G(k; t, t') = 0$$

$$[t \geq t', \quad G(k; t', t') = 1], \tag{4.4}$$

$$\left(\frac{\partial}{\partial t} + 2\nu k^2\right) U(k; t, t) = 2S(k; t, t), \tag{4.5}$$

where

$$S(k; t, t) = \pi k\lambda^2 \iint_\triangle pqdpdq\, \theta_{kpq}(t)$$

$$\times [a(k, p, q)U(p; t, t)U(q; t, t) - b(k, p, q)U(k; t, t)U(q; t, t)], \tag{4.6}$$

$$\eta(k; t) = \pi k\lambda^2 \iint_\triangle pqdpdq\, b(k, p, q)\theta_{kpq}(t)U(q; t, t), \tag{4.7}$$

The form of Markovianized model closest to DIA is obtained by setting

$$\theta_{kpq}(t) = \int_0^t G(k; t, s)G(p; t, s)G(q; t, s)ds. \tag{4.8}$$

Differentiation of (4.8) and the use of (4.3) yield

$$d\theta_{kpq}(t)/dt =$$

$$1 - [\nu(k^2 + p^2 + q^2) + \eta(k; t) + \eta(p; t) + \eta(q; t)]\theta_{kpq}(t), \tag{4.9}$$

and (4.8) gives the initial value $v_{kpq}(0) = 0$. If the modes k, p, q are in a steady state, the solution of (4.9) is

$$\theta_{kpq}(t) = 1/[\nu(k^2 + p^2 + q^2) + \eta(k,t) + \eta(p,t) + \eta(q,t)], \qquad (4.10)$$

which yields an energy balance equation identical with that proposed by Edwards.[36]

The formula (4.8) fixes the characteristic times $\theta_{kpq}(t)$ as Eulerian decorrelation times, like those in the original DIA. The problems with RGI therefore remain. Several Markovianized models have been proposed that make the $\theta_{kpq}(t)$ effective Lagrangian decorrelation times. They give energy-balance equations that exhibit RGI and consequently yield (3.2) in the inertial range. An "eddy-damped-quasi-normal-Markovian" (EDQNM) model is obtained by assuming (4.10) at all t with

$$\eta(k,t) = c[k^3 E(k,t)]^{1/2}, \qquad (4.11)$$

where c is an $O(1)$ constant.[11,50,57] The right side of (4.11) may be interpreted as the inverse of an eddy circulation time associated the velocity field in, say, an octave band centered on k. This choice of $\eta(k,t)$ thus eliminates the sweeping of inertial-range structures by the energy-range velocity field. The resulting two-time functions $U(k;t,t')$ and $G(k;t,t')$ must be interpreted as approximate Lagrangian covariance and response functions.

The "test-field model" (TFM) achieves RGI, and a $-5/3$ inertial-range spectrum, by tying $\theta_{kpq}(t)$ to distortion effects due to pressure and viscous forces.[36,37] The decorrelation time associated with pressure forces is estimated, in a purely Eulerian frame, by examining the rate at which compressive velocity would be built up if the pressure forces were removed from the NS equation. The TFM equations for $\theta_{kpq}(t)$ involve characteristic inverse times both for the actual incompressible (solenoidal) velocity field and the fictitious compressive "test field" so created. The final formulas are

$$\eta^S(k;t) = \pi k \lambda^2 \iint_\triangle pq\,dp\,dq\, b'(k,p,q)\theta'_{pqk}(t)U(q;t,t), \qquad (4.12)$$

$$\eta^C(k;t) = 2\pi k \lambda^2 \iint_\triangle pq\,dp\,dq\, b'(k,p,q)\theta'_{kqp}(t)U(q;t,t), \qquad (4.13)$$

$$d\theta_{kpq}(t)/dt =$$
$$1 - [\nu(k^2 + p^2 + q^2) + \eta^S(k;t) + \eta^S(p;t) + \eta^S(q;t)]\theta_{kpq}(t), \qquad (4.14)$$

$$d\theta'_{kpq}(t)/dt =$$
$$1 - [\nu(k^2 + p^2 + q^2) + \eta^C(k;t) + \eta^S(p;t) + \eta^S(q;t)]\theta'_{kpq}(t), \qquad (4.15)$$

where η^S and η^C are associated with the solenoidal and fictitious compressive velocity fields, $\theta'_{kpq}(t)$ is an auxiliary time parameter and

$$b'(k,p,q) = \tfrac{1}{2}(1 - y^2)(1 - z^2). \qquad (4.16)$$

Markovianized models based on (4.1) guarantee realizability of $U(k; t, t')$ provided that whatever equations are used to evolve $\theta_{kpq}(t)$ keep this quantity positive. They are also relatively economical to integrate because none of the terms in the differential equations involve integrals over past time. But the almost total freedom in choosing $\theta_{kpq}(t)$ makes the Markovianized models less than deductive. The TFM model has been used to predict error growth.[48,49]

4.2 *Renormalized Perturbation Theory*

A formal perturbation expansion of $\mathbf{u}(\mathbf{x}, t)$ may be generated staightforwardly as an infinite functional power series in λ by iterative treatment of (2.1). The coefficient of λ^n is a functional polynomial of degree $n + 1$ in $\mathbf{u}^0(\mathbf{x}, t)$, which is defined as the solution of the linearized equation obtained by setting $\lambda = 0$. If the initial field $\mathbf{u}(\mathbf{x}, t)$ is Gaussian, then $\mathbf{u}^0(\mathbf{x}, t)$ is Gaussian and any moment of $\mathbf{u}^0(\mathbf{x}, t)$ may be expressed as a function of covariances of $\mathbf{u}^0(\mathbf{x}, t)$. Consequently, any moment of $\mathbf{u}(\mathbf{x}, t)$ may be expressed *via* the functional power series as an infinite functional power series[39,44] in the covariance of the zeroth-order field $\mathbf{u}^0$.

Expansion in powers of λ is essentially expansion in powers of Reynolds number, and, at best, useful convergence might be expected only if the Reynolds number is small. The actual convergence properties are probably worse than this. It is plausible that the expansion of $\mathbf{u}(\mathbf{x}, t)$ has a finite radius of convergence in λ in a typical realization. But prototype systems that do exhibit this kind of convergence give zero radius of convergence for the λ expansions of moments because the averaging over the initial Gaussian distribution introduces realizations in which the radius of convergence is indefinitely small.[44] In the absence of contrary evidence, one must therefore assume that the radius of convergence λ of moments like $U(k; t, t')$ is zero.

The formal expansion just described (primitive perturbation expansion) may be "line-renormalized" by a variety of techniques. The result is new formal λ expansions for moments of $\mathbf{u}(\mathbf{x}, t)$ in which the coefficient of λ^n is a functional polynomial in the covariance scalar $U(k; t, t')$ and infinitesimal response function $G(k; t, t')$ of the actual field $\mathbf{u}(\mathbf{x}, t)$. These renormalized expansions appear more physical because they no longer refer to the linearized field $\mathbf{u}^0(\mathbf{x}, t)$.[44] However, the convergence in λ at high orders does not appear to be qualitatively improved over the primitive expansion.

The lowest-order terms in the line-renormalized expansions for the quantities $S(k; t, t')$ and $\eta(k; t, s)$ that appear in (2.10) and (2.11) are precisely the DIA expressions (2.12) and (2.13). As noted in Sec. 5, higher terms in the expansions for the RC model vanish. Thus truncation of the line-renormalized expansions at the lowest order has a special significance and exhibits the assured consistency properties of the RC model. In general, higher truncations of the renormalized expansions lead to unphysical, unbounded results.[30] It is possible to obtain reasonable-appearing higher approximations for $S(k; t, t')$ and $\eta(k; t, s)$ by Padé approximants and other techniques. In general, this is a blind procedure because the analyticity properties of the line-renormalized expansions are badly known.[44]

Both the primitive and the line-renormalized perturbation expansions for $S(k; t, t')$ and $\eta(k; t, s)$ have the following consistency properties in each order in λ: they preserve the quadratic inviscid constants of motion of the NS equation, the equipartition in absolute statistical equilibrium when the inviscid system is truncated in wavenumber, and a fluctuation-dissipation relation between $U(k; t, t')$ and $G(k; t, t')$ in that equilibrium. The primitive expansion preserves RGI in each order in λ, but the line-renormalized expansion does not. This invariance is violated not only in the lowest-order term that gives DIA but in every successive term. Invariance under RGT is recovered only if the formal series can be summed.

A higher renormalized expansion has also been constructed that involves vertex renormalizations. Again, this expansion keeps the conservation and absolute-equilibrium properties in each order but violates RGI in each order. Elegant, formally closed, functional formulations of the renormalized expansions have been presented; they show the same properties that are exhibited order by order in the expansions.[53,59−61]

4.3 Lagrangian Modifications of DIA

The DIA and, in fact, the entire line-renormalized perturbation expansion of which it represents the leading term, can be systematically modified so that the analysis is based on Lagrangian velocity statistics.[14,23−25,32,33,39] The result is that RGI is restored in each order in λ. The price for this is loss of the RC model representation, lack of uniqueness of the modification algorithm, and analytical complication. A DIA modified in this way yields the Kolmogorov inertial range (3.2) and, without the benefit of adjustable parameters, gives excellent quantitative spectral fits over both the inertial-range and dissipation-range of high-Reynolds-number experimental data.[33] Stochastic models that generalize DIA and exhibit RGI without appeal to Lagrangian velocities can be constructed by the decimation methods discussed in Sec. 5.[41]

Statistics constructed from the generalized velocity field $\mathbf{u}(\mathbf{x}, t|s)$ that obeys (3.6) can distinguish between convective decorrelation and distortion due to straining and provide a basis for approximations and expansions that preserve RGI. The generalized velocity covariance $\langle u_i(\mathbf{x}, t|s)u_j(\mathbf{x}', t'|s')\rangle$ is determined in isotropic, homogeneous flow by a spectral scalar $U(k; t|s; t'|s')$. Formal expansions involving the generalized field can exploit the fact that the linearized field $\mathbf{u}^0(\mathbf{x}, t|s)$, the solution of (2.1) and (3.6) with $\lambda = 0$, is independent of the labeling time t: $\mathbf{u}^0(\mathbf{x}, t|s) = \mathbf{u}^0(\mathbf{x}, s|s) = \mathbf{u}^0(\mathbf{x}, s)$. This means that, in the primitive perturbation expansions, each amplitude factor $\mathbf{u}^0(\mathbf{x}, s)$ may be replaced by $\mathbf{u}^0(\mathbf{x}, t'|s)$, where t' is to be determined, without changing the value. Such changes enable the systematic construction of new line-renormalized expansions in which all the integrals over past time (history integrals) are evaluated along fluid-element trajectories instead of at fixed points in space.

The lowest-order truncation of these "Lagrangian-history" line-renormalized expansions gives an altered DIA, the "Lagrangian-history DIA" (LH-DIA) in which energy transfer is expressed in terms of Lagrangian covariances. Every G and U function that appears the equal-time triple moment $S(k; t, t)$, as given by (2.12), is replaced by the Fourier transform of a Lagrangian function:

$$U(k; t, s) \to U(k; t|t; t|s), \quad G(k; t, s) \to G(k; t|t; t|s), \qquad (4.17)$$

where $G(k; t|s; t'|s')$ is a mean infinitesimal response function associated with the generalized velocity field. Eqs. (2.10) and (2.11) are augmented by evolution equations that determine $U(k; t|s; t'|s')$ and $G(k; t|s; t'|s')$. There is also an abridged form of LHDIA (ALHDIA) that yields closed equations solely for the restricted set of Lagrangian functions $U(k; t|t; t|s)$ and $G(k; t|t; t|s)$ and does not involve Eulerian two-time functions at all. The ALHDIA equations present the same level of difficulty to integrate numerically as the original DIA equations, while the full LHDIA equations constitute a much bigger task.

Both the LHDIA and the ALHDIA equations are invariant under RGT. This is also true order-by-order of the full line-renormalized expansions of which LHDIA and ALHDIA represent the leading terms. At the same time, these expansions retain the order-by-order consistency properties of the original Eulerian expansions: detailed energy conservation and absolute-equilibrium solutions that display correct fluctuation-dissipation relations. What is lost in the Lagrangian reworking is the model representation (2.5) of the DIA which is so important in assuring realizability properties of $U(k; t, t')$. No evidence has appeared that solutions of the LHDIA or ALHDIA equations violate realizability, but nevertheless these approximations have lost a principal conceptual strength of the original DIA.

Numerical integrations of the ALHDIA equations have been carried out for isotropic turbulence at infinite Reynolds number.[33] The results give an excellent quantitative fit to the inertial-range and dissipation-range spectrum of sea-water measurements carried out by Grant, Stewart, and Moilliet.[15] See Figs. 6 and 7. The predicted value for the Kolmogorov constant in (3.2) is $C = 1.77$. The ALHDIA and LHDIA equations give physically plausible qualitative results, but poorer quantitative predictions, in a number of other problems where there is a wide range of excited wavenumbers: advection of a passive scalar, where they yield both the $k^{-5/3}$ inertial-range spectrum and Batchelor's k^{-1} range; Burgers-equation turbulence, where they yield the k^{-2} inertial range; Alfvén-wave turbulence, where they yield the $k^{-3/2}$ inertial-range spectrum.[44] In all but the last of these applications, the original DIA gives qualitatively flawed predictions because of spurious convection effects.

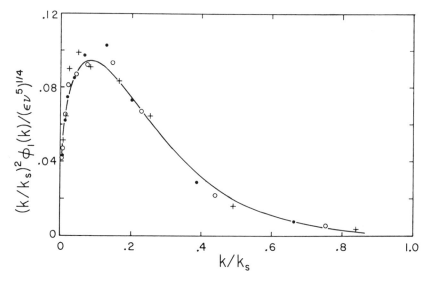

Fig. 6. Normalized "one-dimensional" dissipation spectrum $(k/k_s)^2 \phi_1(k)(\epsilon \nu^5)^{1/4}$ according to ALHDIA equations at infinite $\mathcal{R}_\lambda$ (curve) and tidal channel data of Grant, Stewart & Moilliet[15] at $\mathcal{R}_\lambda \approx 3000$. From Kraichnan.[33]

A number of varieties of Lagrangian closures have been proposed that differ in varying degrees from the LHDIA closure. Kaneda and his coworkers have described a formalism in which the evolution equations for $\mathbf{u}(\mathbf{x}, t|s)$ are simplified by expressing them with the aid of a passive scalar field advected by the turbulence.[14,23−25] This formalism gives rise to Markovianized closures. It is also possible to form Lagrangian-history closures in which the Lagrangian strain field rather than the Lagrangian velocity field is fundamental.[19,45] These closures seem to give better quantitative results than ALHDIA for the spectrum of a passive scalar field advected by a random velocity field.

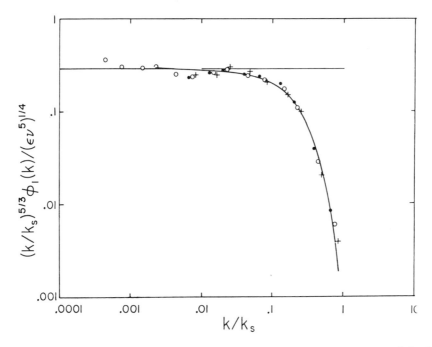

Fig. 7. The data of Fig. 6 presented in log-log form. The spectrum is now weighted by $(k/k_s)^{5/3}$ so that $\phi_1(k) \propto k^{-5/3}$ is a horizontal line.

4.4 Renormalization-Group Approximations

A number of investigators have used renormalization-group (RG) methods to form closures for isotropic turbulence. The various approaches have had three features in common: they are perturbative; they eliminate from the dynamical equations the modes that lie in successive thin shells in k space; in addition to expansion in λ they invoke a power-law external forcing whose exponent is expanded about a value for which the statistical dynamics take a simple form ("epsilon expansion").[40] The closures thus obtained can be expressed in terms of the RC model and associated DIA equations, if the latter are constructed for simplified dynamical problems in which the interactions among wavevector modes are modified from their form in the NS equation. This reduction will be illustrated here for the RG theory of the inertial range originated by Yakhot and Orszag.[9,72,73]

In the Yakhot-Orszag theory, successive bands of modes are eliminated from the high-k end of the inertial range. The dynamics of the band being eliminated are controlled by a dynamical damping due to the modes already eliminated. The interaction of this band with much smaller wavenumbers then is evaluated by second-order perturbation analysis and is found to take the form of an incremental dynamical damping exerted on the smaller wavenumbers. From this analysis, a self-consistent model of the entire non-linear dynamics is constructed in which a mode of given wavenumber experiences dynamical damping exerted by all modes of higher wavenumber. Energy conservation by the nonlinear interactions is maintained: in the representation of the nonlinear interaction, the dynamical damping exerted by a band is accompanied by an effective random forcing that pumps energy into the band at the rate required to maintain energy conservation.[42]

The only interactions that actually enter explicitly into the analysis are those between a given wavenumber and a pair of much larger wavenumbers (distant interactions). Interactions among wavenumbers of similar magnitude (local interactions) are evaluated by extrapolation from the distant interactions. The sweeping effect of energy-range velocity upon inertial-range wavenumbers is omitted from the analysis. The technical means by which both the extrapolation and the omission of sweeping happen is the epsilon expansion: The properties of the Kolmogorov $k^{-5/3}$ inertial range are obtained by an expansion in power-law exponent about a reference regime in which the spectrum goes as k^{+1}. In this regime, there are two essential physical differences from the Kolmogorov case: the energy is dominated by the high-wavenumber end of the spectrum, instead of by the low-wavenumber end, and the eddy-damping felt by modes in the power-law range is dominated by contributions from the high-k end of the power-law range, instead of by local contributions.

The dynamics obtained by the RG analysis just described can be obtained from a RC model, and associated DIA approximation, for an altered system in which the the the interactions of wavevector triads in the NS equation are altered so that sweeping effects of low wavenumbers are eliminated while eddy damping remains.[75] The DIA approximation for the altered NS equation exhibits no effect of energy-range sweeping on inertial-range $U(k;t,t')$ and $G(k;t,t')$ and consequently yields (3.2). The change in (2.10)–(2.16) is solely that the coefficients $a(k,p,q)$ and $b(k,p,q)$ are altered. Two further approximations then express the extrapolation-approximation of local interactions characteristic of the Yakhot-Orzsag RG treatment: (a) the $a(k,p,q)$ and $b(k,p,q)$ coefficients are additionally altered to the form they take when one of the three wavenumbers is much smaller than the other two; (b) it is assumed, in the Langevin equation (2.15), that the decorrelation time of $\eta(k;t,s)$ is much smaller than the characteristic time of $u_i(\mathbf{k},s)$, so that $u_i(\mathbf{k},s)$ may be taken outside the s integral and replaced by $u_i(\mathbf{k},t)$.

The triad interactions of all modes are treated simultaneously in the DIA modeling of RG. There is no need to do the band-by-band elimination in order to obtain the final results of the RG analysis. The same qualitative results can be obtained by a very simple *ad hoc* analysis in which modeling of the nonlinear interaction by random force and eddy-viscosity are assumed at the outset, and the two are related by requiring energy conservation.[42] Again there is no need to do the band-by-band elimination.

It can be seen by the preceding description that The Yakhot-Orszag RG closure involves several drastic approximations, in particular the treatment of local interactions by extrapolation from the dynamics of distant interactions. Nevertheless, the final prediction $C = 1.62$ for the constant in (3.2) is good. Part of the reason for this success may be that a broad class of closures of the DIA type exhibits remarkable insensitivity of C to the dynamics of local interactions. The local interactions have two counteracting effects: a contribution to energy transport through the inertial range and a contribution to the effective inverse correlation times of inertial-range modes. The first effect tends to decrease the value of C while the second effect tends to increase it.[42]

5. Decimation Models

One might hope that closed approximations above DIA could be constructed by giving to the coupling coefficients ϕ_{nrs} in (2.5) or (2.9) values that somehow lie between the totally random symmetric values of the RC model and the value $\phi_{nrs} = M\delta_{nr}\delta_{ns}$ that gives back the original NS dynamics (2.3). Such models have not materialized. However, it is possible to imbed DIA in a set of systematically improved approximations by constructing stochastic models of a different kind, again starting from the collection of NS systems (2.3).[41,76,79]

The collection of M systems may be described by a set of M statistically symmetrical collective fields that are statistically orthogonal linear combinations of the $\mathbf{u}^n(\mathbf{x}, t)$. The models are then constructed by a several stage process: First the collective fields are divided in an explicit set, with $S \leq M$ members and an implicit set containing all the rest of the collective fields. Only the explicit set of field amplitudes is followed in detail. Second, the nonlinear terms representing the action of the implicit set in the equations of motion of the explicit set are considered to be forcing terms that are known only statistically. Third, the statistics of these forcing terms are systematically constrained toward their true values by moment equations that express the statistical symmetry between the explicit amplitudes and the implicit amplitudes. This permits a bootstrap procedure in which the dynamics of the explicit amplitudes is followed as they interact with each other amid the sea of interactions with the implicit modes.

The collective fields may be defined by:

$$\mathbf{u}^{\alpha}(\mathbf{x},t) = M^{-1/2} \sum_{n} e^{i2\pi\alpha n/M} \mathbf{u}^{n}(\mathbf{x},t)$$

$$[\alpha = 0, \pm 1, \pm 2, ..., \pm (M-1)/2], \tag{5.1}$$

where Latin superiors label individual fields in the collection, as before, Greek superiors label the collective fields, and the collection size M is taken odd. The orthogonality identities

$$\sum_{\alpha} e^{i2\pi\alpha(n-m)/M} = M\delta_{nm}, \quad \sum_{n} e^{i2\pi(\alpha-\beta)n/M} = M\delta_{\alpha\beta}, \tag{5.2}$$

make (5.1) equivalent to

$$\mathbf{u}^{n}(\mathbf{x},t) = M^{-1/2} \sum_{\alpha} e^{-i2\pi\alpha n/M} \mathbf{u}^{\alpha}(\mathbf{x},t). \tag{5.3}$$

With the aid of (5.1)–(5.3), Eqs. (2.3) and (2.4) for the collection of statistically identical NS flow systems can be transformed to the following equations for the collective fields:

$$\left(\frac{\partial}{\partial t} - \nu\nabla^2\right)\mathbf{u}^{\alpha}(\mathbf{x},t) =$$

$$-\lambda M^{-1/2} \sum_{\beta+\gamma=\alpha} \mathbf{u}^{\beta}(\mathbf{x},t)\cdot\nabla\mathbf{u}^{\gamma}(\mathbf{x},t) - \nabla p^{\alpha}(\mathbf{x},t), \tag{5.4}$$

$$\nabla\cdot\mathbf{u}^{\alpha} = 0. \tag{5.5}$$

The exact NS dynamics are unaltered by this transformation. The description of the M statistically identical flow systems by the collective fields is formally analogous to the description of a spatially homogeneous dynamical system, with finite coherence lengths, by spatial Fourier modes in a cyclic box. Like the Fourier modes of a homogeneous system, the collective fields yield nonvanishing moments only if the labeling superscripts add to zero. For example,

$$\langle\mathbf{u}^{\alpha}(\mathbf{x},t)\mathbf{u}^{\beta}(\mathbf{x}',t')\rangle = \delta_{\alpha+\beta}\langle\mathbf{u}^{n}(\mathbf{x},t)\mathbf{u}^{n}(\mathbf{x}',t')\rangle, \tag{5.6}$$

where $\delta_{\mu} = 1$ if $\mu = 0$, $\delta_{\mu} = 0$ otherwise, and n denotes any one system from the collection of statistically identical systems. Eq. (5.6) illustrates that statistical properties of individual collective fields are independent of α. In general, the ensemble average of any product of collective field amplitudes with a set of labeling superscripts α, β, ... that adds to zero depends only on how this set decomposes into zero-sum subsets; the value is otherwise independent of the particular values of α, β, ...

RC models may be constructed by altering (5.4); if fact, this was the original method of construction. To do this, factors $\phi_{\alpha\beta\gamma}$ are inserted after the summation sign in (5.4). They take the values $+1$ and -1 at random subject to the symmetry requirements

$$\phi_{\alpha\beta\gamma} = \phi_{\alpha\gamma\beta} = \phi_{\beta\alpha,-\gamma} = \phi_{-\alpha,-\beta,-\gamma}, \tag{5.7}$$

which maintain energy conservation and the reality of all the $\mathbf{u}^n(\mathbf{x}, t)$. The RC model so constructed is not identical with that formed directly from (2.3) in Sec. 3, but in the limit $M \to \infty$ both models yield the same statistics for each $\mathbf{u}^n(\mathbf{x}, t)$ and, in particular, the same DIA equations.

The present Sec. is concerned with the construction of a different kind of model: the decimation (DEC) models which exploit, in a different way, the statistical symmetry of the collection of systems. The DEC models are constructed by dividing the total set of M collective fields into an explicit, or sample, subset with S members and an implicit subset with $M - S$ members. If α is in the explicit subset, then $-\alpha$ is also. The equations of motion for the explicit subset are written as

$$\left(\frac{\partial}{\partial t} - \nu\nabla^2 \right) \mathbf{u}^\alpha(\mathbf{x}, t) =$$

$$-\lambda M^{-1/2} \sum_{\substack{\beta+\gamma=\alpha}}^{S} \mathbf{u}^\beta(\mathbf{x}, t) \cdot \nabla \mathbf{u}^\gamma(\mathbf{x}, t) - \nabla p^\alpha(\mathbf{x}, t) + \mathbf{q}^\alpha(\mathbf{x}, t) \quad (\alpha \in S),$$

$$\tag{5.8}$$

where the S-sum is only over $\beta - \gamma$ pairs that lie in the explicit subset and $\mathbf{q}^\alpha(\mathbf{x}, t)$ then represents all the terms in the sum in (5.8) such that β and/or γ are outside the explicit set. So far the exact dynamics are undisturbed. The models are now constructed by considering $\mathbf{q}^\alpha(\mathbf{x}, t)$ to be a stochastic field constrained only by statistical relations that express the underlying statistical symmetry among the collective fields. In this way, moments, or other statistics, of $\mathbf{q}^\alpha(\mathbf{x}, t)$ are expressed in terms of statistics of the explicit fields alone. The statistical determination replaces determination of $\mathbf{q}^\alpha(\mathbf{x}, t)$ from the actual equations of motion for the implicit collective fields.

The model construction is best illustrated by example. Let the explicit set consist of the single member $\alpha = 0$. Then (5.8) becomes simply

$$\left(\frac{\partial}{\partial t} - \nu\nabla^2 \right) \mathbf{u}^0(\mathbf{x}, t) =$$

$$-\lambda M^{-1/2} \sum_{\beta} \mathbf{u}^\beta(\mathbf{x}, t) \cdot \nabla \mathbf{u}^{-\beta}(\mathbf{x}, t) - \nabla p^\alpha(\mathbf{x}, t) + \mathbf{q}^0(\mathbf{x}, t). \tag{5.9}$$

[The collective field $\mathbf{u}^0(\mathbf{x}, t)$ in (5.9) must not be confused with the linearized field invoked in Sec. 4.] The statistical symmetry among the collective fields implies

$$\langle \mathbf{q}^0(\mathbf{x},t)\mathbf{u}^0(\mathbf{x}',t')\rangle = (M-1)\langle \mathbf{u}^0(\mathbf{x}',t')\mathbf{u}^0(\mathbf{x},t)\cdot\nabla\mathbf{u}^0(\mathbf{x},t)\rangle. \qquad (5.10)$$

This is because the average on the left of (5.10) consists of $M-1$ terms each equal to the average on the right side, according to the statistical symmetry.

The constraint (5.10) is of fundamental importance because for $t = t'$ it constrains $\mathbf{q}^0$ so that the total nonlinear interaction [$\mathbf{q}(\mathbf{x},t)$ plus the explicit sum in (5.9)] conserves mean energy. Clearly this single moment constraint is insufficient to fully determine the statistics of $\mathbf{q}^0$. But the determination can be made unambiguous by requiring that $\mathbf{q}^0(\mathbf{x},t)$ have the minimum variance consistent with (5.9) and (5.10). In the limit $M \to \infty$, it can be shown that this choice yields precisely the DIA, in the form of the Langevin equation (2.15). The amplitude $u_i(\mathbf{k};t)$ in (2.15) now represents the Fourier transform of $\mathbf{u}^0(\mathbf{x},t)$ while the transform of $\mathbf{q}^0(\mathbf{x},t)$ (with pressure contribution incorporated) is

$$q_i^0(\mathbf{k};t) = b_i(\mathbf{k};t) - \int_0^t \eta(k;t,s)u_i(\mathbf{k};s)ds. \qquad (5.11)$$

Thus (2.15) can be interpreted as the result of realizing $\mathbf{q}^0(\mathbf{x},t)$ in the limit $M \to \infty$ with the least variance permitted by the symmetry constraint (5.10).

The advantage of constructing DIA *via* the DEC model is that systematically improved models can be constructed by adjoining higher statistical symmetry constraints to (5.10). These constraints are all obeyed exactly in the actual NS dynamics. If all possible symmetry constraints are imposed on $q^0(\mathbf{x},t)$, the resulting model reproduces all NS statistics. A model logically one step above DIA is obtained by adjoining to (5.10) the constraints that express $\langle \mathbf{q}^0(\mathbf{x},t)\mathbf{u}(\mathbf{x}',t')\mathbf{u}(\mathbf{x}'',t'')\rangle$ and $\langle \mathbf{q}^0(\mathbf{x},t)\mathbf{q}^0(\mathbf{x}',t')\rangle$ in terms of moments of $\mathbf{u}^0(\cdot,\cdot)$ alone. It has been shown that these constraints enforce invariance of energy transfer under RGT.[41]

Since the DEC model obtained by imposing only (5.10) yields the DIA, it corresponds to truncating the line-renormalized expansion for triple moments at the lowest term. The higher DEC models do not appear to have any simple correspondence to line- or vertex-renormalized perturbation expansions. Since the imposed constraints are always ones obeyed in the exact NS dynamics, successively higher DEC models are expected to yield statistics that converge to exact NS statistics. This is in contrast to the divergence of RPT.

6. Stochastic Modeling for Probability Distributions

The stochastic models so far described in this paper all are models of dynamics: the actual dynamics is replaced by altered equations of motion from which exact values of certain statistics can be obtained. Models of this kind, in particular those related to DIA, have been successful in predicting second- and third-order moments of the velocity field; they have had limited success in predicting departures of fourth-order moments from Gaussian values; and they have totally failed to capture the intermittency of small scales. There is little reason to believe that such models will be of value in approximating the shapes of probability distribution functions[62] (PDF's) of the amplitudes of velocity and its spatial derivatives.

A basic shortcoming of the RC model in handling PDF's can be seen immediately from the equation of motion (2.5). In the limit $M \to \infty$, necessary to get analytical results for the model, each field $\mathbf{u}^n(\mathbf{x}, t)$ is coupled to an infinite number of other fields. Arguments based on the central limit theorem then suggest that the statistics of each individual $\mathbf{u}^n(\mathbf{x}, t)$ are Gaussian in the limit. This can be verified. The joint PDF of all the $\mathbf{u}^n(\mathbf{x}, t)$ of course is not Gaussian. Thus, the model values for triple moments, like $S(k; t, t')$ in (2.10), are nonzero, but they are obtained by summing over all the system labels, and they contain the random ϕ coefficients. There does not appear to be any simple way to construct, from the model, a non-empty approximation to PDF's associated with a single field — for example, the probability that the velocity field amplitude at a single space-time point has a given value. Related difficulties arise when low-order decimation models are asked to provide approximations to PDF's.

There is an alternative approach: model the stochastic velocity field itself in x space and apply the exact dynamical equations to the model field. Current work suggests that valid and workable approximations to PDF's can be constructed in this way and that this approach offers powerful, nonperturbative tools for handling strong turbulence.[6,8] The stochastic models to be discussed here are constructed by nonlinear mappings that carry a multivariate Gaussian "reference field" into a field that can be wildly non-Gaussian. Closure approximations are obtained by exploiting the known statistics of the Gaussian reference field. The moment closures discussed already in this paper are based on the assumption of polynomial nonlinearity, in particular quadratic nonlinearity. But the mapping closures to be discussed now are substantially indifferent to the form of the nonlinearity; the latter need not be algebraic. For this reason, the mapping closures are well suited for application to compressible flow with real equations of state and to other problems that are daunting to moment closures.

The modeling by mapping will be illustrated here by application to problems simpler than NS dynamics: relaxation of a scalar field under the heat equation, evolution of a reactive and diffusive scalar field,[6] and the initial development of intermittency in Burgers turbulence.[8] The examples discussed here invoke statistical homogeneity, but the mapping method is also applicable to inhomogeneous flows in any geometry.

Consider first a scalar field, homogeneously and isotropically distributed in D dimensions, that obeys

$$\partial\psi(\mathbf{x},t)/\partial t = Q(\psi(\mathbf{x},t)) + \kappa\nabla^2\psi(\mathbf{x},t), \qquad (6.1)$$

where κ is the molecular diffusivity and $Q(\psi)$ is some local nonlinear function of ψ that represents a one-species chemical reaction. Eq. (6.1) presents two kinds of closure problem. First, the term in $Q(\psi)$ couples moments of different orders. If Q is a polynomial in ψ (which it need not be), this closure problem is of the kind that can be addressed by stochastic models like those already discussed in this paper. Second, the linear terms in κ couple ψ to its second-order space derivatives. The latter problem exists already for the heat equation ($Q = 0$), which is linear. The coupling of space derivatives of different order is not a problem if one's objective is evolution equations for moments of given order. However, it is a crucial problem if closure approximations are sought for quantities like $\mathcal{P}(\psi,t)$, defined as the probability distribution of the scalar field amplitude at a single point.

The second closure problem can be addressed by constructing stochastic models that express time-dependent, nonlinear mappings of a multivariate-Gaussian reference field $\psi_0(\mathbf{x})$. The basic form is[6]

$$\psi = X(\psi_0,t) \quad [\partial X/\partial\psi_0 > 0]. \qquad (6.2)$$

An end objective is to produce valid approximations for $\mathcal{P}(\psi,t)$. The ability to form closures from the mapping (6.2) rests on the fact that all statistics of the Gaussian reference field $\psi_0(\mathbf{x})$ have simple explicit forms. As a consequence, the mappings generate accessible statistical representations of $\nabla^2\psi$ as a function of ψ

Eq. (6.2) is a single-valued, single-point transformation; if $\psi_0(\mathbf{x})$ has the value s at a particular $\mathbf{x}$, then $\psi(\mathbf{x},t)$ has the value $X(s,t)$. This mapping can represent an essentially arbitrary univariate PDF $\mathcal{P}(\psi,t)$. $\mathcal{P}(\psi,t)$ is related to

$$\mathcal{P}_0(\psi_0) = (2\pi)^{-1/2}\langle\psi_0^2\rangle^{-1/2}\exp(-\tfrac{1}{2}\psi_0^2/\langle\psi_0^2\rangle), \qquad (6.3)$$

the PDF of the reference field ψ_0, by

$$\mathcal{P}(\psi,t) = \mathcal{P}_0(\psi_0)[\partial X(\psi_0,t)/\partial\psi_0]^{-1}. \qquad (6.4)$$

Here are two examples. If $X(\psi_0,t) = \psi_0^3$, then

$$\mathcal{P}(\psi,t) = \tfrac{1}{3}(2\pi)^{-1/2}|\psi|^{-2/3}\langle\psi_0^2\rangle^{-1/2}\exp(-\tfrac{1}{2}|\psi|^{2/3}/\langle\psi_0^2\rangle). \tag{6.5}$$

This is an intermittent PDF with an infinite cusp at $\psi = 0$. The second example is

$$X(\psi_0,t) = \psi_0/(1 + \psi_0^2/a^2)^{1/2}, \tag{6.6}$$

where a is a (possibly time dependent) positive parameter. Eq. (6.6) confines ψ to the interval $(-a, a)$ and $\mathcal{P}(\psi)$ has cusps at $\pm a$. In the limit $a/\langle\psi_0^2\rangle^{1/2} \to 0$, the cusps become δ-functions and (6.6) is equivalent to simple clipping of the process $\psi_0(\mathbf{x})$ at the level $|\psi_0| = a$. If a term a is added to the right side of (6.6), the support of ψ is shifted to the interval $(0, 2a)$.

The evolution of $\mathcal{P}(\psi, t)$ is determined by (6.1). Q, on the right side, is a statistically sharp function of ψ, but $\nabla^2\psi$ is not. However, it is easy to see that the only part of $\nabla^2\psi$ that contributes to the evolution of $\mathcal{P}(\psi, t)$ is $[\nabla^2\psi]_{C:\psi}$, defined at any point $\mathbf{x}$ as the ensemble mean of $\nabla^2\psi(\mathbf{x}, t)$ over the full multivariate distribution of the field $\psi(\cdot, t)$, conditional on a given value of $\psi(\mathbf{x}, t)$. Thus

$$V(\psi) = Q(\psi) + \kappa[\nabla^2\psi]_{C:\psi} \tag{6.7}$$

is the effective mean flow velocity in ψ space that evolves $\mathcal{P}(\psi, t)$:

$$\frac{\partial\mathcal{P}(\psi)}{\partial t} + \frac{\partial}{\partial\psi}[\mathcal{P}(\psi)V(\psi,t)] = 0, \tag{6.8}$$

The unfolding of the mapping (6.2) will reproduce the mean flow $V(\psi, t)$ as a function of t if

$$\frac{\partial X}{\partial t} = Q(\psi) + \kappa[\nabla^2\psi]_{C:\psi}, \tag{6.9}$$

$[\nabla^2\psi]_{C:\psi}$ is easily evaluated from (6.2), thereby closing the equations above: Chain differentiation of (6.2) yields

$$\nabla\psi = \nabla\psi_0\partial X/\partial\psi_0, \tag{6.10}$$

$$\nabla^2\psi = \nabla^2\psi_0\partial X/\partial\psi_0 + |\nabla\psi_0|^2\partial^2 X/\partial\psi_0^2. \tag{6.11}$$

Now, since $\psi_0(\mathbf{x}, t)$ is multivariate-Gaussian, the joint PDF of ψ_0, $\nabla\psi_0$ and $\nabla^2\psi_0$ at a point $\mathbf{x}$ is fully determined by the covariances of these quantities. It follows from homogeneity that $\langle\psi_0\nabla\psi_0\rangle = \langle\nabla\psi_0\nabla^2\psi_0\rangle = 0$, so that ψ_0 and $\nabla\psi_0$ are statistically independent, and that

$$[\nabla^2\psi_0]_{C:\psi 0} = -\psi_0\langle|\nabla\psi_0|^2\rangle/\langle\psi_0^2\rangle. \tag{6.12}$$

Since (6.2) is a single-valued transformation, conditional mean with respect to ψ_0 is equivalent to conditional mean with respect to ψ. Then (6.10)–(6.12) yield

$$\langle|\nabla\psi|^2\rangle = \langle|\nabla\psi_0|^2\rangle\langle(\partial X/\partial\psi_0)^2\rangle, \qquad (6.13)$$

$$[\nabla^2\psi]_{C:\psi} = \langle|\nabla\psi_0|^2\rangle\left(-\frac{\psi_0}{\langle\psi_0^2\rangle}\frac{\partial X}{\partial\psi_0} + \frac{\partial^2 X}{\partial\psi_0^2}\right), \qquad (6.14)$$

Eqs. (6.9) and (6.14) give the final closure equation

$$\frac{\partial X}{\partial t} = Q(X) + \kappa\langle|\nabla\psi_0|^2\rangle\left(-\frac{\psi_0}{\langle\psi_0^2\rangle}\frac{\partial X}{\partial\psi_0} + \frac{\partial^2 X}{\partial\psi_0^2}\right). \qquad (6.15)$$

The transformation X defines a trajectory in ψ space. It must be emphasized that this trajectory does *not* represent the ψ history of a typical fluid particle that starts with $\psi = \psi_0$. Instead, X is an effective mean trajectory that yields self-consistent evolution of $\mathcal{P}(\psi, t)$. It is not actually necessary to solve (6.8) since (6.2) implies

$$\mathcal{P}(\psi) = \mathcal{P}_0(\psi_0)[\partial X/\partial\psi_0]^{-1}. \qquad (6.16)$$

Eq. (6.2) replaces whatever may be the true multivariate PDF at each t with a tractable multivariate PDF obtained by local distortion of the field $\psi_0(\mathbf{x})$. In each interval dt, the evolution of current distorted field is treated exactly. Therefore the closure is expected to exhibit the empirically observed tendency of diffusion to relax non-Gaussian $\mathcal{P}(\psi)$ toward Gaussian form.[62] This physics shows clearly in (6.15). The two κ terms describe, respectively, an outgoing wave in ψ space and diffusive smoothing. The wave motion expresses the general damping of excitation while the diffusive term gives relaxation toward Gaussian statistics; $\partial^2 X/\partial\psi_0^2$ vanishes if ψ is Gaussian.

The relaxation toward Gaussian statistics has previously proved hard to capture in closures. The underlying reason is that molecular diffusion makes $\mathcal{P}(\psi, t)$ shrink toward the origin, which is essentially a *negative* diffusion in ψ space. Negative diffusion is a highly unstable phenomenon, and this is reflected by difficulty in constructing realistic closures.[62]

The relaxation property of (6.15) is clearly exhibited by an analytical solution found by Pope[63] for the case $Q = 0$:

$$X(\psi_0, t) = C_0^{1/2}\, \text{erf}\,[a(t)\psi_0/C_0^{1/2}], \qquad (6.17)$$

$$[a(t)]^2 = \tfrac{1}{2}[\exp(2\kappa t C_1/C_0) - 1]^{-1}. \qquad (6.18)$$

where $C_0 = \langle\psi_0^2\rangle$ and $C_1 = \langle|\nabla\psi_0|^2\rangle$. At $t = 0$, the PDF $\mathcal{P}(\psi, t)$ of $\psi = X(\psi_0, t)$ consists of δ–function peaks at $\psi = \pm C_0^{1/2}$. As t increases, $\mathcal{P}(\psi, t)$ relaxes to eventual Gaussian form through a series of intermediate shapes that closely reproduce those reported in simulations by Eswaran & Pope.[10] During the relaxation, $\mathcal{P}(\psi, t)$ vanishes for $|\psi|$ greater than the initial extremal value $C_0^{1/2}$. This reproduces an exact property of the heat equation.

Despite its good physics, (6.15) is a very limited approximation. It takes no account of advective stretching and, more generally, ignores changes in $\langle|\nabla\psi|^2\rangle$ not associated with local rescaling of ψ via (6.2). This deficiency is addressed by the higher mapping closure introduced in the discussion of Burgers' equation that follows. Alternatively, external information about second-order moments can be used, if available, to improve the closure: if $\langle|\nabla\psi|^2\rangle/\langle\psi^2\rangle$, a measure of effective dissipation-scale size, is known, then $\langle|\nabla\psi_0|^2\rangle$ can be reset after each step dt so that (6.2) and (6.13) give this $\langle|\nabla\psi|^2\rangle/\langle\psi^2\rangle$. Eq. (6.1) and homogeneity imply $d\langle\psi^2\rangle/dt = 2(\langle\psi Q(\psi)\rangle - \kappa\langle|\nabla\psi|^2\rangle)$. Thus the resetting gives exact evolution of $\langle\psi^2\rangle$ if $Q = 0$, in which case it simply rescales t in (6.15)

Fig. 8 compares the evolution of $\mathcal{P}(\psi, t)$ obtained by direct computer simulation of (6.1) with an integration of (6.15) for $D = 1$, $Q(\psi) = -\alpha\psi|\psi|$, and Gaussian $\mathcal{P}(\psi, t = 0)$ $(X(\psi_0, 0) = \psi_0)$.[6] The simulations were started from realizations of ψ at 10^5 points unit-spaced over a cyclic line segment, with multivariate Gaussian statistics and a Gaussian-shaped wavenumber spectrum. In the integration of (6.15), $\langle|\nabla\psi_0|^2\rangle$ was reset after each time step so that the closure gave the same value of $\langle|\nabla\psi|^2\rangle/\langle\psi^2\rangle$ as the simulation. Eq. (6.15) would have overestimated the rate of diffusive effects if the resetting had not been performed. Also shown are the exact evolution for $\kappa = 0$ and the prediction of the widely-used Gaussian closure [cf. (6.12)]

$$[\nabla^2\psi]_{C:\psi} = -\psi\langle|\nabla\psi|^2\rangle/\langle\psi^2\rangle. \tag{6.19}$$

For that closure also, $\langle|\nabla\psi|^2\rangle/\langle\psi^2\rangle$ was given the simulation value after each time step.

The results illustrate the crucial role played by the relaxation property of (6.15). Under the Gaussian closure, diffusion preserves the shape of $\mathcal{P}(\psi, t)$. The difference between its prediction and the curve for $\kappa = 0$ is a measure of quenching of the reaction by diffusive reduction of amplitudes. Clearly this represents only a small part of the total effect of the κ term in (6.1)

The construction of higher mapping closures can be illustrated by an application to Burger's equation[4,8]

$$\mathcal{D}\psi/\mathcal{D}t = \eta\psi_{xx}, \tag{6.20}$$

where, for consistency with preceding notation, $\psi(x, t)$ is the the one-dimensional velocity field, $\mathcal{D}/\mathcal{D}t \equiv \partial/\partial t + \psi\partial/\partial x$ and η is kinematic viscosity. Differentiation of (6.20) with respect to x yields

$$\mathcal{D}\xi/\mathcal{D}t = -\xi^2 + \eta\xi_{xx}, \tag{6.21}$$

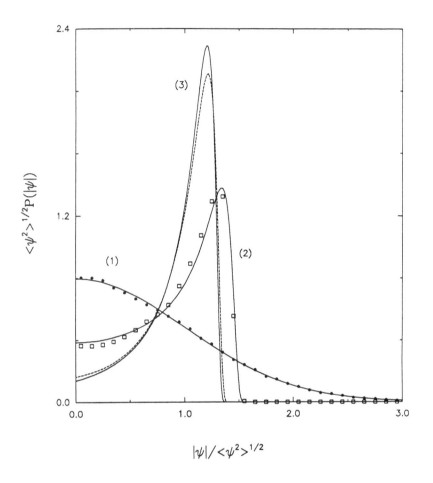

Fig. 8. Diffusive reaction with $Q = \psi|\psi|$, $\kappa = 2.5$ and initial values $\langle\psi^2\rangle = 1$, $\langle\xi^2\rangle = 0.02$. At $t = 4$ (shown), the evolved values were $\langle\xi^2\rangle/\langle\psi^2\rangle = 0.01349$, $\langle\psi^2\rangle = 0.02339$ (simulation), $\langle\psi^2\rangle = 0.02332$ (by (6.15)), while $\langle\xi_0^2\rangle$ was reset from an initial value 0.02 to 0.0101. Curve 1, initial Gaussian PDF; curve 2, evolved PDF according to (6.15); curve 3, exact evolved PDF if $\kappa = 0$; dashed curve, evolved PDF according to Gaussian closure (6.19). The solid (open) data points are the initial (evolved) simulation values. From Chen, Chen & Kraichnan.[6]

where $\xi \equiv \psi_x$. Burger's equation describes an infinitely compressible flow, in which the density $\rho(x,t)$ obeys

$$\mathcal{D}\rho/\mathcal{D}t = -\xi\rho, \tag{6.22}$$

The initial condition $\rho(x,0) = 1$ will be assumed here.

The dominant behavior under Burgers' equation is the steepening of velocity gradients into viscosity-limited shocks.[4] During this process an initially multivariate Gaussian field ψ retains a nearly Gaussian univariate distribution of ψ while the distribution of the gradient ξ becomes highly intermittent. Such behavior can be tracked *via* the single-point joint PDF $\mathcal{P}(\psi, \xi, t)$. Mapping closures are most naturally applied to this problem if $\mathcal{P}(\psi, \xi, t)$ is constructed by sampling *along particle trajectories*. To do this, $\mathcal{D}/\mathcal{D}t$ in (6.19)–(6.21) can be read as $\partial/\partial t$. But, since the flow is compressible, an initially uniformly spaced set of trajectories will tend to migrate into the shocks, so that the Lagrangian PDF $\mathcal{P}_L(\psi, \xi, t)$ so computed will differ from the Eulerian PDF $\mathcal{P}_E(\psi, \xi, t)$ measured in a fixed coordinate system. The relation between the two PDF's is

$$\mathcal{P}_E(\psi, \xi, t) = N_{EL}\mathcal{P}_L(\psi, \xi, t)/\rho(\psi, \xi, t), \qquad (6.23)$$

where N_{EL} normalizes $\mathcal{P}_E(\psi, \xi, t)$ to unity and $[\rho(\psi, \xi, t)]_{C:\psi\xi}$ is the mean of $\rho(\psi, \xi, t)$ conditional on given ψ and ξ.

The compressive character of Burgers' equation makes it natural to adjoin to the point transformation X a further transformation that distorts the space in which the reference field ψ_0 lives. To do this, let the reference Gaussian field now be taken as $\psi_0(z)$ where z is a reference coordinate, related to the laboratory coordinate x by

$$\partial z/\partial x = J(\psi_0, \xi_0, t), \qquad (6.24)$$

with $\xi_0 \equiv \psi_{0z}$. J is a squeezing or stretching factor that is to be determined by closure. Then chain differentiation yields

$$\xi = \xi_0 J(\psi_0, \xi_0)\partial X/\partial\psi_0 \equiv Y(\psi_0, \xi_0). \qquad (6.25)$$

and the joint PDF of ψ and ξ is

$$\mathcal{P}_E(\psi, \xi) = N\mathcal{P}_0(\psi_0)Q_0(\xi_0)\left[\frac{\partial X}{\partial\psi_0}\frac{\partial Y}{\partial\xi_0}\right]^{-1}\frac{1}{J(\psi_0, \xi_0)}, \qquad (6.26)$$

where $Q_0(\xi_0)$ is the Gaussian PDF of ξ_0, and $N(t)$ normalizes $\mathcal{P}(\psi, \xi)$ to unity. Recall that amplitude and gradient of a homogeneous, Gaussian field are statistically independent. The square bracket in (6.26) is the Jacobian of the transformation between ψ_0, ξ_0 and ψ, ξ while the $1/J$ factor represents the weighting of probability associated with squeezing and stretching. If J is independent of ψ_0 then the marginal PDF $\mathcal{P}_E(\psi)$ is independent of J. If, in addition, $X = \psi_0$, then $\mathcal{P}_E(\psi) = \mathcal{P}_0(\psi_0)$.

In some applications the closure (6.15) can be improved, at low cost, by including a degenerate form of the J transformation in which J depends solely on time. The analysis to follow is for the case $Q \equiv 0$. It is easily extended to general Q. If $J = J(t)$, then $N(t) = J(t)$, a factor J appears in the right side of (6.10) and a factor J^2 appears in the right sides of (6.11) and (6.12). Eq. (6.15) becomes

$$\frac{\partial X}{\partial t} = \kappa J^2 \langle |\nabla \psi_0|^2 \rangle \left(-\frac{\psi_0}{\langle \psi_0^2 \rangle} \frac{\partial X}{\partial \psi_0} + \frac{\partial^2 X}{\partial \psi_0^2} \right) \tag{6.27}$$

$J(t)$ is determined by the homogeneous balance equation

$$d\langle \xi^2 \rangle / dt = -2\kappa \langle (\nabla^2 \psi)^2 \rangle, \tag{6.28}$$

which follows from (6.1), upon differentiation and partial integration. If (6.10) and (6.11), with the J factors included, are substituted into (6.28), the resulting equation for J^2 is

$$\frac{dJ^2}{dt} + J^2 \frac{d\ln I}{dt} = -2\kappa J^4 \left[\frac{C_2}{C_1} + \frac{K_1}{C_1 I} \left\langle \left(\frac{\partial^2 X}{\partial \psi_0^2} \right)^2 \right\rangle \right]. \tag{6.29}$$

where

$$I = \left\langle \left(\frac{\partial X}{\partial \psi_0} \right)^2 \right\rangle, \tag{6.30}$$

$$C_0 = \langle \psi_0^2 \rangle, \quad C_1 = \langle |\nabla \psi_0|^2 \rangle,$$

$$C_2 = \langle (\nabla^2 \psi_0)^2 \rangle, \quad K_1 = \langle |\nabla \psi_0|^4 \rangle, \tag{6.31}$$

and $\langle \rangle$ in (6.29) and (6.30) denotes average over $\mathcal{P}_0(\psi_0)$.

An initially multivariate Gaussian distribution remains multivariate Gaussian under the heat equation ($Q = 0$). This remains true under either of the closures (6.15) or (6.27)–(6.31). The appropriate initial transformation is simply $X(\psi_0, t = 0) = \psi_0$. Closure (6.27)–(6.31) has the additional property that the evolution of $\mathcal{P}(\psi, t)$ (decay of the Gaussian amplitudes) then is exactly correct, provided that the wavenumber spectrum of ψ has the self-preserving form

$$\Psi(k, t) \propto k^n \exp[-\tfrac{1}{2} b(t) k^2] \quad (n > -1). \tag{6.32}$$

In this case, (6.27)–(6.31) reduce to

$$X(\psi_0, t) = r(t)\psi_0, \quad dr/dt = -\kappa J^2 r C_1 / C_0, \tag{6.33}$$

$$\frac{dJ^2}{dt} = -2\kappa J^4 \left(\frac{C_2}{C_1} - \frac{C_1}{C_0} \right), \tag{6.34}$$

In the application to Burgers' equation, J must, at the least, depend on the rate of strain ξ as well as t. The analysis is greatly simplified by the approximation that $P_E(\psi)$ is Gaussian and that ψ and ξ are statistically independent. The first part of the assumption is excellently supported by direct computer simulation of Burgers' equation. The second part is more suspect and is not wholly self-consistent within the closure method. The error in these approximations can be estimated *a posteriori* within the closure scheme, but this analysis will not be discussed here. Thus it will be assumed that

$$P_E(\psi, \xi) = P_E(\psi) Q_E(\xi), \tag{6.35}$$

$$\psi = X(\psi_0, t) = r(t)\psi_0, \quad \xi = Y(\xi_0, t) = r(t)\xi_0 J(\xi_0). \tag{6.36}$$

Here $r(t)$ measures the decay of velocity amplitudes under viscosity and the second part of (6.36) follows from (6.25). The energy-balance equation

$$d\langle \psi^2 \rangle / dt = -2\eta \langle \xi^2 \rangle \tag{6.37}$$

yields

$$dr/dt = -r\eta \langle \xi^2 \rangle / \langle \psi_0^2 \rangle. \tag{6.38}$$

Now only $[\xi_{xx}]_{C:\xi}$, the conditional mean at given ξ is needed. To obtain closure, the requirement is made that, over each dt, the stretching factor J change so that the change in $P_E(\psi, \xi)$ given by (6.26) and (6.35)–(6.38) matches the change under the equations of motion. This may be assured as follows: The reduced Liouville equation for the marginal PDF $Q_E(\xi)$ implied by (6.21) is

$$\frac{\partial Q_E(\xi)}{\partial t} + \frac{\partial}{\partial \xi} \left(\left[\frac{D\xi}{Dt} \right]_{C:\xi} Q_E(\xi) \right) = \xi Q_E(\xi), \tag{6.39}$$

where the divergence term on the right side expresses the effects of transformation from Lagrangian to Eulerian coordinates.

On the other hand, (6.26), (6.35) and (6.36) give

$$Q_E(\xi) = Q_0(\xi_0) \left(\frac{\partial Y}{\partial \xi_0} \right)^{-1} \frac{N}{J}. \tag{6.40}$$

This implies the reduced Liouville equation

$$\frac{\partial Q_E(\xi)}{\partial t} + \frac{\partial}{\partial \xi} \left(\frac{\partial Y}{\partial t} Q_E(\xi) \right) = \alpha(\xi) Q_E(\xi), \tag{6.41}$$

where

$$\alpha(\xi) = \frac{\partial}{\partial t} \ln \left(\frac{N}{J} \right). \tag{6.42}$$

The right side of (6.41) expresses the effects of squeezing and stretching by the J transformation. Subtraction of (6.39) from (6.41) and integration over ξ now yields

$$\frac{\partial Y}{\partial t} = -\xi^2 + \eta [\xi_{xx}]_{C:\xi} + \frac{1}{Q_E(\xi)} \int_{-\infty}^{\xi} [\alpha(\xi') - \xi'] Q_E(\xi') d\xi', \tag{6.43}$$

where the constant of integration is chosen to give finite results at $\pm\infty$. In this connection, note that homogeneity and the definition of N imply

$$\int_{-\infty}^{\infty} \xi Q_E(\xi)d\xi = \int_{-\infty}^{\infty} \alpha(\xi)Q_E(\xi)d\xi = 0. \qquad (6.44)$$

Eqs. (6.36)–(6.38) and (6.43), lead to the following somewhat complicated equation of motion for J:

$$\frac{\partial J}{\partial t} = -r\xi_0 J^2 - \frac{1}{r\xi_0 Q_E(\xi)} \int_{-\infty}^{\xi} [\alpha(\xi') - \xi']Q_E(\xi')d\xi' \qquad (6.45)$$

$$+\eta(r\xi_0)^{-1}[\xi_{xx}]_{C:\xi} + \eta J\langle(\xi_0 J)^2\rangle/\langle\psi_0^2\rangle$$

and evaluation of $[\xi_{xx}]_{C:\xi}$ by the techniques that gave (6.15) or (6.27) gives

$$[\xi_{xx}]_{C:\xi} = -r\xi_0 k_d^2 \left(J^3 + \frac{\xi_0}{3}\frac{\partial J^3}{\partial \xi_0}\right) + rC_2 \left(\frac{\partial J^3}{\partial \xi_0} + \frac{\xi_0}{2}J\frac{\partial^2 J^2}{\partial \xi_0^2}\right).$$

$$(6.46)$$

with

$$C_2 = \langle(\partial\xi_0/\partial z)^2\rangle, \quad k_d^2 = C_2/\langle\xi_0^2\rangle.$$

The J^2 term in (6.45) comes directly from the $-\xi^2$ term in (6.21). The derivative terms on the right side of (6.46) come from consistent treatment of ξ_{xx} under the space distortion represented by J. The integral term in (6.45) arises from the effect on $P_E(\psi,\xi)$ of the N/J factor in (6.26). This term makes (6.45) an integro-differential equation that must be solved iteratively. The derivative and integral terms play an essential role in shaping the PDF near its maximum.

Figs. 9–14 compare the marginal PDF $Q_E(\xi)$ obtained from solution of the closure (6.26), (6.35), (6.36), (6.38), (6.42), (6.45) and (6.46) with the results of a series of simulations of Burgers' equation.[8] The simulations started from a multivariate-Gaussian distribution of $\psi(x,0)$ and a spectrum of the form (6.32). Each simulation involved 10^5 points unit-spaced on a cyclic line segment. Sets of simulation runs were carried out for two parameter assignments: (Case A) $n = 0$, $\eta = 5$, $C_0 = 1$, $C_1 = 0.005$, $C_2 = 7.5 \times 10^{-5}$; (Case B) $n = 2$, $\eta = 2.5$, $C_0 = 1$, $C_1 = 0.015$, $C_2 = 3.75 \times 10^{-4}$. The same parameter values were used in integrations of the closure equations.

Figs. 9 and 10 show the evolution of velocity-gradient skewness $S = \langle\xi^3\rangle/\langle\xi^2\rangle^{3/2}$. S is a measure of the production of $\langle\xi^2\rangle$ by nonlinear effects. The closure results capture two of the qualitative differences between Cases A and B: the initial rapid rise in S is concave downward in Case A and concave upward in Case B; and, after the initial rapid rise, S continues to increase in Case A but decreases in Case B. However, the closure misses

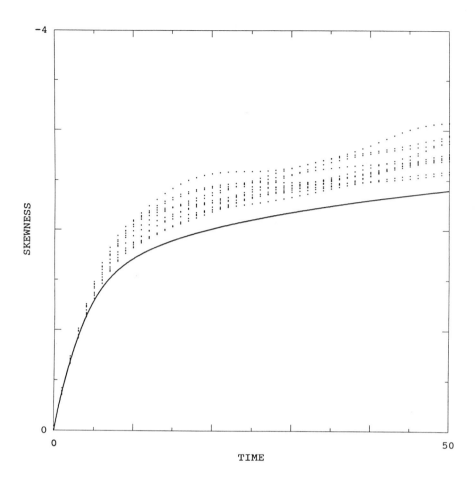

Fig. 9. Velocity-gradient skewness S *vs* time for Case A, as given by 12 simulation runs (data points) and by closure (solid line).

the sharp overshoot exhibited by the Case B simulations. Figs. 11 and 12 show simulation and closure results for $\langle \xi^2 \rangle$ and $\langle \xi_x^2 \rangle / \langle \xi^2 \rangle$ in Case A. The simulation curves are the means of all runs. Again, the closure captures qualitative behavior.

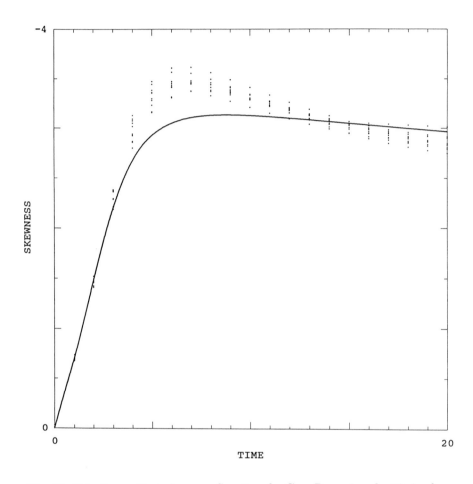

Fig. 10. Velocity-gradient skewness S *vs* time for Case B, as given by 10 simulation runs (data points) and by closure (solid line).

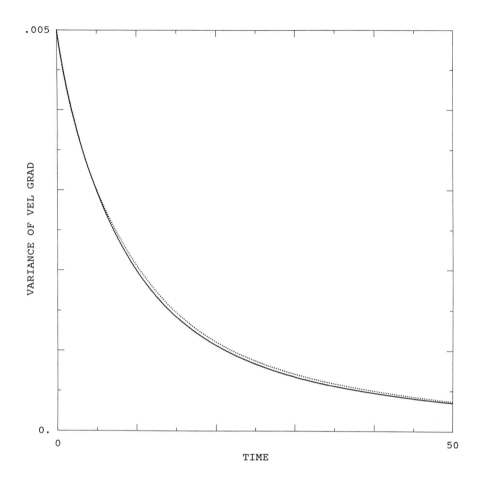

Fig. 11. Velocity-gradient variance *vs* time for Case A, as given by mean of 12 simulation runs (dotted line) and by closure (solid line).

Figs. 13 and 14 show the full PDF's of ξ for Cases A and B at several times of evolution. There is increasing departure from Gaussian shape with time in both Case A and Case B. Case B shows stronger departure at given values of S and, in particular, displays a marked upward concavity of the skirt. There are no adjustable parameters or functions in the closure.

Both the closure and simulation results clearly show an almost exponential tail to $Q(\xi)$ for large negative ξ. This phenomenon is easily deduced from the closure equations. At large negative ξ_0, the dominant terms in (6.45) are the first term on the right side, which is proportional to J^2, and

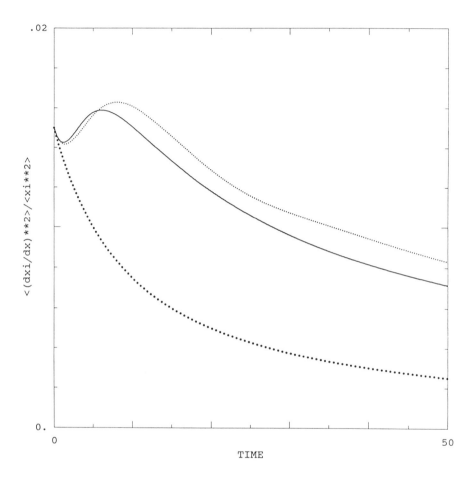

Fig. 12. The quantity $\langle \xi_x^2 \rangle / \langle \xi^2 \rangle$ (a measure of dissipation of velocity-gradient variance) *vs* time for Case A, as given by mean of 12 simulation runs (light dotted line) and by closure (solid line). The heavy dotted line shows what the evolution would be if only viscosity acted.

the first parenthesis in expression (6.46) for $[\xi_{xx}]_{C:\xi}$, which is proportional to J^3 [by (6.47) below, J becomes $\propto |\xi_0|$ so the two terms in the parenthesis give equal contributions]. A quasi-equilibrium is quickly reached at large negative ξ_0 in which these two terms very nearly cancel each other:

$$ r|\xi_0|J^2 \approx 2\eta k_d^2 J^3, \quad J \approx \tfrac{1}{2} r|\xi_0|/\eta k_d^2 \quad (\xi_0 < 0, \xi_0^2 \gg \langle \xi_0^2 \rangle). \qquad (6.47) $$

This implies $|\xi| \approx \tfrac{1}{2}(r\xi_0)^2/(\eta k_d^2)$ and, by (6.25) and (6.26),

$$Q(\xi) \approx \frac{N}{2\pi\langle\xi_0^2\rangle^{1/2}} \left(\frac{\eta k_d^2}{r|\xi|}\right) \exp\left[-\frac{\eta k_d^2|\xi|}{r^2\langle\xi_0^2\rangle}\right], \quad (\xi < 0, \xi^2 \gg \langle\xi_0^2\rangle). \quad (6.48)$$

Under Burgers' equation, the fluid density becomes highly concentrated in the shock regions. The zero-velocity points in the shocks are accumulation points for fluid particles. Unlike the velocity field, the fluid density is not diffused by viscosity. Consequently the density PDF becomes extremely intermittent.

Eq. (6.22) gives the evolution of fluid density along the particle paths. The latter, considered as trajectories in ξ, are well approximated at early times of evolution by the trajectories induced under the J mapping. Later, when viscosity is more important, the two kinds of trajectory can be significantly different. This makes some complication in the analytical treatment of the density PDF. Manipulations like those that lead to (6.43) yield the following equation for the effective evolution of ρ along a trajectory with fixed ξ_0:

$$\frac{\partial\gamma}{\partial t} = -\xi + \frac{1}{P_E(\gamma)} \int_{-\infty}^{\xi} [\alpha(\xi') - \xi']Q_E(\xi')d\xi'. \quad (6.49)$$

Here $\gamma = \ln\rho$ and $P_E(\gamma)$ is the Eulerian PDF of γ. The integral term in (6.49), like those in (6.43) and (6.45), corrects the Lagrangian equation of motion for differences between particle trajectory and mapping trajectory. The integral is the same quantity in all three cases.

Fig. 15 shows a log-log plot of $P_E(\rho) = P_E(\gamma)/\rho$ for simulation and closure for Case A at $t = 10$. The most striking feature is the power-law tail at large ρ, which indicates that moments of $P_E(\rho)$ do not exist above some critical order. This behavior can be understood in a simple way. The qualitative behavior of γ at large negative ξ is controlled by the $-\xi$ term in (6.49). Since ξ rapidly reaches quasi-equilibrium at large negative ξ, the qualitative behavior of γ in the tail is then $\gamma \approx |\xi|t$. Thus the PDF of γ has an exponential-like tail resembling that of ξ, and the power-law tail for ρ follows immediately.

The preceding closure for Burgers' equation can be used as a model for the construction of an heuristic closure for the PDF of velocity gradients in incompressible NS turbulence.[80] Suppose that the initial velocity field is multivariate-Gaussian, so that the components of the rate-of-strain tensor also are Gaussian. The heuristic model equation for the evolution of a strain component s is

$$s = J(s_0)s_0, \quad (6.50)$$

$$\partial J/\partial t = |s_0|J^2 - \eta k_d^2 J^3. \quad (6.51)$$

Here s_0 is the initial, Gaussianly distributed, value of s, J is an overall effective straining ratio for regions of the field with given s_0, η is kinematic

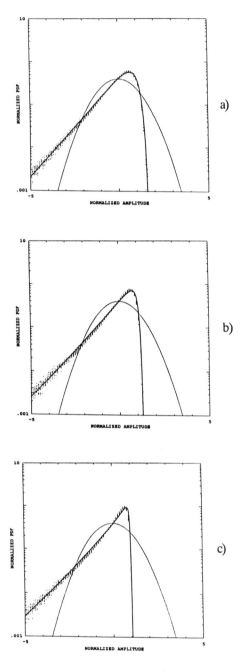

Fig. 13. (a) Normalized velocity-gradient PDF $\langle \xi^2 \rangle^{1/2} Q_E(\xi)$ *vs* $\xi/\langle \xi^2 \rangle^{1/2}$ for Case A, as given by closure (solid line) and 12 simulation runs (data points) at $t = 10$. Also shown is a normalized Gaussian PDF for reference (dotted line). Same as (a), but for $t = 20$. (c) Same as (a), but for $t = 40$.

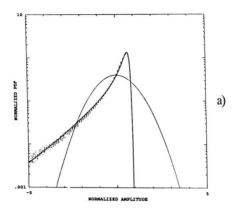

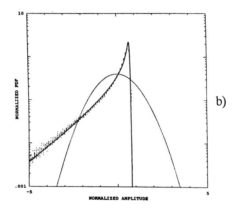

Fig. 14. (a) Normalized velocity-gradient PDF $\langle \xi^2 \rangle^{1/2} Q_E(\xi)$ *vs* $\xi/\langle \xi^2 \rangle^{1/2}$ for Case B, as given by closure (solid line) and 10 simulation runs (data points) at $t = 10$. Also shown is a normalized Gaussian PDF for reference (dotted line). (b) Same as (a), but for $t = 20$.

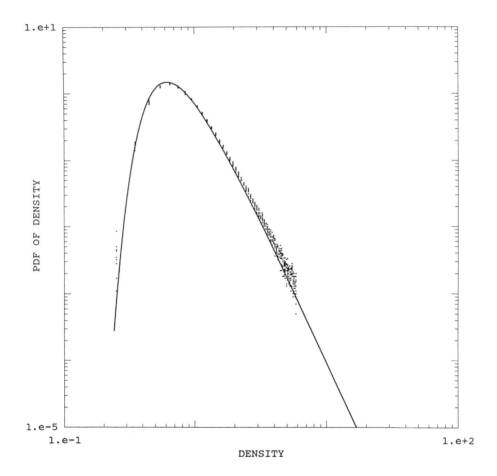

Fig. 15. Pdf $P_E(\rho)$ of fluid density ρ for Case A, as given by closure (solid line) and 12 simulation runs (data points) at $t = 10$.

viscosity and the parameter k_d is a characteristic dissipation wavenumber. $J > 1$ represents the intensification of gradients.

The two terms on the right side of (6.51) correspond to the J^2 and J^3 terms that are asymptotically dominant in the Burgers' closure (6.45). The first term on the right side represents nonlinear straining effects. J appears as square because the nonlinearity is quadratic. Alone, this term would lead to extreme intermittency, and a singularity at finite time for any given value of s_0. The second term on the right side represents viscous decay with a time constant inversely proportional to the square J^2 of the stretching ratio. The third J factor in the viscous term comes from (6.50). The ratio $\langle s_0^2 \rangle^{1/2}/\eta k_d^2$ is an effective Reynolds number. The only other parameter for the model is the normalized evolution time $\langle s_0^2 \rangle^{1/2}t$. The decay of velocity amplitudes [r factor in (6.45)] is ignored here, so the model is properly applicable only over times short compared to overall decay times.

At large enough values of $|s_0|$, (6.51) quickly leads to a near equilibrium in which J grows to make the two terms on the right side balance; thus, at large s,

$$J \approx s_0/(\eta k_d^2), \quad |s| \approx s_0^2/(\eta k_d^2), \quad J \approx |s|^{1/2}/(\eta k_d^2)^{1/2} \qquad (6.52)$$

The PDF of s is

$$P(s) = P_0(s_0)\frac{\partial s_0}{\partial s}, \qquad (6.53)$$

where

$$P_0(s_0) = (2\pi \langle s_0^2 \rangle)^{-1/2}\exp(-\tfrac{1}{2}s_0^2/\langle s_0^2 \rangle) \qquad (6.54)$$

is the Gaussian PDF of s_0. No weighting factor N/J is is included in (6.53) because the mapping distortions appropriate to straining of gradients in incompressible flow are themeselves incompressive.

Eqs. (6.52) and (6.53) imply that $P(s)$ at large enough $|s|$ has the form

$$P(s) \approx \left(\frac{\eta k_d^2}{8\pi \langle s_0^2 \rangle |s|}\right)^{1/2}\exp\left(-\frac{1}{2}\frac{\eta k_d^2 |s|}{\langle s_0^2 \rangle}\right) \quad \left(|s| \gg \frac{\langle s_0^2 \rangle}{\eta k_d^2}\right) \qquad (6.55)$$

The nearly exponential tail exhibited in (6.55) recalls the exponential-like tails that have been observed in incompressible NS simulations (Refs. 21, 22, 67, 71, 74). Fig. 16 shows the PDF of transverse velocity derivative found in an isotropic-turbulence simulation by Vincent and Meneguzzi[71] at a macroscale Reynolds number of about 1000. Fig. 17 shows the PDF (6.53), evolved at $t = 0.3$ under (6.50) and (6.51) from a start at $t = 0$ with $s = s_0$, $\langle s_0^2 \rangle = 1$, $\eta k_d^2 = 1.5$. The model reproduces not only the broad skirt of Fig. 16 but also the characteristic upward curvature of the skirt. The curvature arises from the $s^{-1/2}$ factor in (6.55).

It is notable not only that the match is possible but that it is obtained with values of the two model parameters that correspond to a Reynolds number and a normalized evolution time both of order unity. This hints that the essential intermittency mechanism in NS dynamics may not require high Reynolds number. The unrealistic sharp peak at the PDF maximum in Fig. 17 is the result of the simplification made in writing (6.51): The derivative and integral terms in (6.45) are important in shaping the PDF near its maximum and these terms are not represented in (6.51)

The differences between Burgers and NS dynamics are substantial. The shock-formation mechanism for velocity-gradient growth in Burgers dynamics is substantially different from the vortex stretching that arises from the

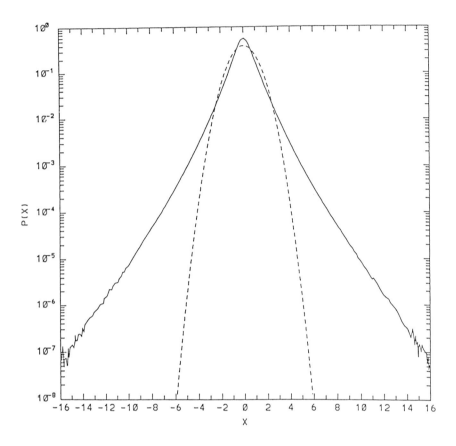

Fig. 16. Normalized PDF of transverse velocity derivative in a simulation of isotropic turbulence at macroscale Reynolds number about 1000. From Vincent and Meneguzzi.[71]

NS equation. A fluid element once caught in a forming shock in the Burgers' case is there forever, while an element of vorticity in the NS case may be subjected to a series of statistically uncorrelated stretchings[38,66] before it is diffused by viscosity. However, the exponential tail characteristic of the heuristic model above is plausible also in NS dynamics. Perhaps only the final stretching, the one that narrows a vortex to the point where viscosity acts strongly on it, is actually important in determining the asymptotic skirts of the PDF. It may be that the details of the path in parameter space followed by a fluid element are relatively unimportant in this respect

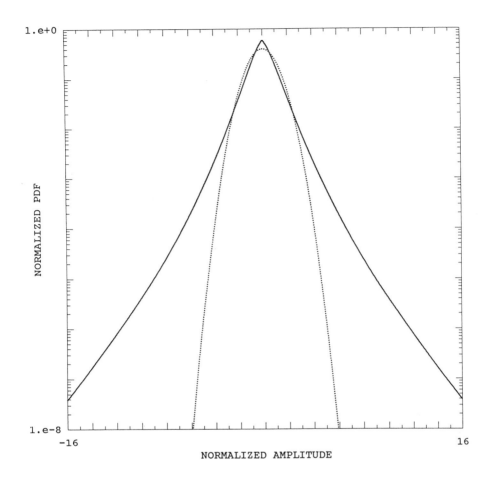

Fig. 17. Normalized PDF $\langle s^2 \rangle^{1/2} P(s)$ for model (2), (3) *vs* normalized rate-of-strain amplitude $s/\langle s^2 \rangle^{1/2}$. See text.

and, instead, the Gaussianly distributed initial gradient and the final viscous relaxation are what count. If this argument is valid, fractal cascade processes do not play the crucial role in fixing the PDF of velocity gradients: the essential ingredients for intermittency already are present at low Reynolds numbers. A qualitative argument to this end was presented some years ago.[34]

The one-point PDF of velocity gradient does not provide the information needed to reconstruct structures in space. Thus it seems possible that the sharply-defined rope-like dissipative structures of high-Reynolds-number turbulence and the relatively structureless fields of low-Reynolds-number turbulence both are consistent with the same form of gradient PDF. If this is so, the presence or absence of clearly-defined dissipative structures may not have a strong effect on the one-point PDF of velocity gradient and, therefore, on the distribution and mean magnitude of dissipation.

Acknowledgements

I am grateful to Hudong Chen for his help in organizing this paper. This work was supported by the National Science Foundation, Division of Atmospheric Sciences, under Grant ATM-8807861 and the Department of Energy under Contract W-7405-Eng-36 with the University of California, Los Alamos National Laboratory.

References

[1] F. Anselmet, Y. Gagne, E. J. Hopfinger and R. A. Antonia, J. Fluid Mech. **140**, 63 (1984).

[2] G. K. Batchelor, *Theory of Homogeneous Turbulence* (Cambridge Univ. Press, 1953).

[3] G. K. Batchelor and A. A. Townsend, Proc. Roy. Soc. (London) **A199**, 238 (1949).

[4] J. M. Burgers, in *Statistical Models and Turbulence*, edited by M. Rosenblatt and C. Van Atta (Springer-Verlag, New York, 1972), p. 41.

[5] B. Castaing, Comptes Rendu (1989), in press.

[6] H. Chen, S. Chen and R. H. Kraichnan, Phys. Rev. Lett. **63**, 2657 (1989).

[7] H. Chen, J. R. Herring, R. M. Kerr and R. H. Kraichnan, Phys. Fluids A, **1**, 1844 (1989).

[8] S. Chen and R. H. Kraichnan, Phys. Fluids A, **2**, to be submitted.

[9] W. P. Dannevik, V. Yakhot and S. A. Orszag, Phys. Fluids **30**, 2021 (1987).

[10] V. Eswaran and S. B. Pope, Phys. Fluid **31**, 506 (1988).

[11] U. Frisch, M. Lesieur and A. Brissaud, J. Fluid Mech. **65**, 145 (1974).

[12] U. Frisch, P. L. Sulem and M. Nelkin, J. Fluid Mech. **87**, 719 (1978).

[13] C. H. Gibson, G. R. Stegen and R. B. Williams, J. Fluid Mech. **41**, 153 (1970).

[14] T. Gotoh, Y. Kaneda and N. Bekki, J. Phys. Soc. Japan **57**, 866 (1988).

[15] H. L. Grant, R. W. Stewart and A. Moilliet, J. Fluid Mech. **12**, 241 (1962).

[16] J. R. Herring in *Frontiers in Fluid Mechanics* (Springer-Verlag, New York, 1985).

[17] — and R. H. Kraichnan, in *Statistical Models and Turbulence*, edited by M. Rosenblatt and C .Van Atta (Springer-Verlag, New York, 1972).

[18] —, J. J. Riley, G. S. Patterson and R. H. Kraichnan, J. Atm. Sci. **30**, 997 (1973).

[19] — and R. H. Kraichnan, J. Fluid Mech. **91**, 581 (1979).

[20] — and O. Métais, J. Fluid Mech. **202**, 97 (1989).

[21] I. Hosokawa, J. Phys. Soc. Japan **58**, 1125 (1989); Phys. Fluids A, **1**, 186 (1989).

[22] I. Hosokawa and K. Yamamoto, J. Phys. Soc. Japan **58**, 20 (1989).

[23] Y. Kaneda, J. Fluid Mech. **107**, 131 (1981).

[24]—, Phys. Fluids **29**, 701 (1986).

[25]—, Phys. Fluids **30**, 2672 (1987).

[26]R. M. Kerr, J. Fluid Mech. **153**, 31 (1985).

[27]—, Phys. Rev. Lett. **59**, 783 (1987).

[28]S. Kida and Y. Murakami, Fluid Dyn. Res. **4**, 347 (1989).

[29]R. H. Kraichnan, J. Fluid Mech. **5**, 497 (1959).

[30]—, J. Math. Phys. **2**, 124 (1961).

[31]—, Phys. Fluids **7**, 1030 (1964).

[32]—. Phys. Fluids **8**, 597 (1965); **9**, 1884 (1966)

[33]—, Phys. Fluids **9**, 1728 (1966).

[34]—, Phys. Fluids **10**, 2081 (1967).

[35]—, Phys. Fluids **13**, 569 (1970).

[36]—, J. Fluid Mech. **47**, 513 (1971).

[37]—, J. Fluid Mech. **47**, 525 (1971).

[38]—, J. Fluid Mech. **62**, 305 (1974).

[39]—, J. Fluid Mech. **81**, 385 (1977).

[40]—, Phys. Rev. A **25**, 3281 (1982).

[41]—. in *Theoretical Approaches to Turbulence*, edited by D. L. Dwoyer, M. Y. Hussaini and R. G. Voight (Springer-Verlag, New York, 1985).

[42]—, Phys. Fluids **30**, 2400 (1987).

[43]—, Complex Systems **1**, 805 (1987).

[44]— in *Current Trends in Turbulence Research*, edited by H. Branover, M. Mond and Y. Unger (American Inst. of Aeronautics, Washington, 1988), p. 198.

[45]— and J. R. Herring, J. Fluid Mech. **88**, 355 (1978).

[46]— and R. Panda, Phys. Fluids **31**, 2395 (1988).

[47]A. Y. Kuo and S. Corrsin, J. Fluid Mech. **56**, 447 (1972).

[48]C. E. Leith, J. Atm. Sci. **28**, 145 (1971).

[49]— and R. H. Kraichnan, J. Atm. Sci. **29**, 1041 (1972).

[50]M. Lesieur, *Turbulence in Fluids* (M. Nijhoff, Dordrecht, 1987).

[51]D. C. Leslie, *Developments in the Theory of Turbulence* (Oxford University Press, 1973).

[52]T. S. Lundgren, Phys. Fluids **10**, 969 (1967).

[53]P. C. Martin, E. D. Siggia and H. A. Rose, Phys. Rev. A **8**, 423 (1973).

[54]J. C. McWilliams, J. Fluid Mech. **198**, 199 (1989).

[55]O. Métais and J. R. Herring, J. Fluid Mech. **202**, 117 (1989).

[56]A. Monin and A. M. Yaglom, *Statistical Fluid Mechanics II*, edited by J. L. Lumley (MIT Press, Cambridge, Mass., 1971).

[57]S. A. Orszag, J. Fluid Mech. **41**, 363 (1970).

[58]— and G. S. Patterson, Phys. Rev. Lett. **28**, 76 (1972).

[59]R. Phythian, J. Phys. A **2**, 181 (1969).

[60]—, J. Phys. A **8**, 1423 (1975).

[61]—, J. Phys. A **10**, 777 (1977).

[62]S. B. Pope, Prog. Energy Combust. Sci. **11**, 119 (1985).

[63]—, private communication

[64]J. Qian. Phys. Fluids **29**, 2165 (1986).

[65]H. A. Rose and P. L. Sulem, J. Physique **39**, 441 (1978).

[66]P. Saffman, Phys. Fluids **13**, 2193 (1970).

[67]E. D. Siggia, J. Fluid Mech. **107**, 375 (1981).

[68]G. I. Taylor, Proc. Roy. Soc. London A, **151**, 421 (1935).

[69]P. Tong and W. I. Goldburg, Phys. Fluids **31**, 2841 (1988).

[70]C. W. van Atta and T. T. Yeh, J. Fluid Mech. **71**, 417 (1975).

[71]A. Vincent and M. Meneguzzi, J. Fluid Mech., submitted (1990).

[72]V. Yakhot and S. A. Orszag, Phys. Rev. Lett. **57**, 1722 (1986).

[73]—, J. Sci. Computing **1**, 3 (1987).

[74]K. Yamamoto and I. Hosokawa, J. Phys. Soc. Japan **57**, 1532 (1988).

[75]S. Chen and R. H. Kraichnan, Physics of Fluids A, **1**, 2019 (1989).

[76]R. H. Kraichnan and S. Chen, Physica D **37**, 160 (1989).

[77]M. Nelkin and M. Tabor, Physics of Fluids A, **2**, 81 (1990).

[78]L. Shtilman and W. Polifke, Physics of Fluids A, **1**, 778 (1989).

[79]T. Williams, E. R. Tracy and G. Vahala, Phys. Rev. Lett. **59**, 1922 (1987).

[80]R. H. Kraichnan, Phys. Rev. Lett. **65**, 575 (1990).

Chapter 2

Rapid distortion theory as a means of exploring the structure of turbulence

J.C.R. Hunt, D.J. Carruthers and J.C.H. Fung

Summary

Turbulence structure is discussed in terms of the different length scales of a turbulent flow—the large scale motions characteristic of the boundary conditions and forcing of the particular flow and, for high Reynolds numbers, the universal small scale motions. Evidence is presented that in shear flows and flows near boundaries the large scale structure of many different turbulent flows is similar. The analysis and understanding of different types and different regions of turbulent flows, and in particular their sensitivity to boundary and initial conditions, is clarified by using the classification of "rapidly changing turbulence" (RCT) and "slowly changing turbulence" (SCT), according to whether the time (T_D) over which fluid particles pass through (or near) changes in the mean flow or boundary conditions is much less or much greater than the characteristic time of the large scales of the turbulence (T_L). It is noted that in all unconfined turbulent flows, the turbulence structure adjusts so that, $T_D \sim T_L$, which implies that some features of the initial conditions, and boundary conditions can persist throughout the flow. A possible physical explanation is suggested.

The linear "Rapid distortion theory" (RDT) of turbulence and its assumptions are briefly reviewed here. It is shown that it is strictly applicable to the former category (RCT). But for some flows the linear theory has slowly changing solutions, or "statistical eigensolutions", and these approximate to certain features of slowly changing turbulent flows (SCT). Examples are given of shear flows and turbulence near boundaries. In addition RDT can be used to simulate some aspects of the large eddy structures of turbulent shear flows.

The style of this paper is informal; for details of some of the results see Hunt & Carruthers (1990) (HC) and Carruthers, Fung, Hunt & Perkins (1990).

1. Introduction

1.1. Turbulence and turbulence research

A useful working definition of a turbulent flow is that it contains motions
with many time scales and length scales which are random in the sense there
is zero probability of any flow variable having a particular value and there
is zero energy in any one particular frequency or wavenumber (i.e. continuous and finite probability density functions and spectra). The difference
between turbulence and weak spatio-temporal chaos is that, if H is the
imposed scale of the flow in turbulence, there is significant random motion
on length scales much less than H. This implies that the Reynolds number
Re is large; but only if Re is very large (the magnitude depending on the
flow, as shown in §4) are the strain rates and vorticity of these small scale
motions greater than that of the large scale motions. Usually these smallest
scale motions are intermittent.

There are different ways of understanding turbulence, and different
information is required for different kinds of practical or scientific problem
involving turbulent flows. Does one want purely statistical information, such
as probability distributions or moments of these distributions (e.g. means
and covariances), or does one want to know the typical forms of eddies, or
"structures", in the turbulent flows? A third possible way of describing a
turbulent flow is in terms of its "state" – is it tending to some statistical
equilibrium or a state in which some global property has some stationary
value (c.f. Malkus 1956)? In most discussions of turbulence, including this
one, all these aspects are considered, but usually with a particular emphasis
on one or other aspect. (These methodological or philosophical questions
were raised in the very interesting collection of essays edited by Favre *et
al.* (1988); see also the review by Hunt (1990).)

In the next sections of the introduction (1.2, 1.3) we provide a context for our later analysis of the structure of turbulence in high and low
Reynolds turbulence, and we review theoretical methods for investigating
the eddy structure of turbulence, and briefly mention recent developments
in the kinematic and probabilistic definitions of the significant kinds of eddy
motion in turbulent flows.

In §2 we discuss the boundary conditions and initial conditions that
define different turbulent flows. This gives rise to an interesting distinction
between different types of turbulent flow and different problems involving
turbulent flows, according to whether the natural time scale (T_L) of the
turbulence is large or small compared with the imposed time scale (T_D)
during which fluid particles are in the domain of interest. This distinction
provides a rational basis for using different kinds of approximate model and
it also helps our understanding. In §3 RDT is introduced, and in §4 RDT
is used to demonstrate certain features of turbulence structure. Finally,
conclusions are presented in §5.

1.2. Do turbulent flows have a generic structure?

A basic theme of turbulence research, from L.F. Richardson, G.I. Taylor and A.N. Kolmogorov onwards, has always been the search for statistical properties and eddy structures that might generally occur in wide classes of turbulent flow. One might expect that such generality should exist from naive analogies between random eddies and the motion of molecules in gases, or, at a superficial theoretical level, between the chaotic solutions of the nonlinear equations of fluid motion and the equilibrium state of a system in statistical physics. In fact turbulence research has shown that a completely universal structure of turbulence does not exist, mainly because there are large scales in turbulent flows, with significant energy, which are of the same scale as that of the inhomogeneities of the flow; consequently they are determined by the boundary and initial conditions, and cannot be universal. However some common or generic features of turbulence have been discovered and have led to useful models of turbulence and better understanding; so the search continues!

Both the analogies of molecular motion and statistical physics can be justified for the small scale motions in turbulent flows at high Reynolds number

$$Re = u_0 L/\nu \gtrsim 10.$$

where u_0 is the rms velocity and L the integral sale of the turbulence, and ν is the kinematic viscosity. Then the dynamical interactions of the smallest eddies (with length and velocity scales ℓ and $u(\ell)$ respectively) is mainly with other small eddies, because the strain rates produced by these small eddies $(u(\ell)/\ell)$ are much larger than those of the large scale eddies (u_0/L) (see figure 1). The small scales reach a state of statistical equilibrium, approximately independent of the large scales, which implies that there is a universal form for the local energy spectrum $(E(k))$; this is only known definitely for the inertial range of $k \sim \ell^{-1}$, where $u(\ell)\ell/\nu \gg 1$ and

$$E(k) = \alpha_k \varepsilon^{\frac{2}{3}} k^{-\frac{5}{3}}, \qquad (1.1)$$

where α_k is the Kolmogorov constant of order one (Kolmogorov 1941, Batchelor 1953, Townsend 1976). Wyngaard & Cote (1972) used their atmospheric measurements to show that in a shear flows $U(y)$ at high Reynolds number $E(k)$ is approximately unchanged. Using their data and dimensional arguments they also showed that weak perturbations from isotropy must also have a universal form; the one dimensional cospectrum of the Reynolds shear stress is given by

$$C_0(k) = c |dU/dy| \varepsilon^{\frac{1}{3}} k^{-\frac{7}{3}}. \qquad (1.2)$$

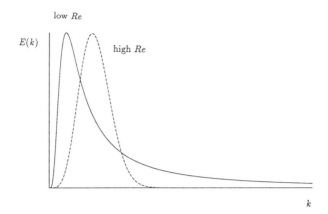

Fig. 1 The difference between spectra of energy $E(k)$ of high and low Reynolds numbers in the various wavenumber ranges.

(The approximation of local isotropy of the small scales (even in shear flow) is the one of the key assumption in Renormalisation Group theory of turbulence (Yakhot & Orszag 1986).)

It is clear from the above argument that for low Reynolds number turbulence ($200 > uL/\nu > 20$), where the strain rates for the small scales are much less than those produced by the large scales (e.g., Domaradzki & Rogallo 1990), the structure of the small scale turbulence cannot have a universal form, as indeed is shown by the spectra in numerical simulations (e.g. Rogers & Moin 1987) and laboratory measurements of grid turbulence without and with shear (e.g. Comte-Bellot & Corrsin 1966, Champagne, Harris & Corrsin 1970), where for example $E(k)$ changed from being proportional to $\exp\left(-k^2 L^2\right)$ to k^{-2}.

Neither the molecular motion nor the statistical physics analogies are plausible reasons why the larger scale eddy motions should have a similar structure in different turbulent flows with different boundary conditions and different forcing, because the larger scales span the flow and are directly affected by the same factors that affect the mean flow. Nevertheless there is much experimental evidence that within certain classes of flow there are common features as defined by (a) statistical quantities, (b) approximate equations for certain statistical quantities and (c) eddy structure.

For example in shear flows $U(y)$ away from boundaries or the edges of the shear layers, there are similarities in the statistical quantities such as $-\langle uv \rangle / u'v' = 0.4 \pm 0.1$, (e.g. Townsend 1976 p. 87; Jeandel, Brison & Mathieu 1978) where u' and v' are the rms velocities, or the spectrum for high wavenumbers is

$$E(k) \propto k^{-n}, \qquad \text{where} \qquad 5/3 \lesssim n \lesssim 2 \qquad (1.3)$$

(e.g. Ho & Huerre 1984, Champagne *et al.* 1970). These similarities in structure occur over a wide range of Reynolds' number, in contrast to the variability of the spectra at moderate to low Reynolds number in flows without shear.

Secondly many different types of shear flow have the same generic structure in that their development is described by the same algebraic or differential model equations relating the different statistical moments of the turbulence to each other and to the mean flow. In general, the more complex the equations and the more of them there are, the wider the class of flows that can be modelled. Launder & Spalding (1972) and Lumley (1978) pointed out that these model equations can only be valid over a wide range of different flows if the structure of the turbulence is approximately similar in these flows; so these models may be inaccurate where the turbulence is quite dissimilar to fully developed shear flows, such as where large energetic vortices, produce by a wake of a nearly body, are present in the shear layer (Murakami & Mochida 1988, Zhou 1989).

Thirdly the large eddies in the outer parts of boundary layers, in wakes, in jets and other shear layers have a generic structure once the flows have had time to develop, although there continue to be some important features of the initial eddy structure of these shear flows that persist far downstream. (Hussain 1986, Hayakawa & Hussain 1989, Lesieur & Metais 1989 and Antonia & Bisset 1990).

1.3. Methods for exploring turbulence structure

Two main types of theoretical methods have been used for investigating the structure of large eddies in turbulent flows in which the turbulence is being distorted by variations in the mean flow, by body forces or by the imposition of boundary conditions. One approach has been to look for eigensolutions $f_n(x)$ in the perturbations to the mean flow equations in the form

$$\mathbf{u}(\mathbf{x}, t) = \sum_n a_n f_n(\mathbf{x}) \exp(i\sigma_n t), \qquad (1.4)$$

where a_n are the amplitudes, and σ_n (complex) determine the time dependence. The main feature of these eigensolutions (ES) is that they span the whole flow and therefore are characterised by each particular mean flow and boundary conditions. These modes describe the large eddies in free shear layers in their early stages of growth, whether they are natural or forced (e.g. by sound), Ho & Huerre (1984). The use of nonlinear analysis enables larger amplitude eddies to be modelled. (Gaster, Kit & Wygnanski 1985, Liu 1989). These modes do not describe the large eddies in the later stages of free shear layers (e.g. Townsend 1976, Hayakawa & Hussain 1989). However in other kinds of fully developed turbulent flows there are destabilising forces which dominate the turbulence to such an extent that these kind of eigenmodes are the large scale eddies, an important example being the streamwise rolls in the atmospheric boundary layer (Mason & Sykes 1982).

The other approach has also been to examine perturbations to mean flows by using a different mathematical procedure, "Rapid Distortion Theory" (RDT). In this case the perturbations are not forced to be of the form of (1.4), but rather

$$\mathbf{u}(\mathbf{x}, t) = \sum_n a_n f_n(\mathbf{x}, t), \qquad (1.5)$$

which allows for the possibility of some components of velocity to increase and some to decrease with time, and also allows for different rates of growth of disturbances in different parts of the flow. Usually RDT is only applied to linear perturbations. In principle it includes perturbations of the form (1.4), such as the oscillatory perturbations found in the analysis of stratified turbulent flows (Komori *et al.* 1983; Hunt, Stretch & Britter 1988).

It is found that in many shear flows the solutions of the form (1.4) do not grow (i.e., $\mathrm{Im}\sigma_n > 0$) for the particular mean shear profiles $U(y)$, and yet the flows are turbulent. In such flows there are solutions of the form (1.5) which depend only on the the the local value of the mean shear dU/dy; in that case local solutions are constructed of the form

$$f_n(\mathbf{x}, t) = A_n(\mathbf{x}, t) \exp\{i\boldsymbol{\kappa}_n(\mathbf{x}, t).\mathbf{x}\}, \qquad (1.6)$$

where the amplitudes of the *different* components grow or diminish algebraically as a function of $(t\, dU/dy)$. Also in inhomogeneous flow RDT solution can be constructed where the growth rate varies across the flow. These kind of solutions are not described in the framework of (1.4). The initial velocity field and/or the initial spectrum has to be defined. The eddy structure and the statistics derived from the solutions (1.6) depend largely on $t\, dU/dy$ where t is a characteristic time of the flow and to some extent on the initial conditions, (e.g. the degree of anisotropy). These local solutions (1.6), which are independent of the general form of $U(y)$, have many of the properties of the observed generic structure of turbulent flows (Deissler 1968, Townsend 1976, Lee, Kim & Moin 1987, Savill 1987, Landahl 1990, Hunt & Carruthers 1990 and Favre *et al.* 1990). In contrast, the dominant eigen modes of (1.4) are independent of the initial conditions and only depend on the form of the mean flow, boundary conditions and the form of the forcing (e.g., its frequency), and therefore do not describe the generic features of well developed shear flows. It appears that the two methods (ES) and RDT are complementary; each are effective for modelling the eddy structure in different kinds of turbulent flow or different regions of the same flow, or even, perhaps (as in the case of free shear layers) different scales of eddy motion at the same location.

There is an important, and recently discovered variant of the RDT approach which explains how large scale slowly varying eddies can be *forced* by smaller scale eddy motion when the turbulence is anisotropic and uniform (known as the Anisotropic Kinetic Alpha effect, e.g. Sulem *et al.* 1990), or anisotropic and non-uniform. A particular example of the latter case is homogeneous turbulence with scale L near a rigid interface; the distortion of the turbulence is calculated by RDT; if the interface has a long length scale undulation Ω ($\gg L$) a mean eddying motion with length scale Ω and rms velocity U_0 begins to grow, where $U_0 \propto tu_0^2/\Omega$ (Wong 1985).

So far the general nonlinear theories of turbulence, such as EDQNM (Lesieur 1987) or DIA (see Leslie 1973), have been used to calculate the spectra and correlations; they have not given any other information about the eddy structure. These theories can also give more general insight when they predict and explain how the statistical properties of distorted turbulence are affected by different initial conditions, such as Reynolds number, anisotropy, and initial forms of the spectra (e.g., van Haren 1991).

The close analysis and understanding of eddy structures is only possible if there are systematic procedures for defining the locations of the eddies, and the nature of their flow fields. Hitherto there have been several different procedures, (none of which have been generally accepted), based on a few measurements at a particular time; and then a representation of a 'typical' or 'significant' eddy structure is built up from many sets of measurements. (e.g. Lumley 1965, Adrian & Moin 1988, Mumford 1982, Hussain 1986). Despite the differences in the methods, and some differences in the resulting eddy structure, many of the conclusions about the general forms of the eddies are similar. With the availability of Direct Numerical Simulation (DNS) and large Eddy Simulation (LES) (Rogallo & Moin 1984) to compute turbulent flows it is possible to define the eddy structure based on information about the whole flow field at one time. This has given rise to two new approaches; one is based on the strain tensor du_i/dx_j, "eddies" being regions where the vorticity is large compared with the pure strain $(du_i/dx_j + du_j/dx_i)/2$, and the pressure is low (Wray & Hunt 1990). It appears that this approach gives a similar picture for the eddy structure in shear flows (see figure 2), as other methods (see Fung et al. 1991). Another new approach is based on "wavelet" analysis, which gives information about distribution of the significant scales of motion associated with certain regions of the flow, such as eddies (Farge & Holschneider 1990).

Once the eddies (and other significant regions of motion in the flow, such as regions of high strain) are identified, it becomes possible to consider local dynamical analysis, for example by modelling elongated eddy regions by vortex tubes and computing their interaction (e.g. Melander & Hussain 1990), or by considering the interaction between a larger eddy and small scale motions within it (e.g. Kida & Hunt 1989).

(a)

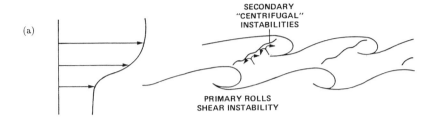

(b)

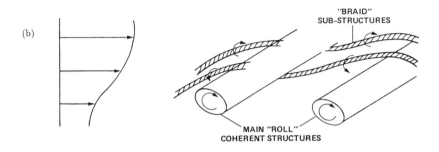

(c)

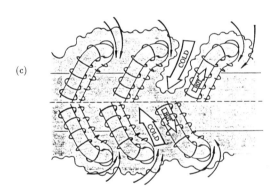

Fig. 2 Coherent structure generated from instabilities: (a), (b) primary and secondary instabilities in a free layer leading to two kinds of coherent structure – 'rolls' and 'braids'. The latter are also known as 'sub-structures'. (a) Linear + nonlinear effects; (b) nonlinear effect + harmonics; (c) fully developed flow where structure are modelled by RDT + eigensolutions.

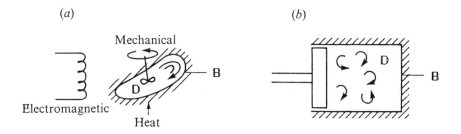

(a)   *(b)*

Mechanical

Electromagnetic

Heat

Fig. 3 Example of a closed domain. (a) Forced mixing with an impeller or body
forces (electromagnetic or buoyancy); (b) a cylinder with a moving piston.

2. Classification of Turbulent Flow Problems

2.1. Classification in terms of boundary conditions

The study and understanding of turbulence always has to be related to
what kind of flow is being considered, because the structure of turbulence
is not the same in all turbulent flows. Also, depending on the flow and its
boundary conditions, the turbulence may be an intrinsic feature of the flow
or may be largely determined by the turbulence on the boundary of the flow
domain. Classification of turbulent flows has, surprisingly, seldom been at-
tempted by turbulence researchers (perhaps because they are attempting to
describe all of turbulence?); but we have found it to be useful in guiding the
choice of appropriate turbulence models, and for relating complex practical
problems in turbulence to simpler and better understood turbulent flows.
This is related to the the approach, advocated by Kline (1981), of analysing
complex flows in terms of "flow zones", each of which may be modelled in
different ways.

We first classify turbulent flows according to three types of boundary
conditions:

2.1.1. *Closed domains* (fig. 3):

In this case the turbulent flow domain $\mathbb{D}$ is confined by surfaces on which the
velocity is zero or is completely specified, such as flow in a heated or stirred
tank, an electromagnetic induction furnace, or by a piston in a closed cylinder. The turbulence is generated by flow instabilities (e.g. shear, buoyancy,
rotational etc.), which along with the direct effects of the boundaries determine the large scale structure of the turbulence. (c.f. Klebanoff *et al.* 1962,
Davidson, Hunt & Moros 1988 and Hunt, Kaimal & Gaynor 1988).

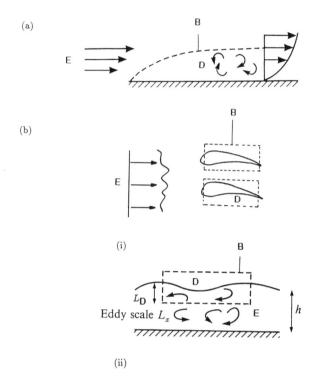

Fig. 4 (a) Open domains and statistical boundary conditions. Non-turbulent flow entering $\mathbb{D}$ by mean flow across $\mathbb{B}$. Turbulence generated in $\mathbb{D}$. (b) (i) Turbulence approaching a row of turbine blades; (ii) Random motion across bounding surface $\mathbb{B}$ advects turbulence into $\mathbb{D}$.

2.1.2. *Open domains* (fig. 4):

In this class, flow enters and leaves $\mathbb{D}$, through the bounding surface $\mathbb{B}$, from or into the external domain $\mathbb{E}$. The flow in $\mathbb{D}$ is determined by the velocity field on $\mathbb{B}$, but there may be significant correlation between the turbulence in $\mathbb{D}$ and $\mathbb{E}$, so the specification of these conditions is often a subtle and particular problem.

(a) No turbulence entering $\mathbb{D}$ from $\mathbb{E}$: in this case the flow entering $\mathbb{D}$ from $\mathbb{E}$ is laminar, and turbulence is generated within $\mathbb{D}$. But where the flow leaves $\mathbb{D}$ back into $\mathbb{E}$ the flow may well be turbulent. An example would be a turbulent boundary layer $\mathbb{D}$ with no free stream turbulence; the flow leaving the layer forms a turbulent wake in $\mathbb{E}$. (For the matching between these regions see Naish & Smith 1988.)

(b) Turbulence enters $\mathbb{D}$ from $\mathbb{E}$, so that turbulence in $\mathbb{D}$ is affected or even determined by the turbulence specified on $\mathbb{B}$: there is an important physical and computational difference between flows where there is a strong mean velocity entering $\mathbb{D}$ (relative to the turbulent eddy ve-

locities), and flows where there is no appreciable mean velocity across
$\mathbb{B}$, because in the latter case turbulent energy is transported into $\mathbb{D}$ by
the action of the eddies, rather than by the mean flow. This implies
a different structure of turbulence on $\mathbb{B}$ in the latter case. In either
case the turbulence transported into $\mathbb{D}$ may or may not significantly
affect the turbulence in $\mathbb{D}$, depending on the local sources of turbulent energy in $\mathbb{D}$, or types of instability. As shown in figure 4(b,i), a
typical example of the former case is turbulence approaching a row of
turbine blades; outside their boundary layers the turbulence is largely
determined by that entering from $\mathbb{E}$, (e.g. Goldstein & Durbin 1980)
whereas the turbulence in the boundary layers depends on an interaction between the incoming turbulence and the local boundary layer
instabilities (e.g. Goldstein 1978). A typical example of the latter case
is turbulence near a density interface, see figure 4(b,ii) (Carruthers &
Hunt 1986).

2.1.3. *Time dependent problems*

Here the complete flow field $\mathbf{u}(\mathbf{x}, t)$ in $\mathbb{D}$, involving mean and random quantities, is specified at an initial moment in time, say t_I, and boundary conditions are also applied on $\mathbb{B}$. The problem is to calculate and understand
how the the flow changes with time. This may be slowly or not at all (in
a statistical sense) if the initial form of the turbulent velocity field and
the boundary conditions define a state that is close to equilibrium. On the
other hand there may be a rapid rate of change if the initial turbulence
does not satisfy the boundary conditions, so that the turbulence has to adjust. Examples of such flows which have been studied experimentally and
theoretically include the change of isotropic turbulence under the actions
of mean straining motions or shear, or anisotropic body forces.

Most basic turbulent flows, and many practical problems, fall into one
or other of these categories. Other flows can be understood in terms of
combinations of these categories. Time dependent flows have proved to be
useful model problems for studying the structure of distorted turbulence in
well developed or even statistically stationary flows; as Lumley remarked
"turbulence is a black box that needs shaking to find out what is inside".
The results of many distortions or "shaking", such as those performed in
their laboratories by Corrsin, Mathieu and Townsend and their colleagues
using ingenious different wind tunnel experiments, form the basis of much
current understanding and modelling (e.g. Launder, Reece & Rodi 1975,
Lumley 1978) and the methodology for further advance.

2.2. Classification according to the turbulence and imposed time scales

We have seen in the previous §2.1 that turbulence in a flow region $\mathbb{D}$ at a
given time may be determined by the turbulence $u_{\mathbb{B}}$ or $u_{\mathbb{I}}$ defined by the

conditions on the bounding surface $\mathbb{B}$ (especially where turbulent energy is transported into $\mathbb{D}$) and conditions at the initial time t_I, or by turbulence generated in the interior of $\mathbb{D}$. (These different sources of turbulence in any given flow could only be identified by considering a number of different conditions on $\mathbb{B}$ and at t_I). In order to understand and model the turbulence $\mathbf{u}(\mathbf{x}, t)$ in the interior of $\mathbb{D}$ it is essential to know the extent to which it depends on the imposed turbulence, relative to the locally generated turbulence. For example is $\mathbf{u}(\mathbf{x}, t)$ statistically correlated with $u_{\mathbb{B}}$ or $u_{\mathbb{I}}$, so that the structure of turbulence in $\mathbb{D}$ depends sensitively on the structure of $u_{\mathbb{B}}$ or $u_{\mathbb{I}}$, or does only the variance of $\mathbf{u}$ ($= u_0^2$) depend on the variance of $u_{\mathbb{B}}$ or $u_{\mathbb{I}}$, and perhaps other factors such as the Reynolds number of the flow? An example of the former situation is the turbulence between the turbine blades which depends on the wake turbulence of upstream blades, but the classic example of the latter situation is the flow in a circular pipe, where turbulence only occurs when the inlet flow contains velocity fluctuations of finite amplitude.

As in other statistical problems in physics, the criterion for the sensitivity to boundary or initial conditions must depend on whether the turbulence time scale T_L, (the intrinsic or "relaxation" time scale of the system) is of the same order as the imposed time scale T_D in the domain. T_D is the time over which the fluid elements are in the domain, where the turbulence may or may not be undergoing distortion.

One measure of the intrinsic time scale of the turbulence in the interior of $\mathbb{D}$, T_L, is the time over which some disturbance to the turbulence structure decays. Since any adjustment of a turbulence flow field involves the transfer of energy between different scales,or "turning over" of eddies (e.g. Kellogg & Corrsin 1980), and since this is also an essential part of the process of the dissipation of turbulent energy, T_L is, secondly, also the time scale for the rate of dissipation of turbulent energy, for given kinetic energy. Thirdly T_L is of the order of the Eulerian or Lagrangian integral time scales of the velocity measured in frames of reference moving with the mean flow or with fluid particles, respectively (see Tennekes & Lumley 1971). At high Re the only scales for determining T_L are the integral scale L and u; hence by dimensional arguments T_L is of order L/u_0, a result which is also consistent with nonlinear theories of turbulence (e.g. Leslie 1973, Lesieur 1987), and with most pictures of large eddies with scale L and velocity u_0.

Rapidly changing turbulence (RCT): Since T_L is the time scale over which the turbulence is correlated for a fluid element, the criterion for the turbulence $\mathbf{u}$ to be statistically correlated with the boundary or initial turbulence, $u_{\mathbb{B}}$ or $u_{\mathbb{I}}$, is that the time that a fluid element spends in $\mathbb{D}$ must be much less than T_L; this criterion is equivalent to specifying that the time scale T_D overwhich the turbulence is changing in $\mathbb{D}$ is much less than the turbulence time scale T_L, i.e. for RCT $T_D \ll T_L$. For an open domain (§2.1.2) of length H with a significant mean velocity advecting turbulence into $\mathbb{D}$, the criterion for RCT is that $H/U_0 \ll L/u_0$, and for a time dependent problem $t - t_I \ll L/u_0$. These are situations for which the linearised RDT is valid and of practical use.

Slowly changing turbulence (SCT): this state is defined to be where the imposed timescale T_D is of the same order or greater than the turbulence timescale T_L, i.e. for SCT, $T_D \gtrsim T_L$. Therefore there can only be a weak correlation between the turbulence in $\mathbb{D}$, and the external or initial turbulence $u_{\mathbb{B}}$, $u_{\mathbb{I}}$.

This point needs some amplification. In many flows the turbulence time scale at the moving location of a fluid element is changing with time, for example in a two-dimensional wake $L \propto T_D^{\frac{1}{2}}$, and $u_0 \propto T_D^{-\frac{1}{2}}$, so that $T_L \sim L/u_0 \propto T_D$. The question to answer is whether T_L is increasing faster or slower than T_D, in order to define whether the turbulence is rapidly or slowly changing.

It is interesting to note that *all* unconfined turbulent flows seem to adjust so that these two time scales remain of the same order i.e. $T_D \sim T_L$. (See table 1, derived from the standard tables of results of length and velocity scales, e.g. Tennekes & Lumley 1971.) This (apparently new) result implies that in all such flows the eddy structure of turbulence must display some sensitivity to boundary or initial conditions throughout the flow. In some shear flows are much more sensitive to initial condition than other (c.f. wakes, Bevilaqua & Lykoudis 1978 and jets, Husain & Hussain 1990).

A possible physical explanation is that in all these flows L increases as the vortical, or eddy, regions in the turbulence (Wray & Hunt 1990) entrain surrounding fluid, and interact and merge with each other, while u_0 decreases as these regions spread and turbulent energy is dissipated. As the eddies grow and merge they preserve some dynamical properties, or "signature", so that their growing time scale T_L is always just increasing at the same rate as the time T_D that fluid elements spend in the flow.

Even for boundary layers which have one confining wall it appears that $T_D \sim T_L$ (if the external flow is not changing), presumably because the large eddies can still grow. However in pipes and channels with surrounding walls the growth of eddies is inhibited and therefore T_L is constant, so that turbulence along a pipe eventually (over about 40 diameters – Sabot & Comte-Bellot 1976) forgets its initial conditions.

In a number of turbulent flows in different regions the turbulence is rapidly changing or slowly changing. In the example of flow over the turbine blades, (length H), near the surface of the blades, the turbulence scale L is small (of the order of the boundary layer thickness δ); so $\dfrac{T_D}{T_L} \sim \dfrac{H}{U_0}\dfrac{u_0}{\delta}$, where U_0 is the mean flow. Thus if $\delta \lesssim H/(U_0/u_0) \lesssim H/20$, $T_D/T_L \gtrsim 1$ and so the boundary layer is a region of SCT. Outside the boundary layer is an RCT region when $T_D/T_L \ll 1$. This concept needs to be understood to model complex turbulent flows where there are significant gradients of turbulence time scale; in particular it helps obviate gross errors such as using

Table 1.

Power of x describing the downstream variation of u_0, L, Re, Turbulent time
scale $T_L = L/u_0$ and time of particle spreads in the flow $T_D = x/(u_0 + U_0)$.

			Powers of x for		
	u_0	L	Re	T_L	T_D
Plane wake	$-1/2$	$1/2$	0	1	1
Self-propelled plane wake	$3/4$	$1/4$	$-1/2$	1	1
Axisymmetric wake	$-2/3$	$1/3$	$-1/3$	1	1
Self-propelled axisymmetric wake	$4/5$	$1/5$	$3/5$	1	1
Mixing Layer	0	1	1	1	1
Plane jet	$-1/2$	1	$1/2$	$3/2$	$3/2$
Axisymmetric jet	-1	1	0	2	2
Plane plume	0	1	1	1	1
Axisymmetric plume	$-1/3$	1	$2/3$	$4/3$	$4/3$
Boundary layer (not power of x)	$\dfrac{1}{\ln x}$	$\dfrac{x}{\ln x}$	$\dfrac{1}{\ln^2 x}$	x	x

'mixing length' models for shear stress in RCT (e.g. Britter *et al.* 1981).

Can RDT contribute to understanding the structure of these examples
of SCT? It seems unlikely, since this theory is valid only for RCT. However
as explained in the next sections, and by Hunt & Carruthers (1990), in certain kinds of RCT, over a time of order T_L, there are features of turbulence
structure which change slowly, or not at all, and are not particularly sensitive to the structure of the turbulence on the boundary $\mathbb{B}$ or at the initial
time t_I. This "eigen" property of certain RCT flows is found, for example,
in shear flows and near rigid boundaries and density interfaces. The nonlinear processes in turbulence act slowly on a time scale greater than T_L, and
are continuously readjusting the turbulence being distorted rapidly, on a
time scale significantly less than T_L, in this case by shear or the boundaries.
Since the structure defined by the linear process is not especially sensitive
to the initial conditions, it means that the linear distortion of turbulence
in shear flow is not very sensitive to the nonlinear processes. Consequently

the structure of turbulence can remain in a near steady state as a result of linear processes. This is broadly the reason (put in a novel form) why the linear theory of RDT can be used to explain and quantify certain general features in the structure of slowly changing turbulent shear flows. But for the reasons given above, if $T_L \sim T_D$ we should also expect certain features of the structure to be peculiar to that particular flow, and dependent on the initial and boundary conditions.

So it is not surprising that the basic free shear flows of turbulence, such as wakes, jets and shear layers should continue to be studied, on the one hand in terms of features that are common to wide classes of these flows, and on the other hand in terms of the particular initial and boundary conditions!

3. Rapid distortion theory—summary of the method

3.1. Linearised equations and approximations

Some of the assumptions and implications of using linearised equations for studying the changes in turbulent flows have been discussed in §1.3 and §2.2. Here we present a summary of the methods used in the RDT approach, which as we explained in §1.3 is usually focussed on aspects of turbulence and turbulent flows which differ from those studied using hydrodynamic stability theory (e.g. Liu 1989).

Consider the random velocity, pressure and vorticity fields $\mathbf{u}^*(\mathbf{x},t)$, p^*, $\boldsymbol{\omega}^*(\mathbf{x},t)$ divided into the ensemble mean and fluctuating components, $\mathbf{u}^* = \mathbf{U}(\mathbf{x},t)+\mathbf{u}$, $p^* = \rho(P+p)$ and $\boldsymbol{\omega}^* = \boldsymbol{\Omega}+\boldsymbol{\omega}$. The ensemble means of $\mathbf{u}$, p and $\boldsymbol{\omega}$ are zero. We now review the estimation of the errors associated with linearisation, both for the large energy-containing scales of the turbulence with a typical r.m.s. velocity $u_o = \left(\frac{1}{3}\langle u_i u_i \rangle\right)^{\frac{1}{2}}$, and integral scale L_x, and for small eddies with velocity scale $u(\ell)$ and length scale ℓ. The typical values of the mean velocity, and its change over a typical length scale of the mean flow $\mathbb{D}$, are U_0 and ΔU_0 respectively.

In this paper the discussion of RDT is restricted to incompressible flows with uniform density and no body forces. But RDT is also a useful technique for estimating how such body forces affect rapidly changing turbulent flows (e.g. Moffatt 1967; Komori et al. 1983).

The governing equations for $\mathbf{u}$ and $\boldsymbol{\omega}$ are:

$$\frac{\partial u_i}{\partial t} + U_j \frac{\partial u_i}{\partial x_j} + u_j \frac{\partial U_i}{\partial x_j} = -\frac{\partial p}{\partial x_i} + \nu \nabla^2 u_i - (\text{NL})_{u_i}, \qquad (3.1a)$$

$$\underbrace{\frac{\partial \omega_i}{\partial t} + U_j \frac{\partial \omega_i}{\partial x_j}}_{(i)} + \underbrace{u_j \frac{\partial \Omega_i}{\partial x_j}}_{(ii)} - \underbrace{\omega_j \frac{\partial U_i}{\partial x_j}}_{(iii)} - \underbrace{\Omega_j \frac{\partial u_i}{\partial x_j}}_{(iv)} = \nu \nabla^2 \omega_i + (\text{NL})_{\Omega_i}, \quad (3.1b)$$

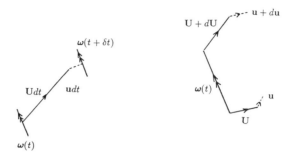

Fig. 5 Change of vorticity produced by the mean flow field $\mathbf{U}$ and the turbulent flow field $\mathbf{u}(\mathbf{x}, t)$. (i) advection by $\mathbf{U}$ and $\mathbf{u}$; (ii) stretching and rotation by spatial *changes* of $\mathbf{U}$ and $\mathbf{u}$ across a vortex line element. In RDT the dashed contribution of $\mathbf{u}$ is neglected.

where
$$\partial u_i / \partial x_i = 0, \qquad \omega_i = \epsilon_{ijk} \partial u_k / \partial x_j, \qquad (3.1c)$$

so that $\partial \omega_k / \partial x_k = 0$.

The physical interpretation of the terms in (3.1b) has been given by various authors (e.g. Tennekes & Lumley 1971, Hunt 1978). The terms (i) and (iii) for the advection and stretching of $\boldsymbol{\omega}$ by the mean flow are important in all flows with a mean velocity. The fourth term (iv) is significant where the mean vorticity exists and can be distorted by velocity fluctuations. The second term (ii) caused by the advection of the mean vorticity by the turbulence is only significant if the mean vorticity is non-uniform (Gartshore, Durbin & Hunt 1983).

The nonlinear terms are

in (3.1a)
$$(\mathrm{NL})_{u_i} = -\left[\frac{\partial(u_i u_j)}{\partial x_j} - \frac{\partial \langle u_i u_j \rangle}{\partial x_j} \right], \qquad (3.2a)$$

and (3.1b)
$$(\mathrm{NL})_{\Omega_i} = -u_j \frac{\partial \omega_i}{\partial x_j} + \omega_j \frac{\partial u_i}{\partial x_j} + u_j \left\langle \frac{\partial \omega_i}{\partial x_j} \right\rangle - \omega_j \left\langle \frac{\partial u_i}{\partial x_j} \right\rangle. \qquad (3.2b)$$

The first term is advection of $\boldsymbol{\omega}$ by the fluctuating vorticity and the second is the stretching of $\boldsymbol{\omega}$ by the fluctuating velocity (see figure 5).

In RDT the linearised velocity and/or vorticity equations (3.1) are primarily used to calculate the two-point moment of the velocity field, $R_{ij}(\mathbf{r}) = \langle u_i(\mathbf{x}) u_j(\mathbf{x}, \mathbf{r}) \rangle$, or the two-point structure function such as $\Delta R_{ii}(\mathbf{x}, \mathbf{r}) = \left\langle (u_i(\mathbf{x}) - u_i(\mathbf{x}, \mathbf{r}))^2 \right\rangle = 2 \left[\langle u_i^2 \rangle - R_{ii}(\mathbf{x}, \mathbf{r}) \right]$. The conditions for linearisation of (3.1b) are quite different if $\boldsymbol{\omega}$ is calculated for the purpose of calculating $\mathbf{u}$ and R_{ii}, as compared with calculating $\langle \omega_i^2 \rangle$. It is necessary to define which scale of the vorticity field contributes to the moments of the velocity field.

Using the Biot-Savart integral (Batchelor 1967, chap. 2), $\Delta R_{ii}(\mathbf{x}, \mathbf{r})$ can be expressed as an integral of $\langle \omega_k(\mathbf{r}')\omega_\ell(\mathbf{r}'') \rangle$ which, for high-Reynolds-number turbulence, can be estimated in terms of the rate of energy dissipation per unit mass, ε. When

$$|(\mathbf{r}' - \mathbf{r}'')| < L_x \qquad \text{and} \qquad \langle \omega_k(\mathbf{r}')\omega_\ell(\mathbf{r}'') \rangle \sim \varepsilon^{\frac{2}{3}} \hat{r}^{-\frac{4}{3}},$$

where $\hat{r} = |(\mathbf{r}' - \mathbf{r}'')|$, and thence, if $|\mathbf{r}| = \ell$,

$$\Delta R_{ii}(\mathbf{x}, \mathbf{r}) \sim \int_0^\ell \varepsilon^{\frac{2}{3}} \hat{r}^{-\frac{4}{3}} \hat{r} \, d\hat{r} \sim \varepsilon^{\frac{2}{3}} \ell^{\frac{2}{3}}. \tag{3.3}$$

This integral shows that, although the vorticity correlation is largest at very small separations of $\hat{r}$, the contribution to ΔR_{ii} from smaller-scale vorticity is comparable with that from lengthscales of the vorticity centred on ℓ.

The contributions to eddies on scale ℓ or $\Delta R_{ii}(\ell)$ from vorticity of different lengthscales could come from vortex sheets separated by L_x or from smooth distributions of vorticity on a scale ℓ, or from both. Flow visualisations and measurements using conditional sampling (e.g. Hussain 1986), the analyses of Moffatt (1984) and Gilbert (1988), and the numerical simulations of two-dimensional high-Reynolds-number vortices by Dritschel (1989) suggest that both forms exist simultaneously because each type of vorticity distribution eventually develops into the other type. However, it does appear that velocity fluctuations on a scale ℓ are primarily associated with smoothly distributed vorticity regions with lengthscale ℓ, so that nonlinear terms like $\omega_j \partial u_i / \partial x_j$ can be estimated as being of order $u^2(\ell)/\ell^2$. Of course if (3.1b) was used to compute the mean-square vorticity, the nonlinear terms would be of order $\langle \omega^2 \rangle$ and greater by a factor of $\mathcal{O}\left(\langle \omega^2 \rangle \ell^2 / u^2\right) = \mathcal{O}(Re)$ if $\ell \sim L$ (Tennekes & Lumley 1971).

The effect of the nonlinear terms in (3.1b) not only has to be estimated over the appropriate lengthscale of the velocity field (ℓ), but also over the period T_D in which the mean distortion is applied. Since these terms are randomly varying in time and space, their effect is reduced. The second nonlinear term in (3.2b) is caused by the stretching of the fluctuating vorticity $\boldsymbol{\omega}$ by the fluctuating velocity, $\mathbf{u}$, and can be estimated for high-Reynolds-number turbulence, using the fact that, on a length scale ℓ, $\partial u_i / \partial x_j$ is of order $\varepsilon^{\frac{1}{3}} \ell^{-\frac{2}{3}}$. Now we can estimate the relative changes in ω produced, in a time T_D, by the linear ($\Delta \omega_{\text{Lin}}$) and nonlinear terms ($\Delta \omega_{\text{NL}}$) compared with the initial vorticity (ω_0). Thus $\Delta \omega_{\text{Lin}}/\omega_0 \sim (\Delta U/L_D) \, T_D$, $\Delta \omega_{\text{NL}}/\omega_0 \sim (u(\ell)/\ell) \, T_D$. Therefore, for the purposes of estimating moments of $u(\ell)$ on a scale ℓ, the criterion for neglecting the nonlinear vortex stretching term $\omega_j \partial u_i / \partial x_j$ is

$$\frac{u(\ell)}{\ell} \sim \varepsilon^{\frac{1}{3}} \ell^{-\frac{2}{3}} \ll \max\left(\frac{\Delta U}{L_D}, \frac{1}{T_D}\right). \tag{3.4}$$

This criterion can also be expressed in terms of the characteristic velocity of the energy-containing eddies u_0. Since, from the inertial-range scaling $u(\ell) \sim u_0(\ell/L_x)^{\frac{1}{3}}$, (3.4) becomes

$$\frac{u_o}{L_x} \left(\frac{\ell}{L_x} \right)^{-\frac{2}{3}} \ll \max \left(\frac{\Delta U}{L_D}, \frac{1}{T_D} \right). \tag{3.5}$$

Thus the two dimensionless parameters which characterise the energy-containing eddies ($\ell \sim L_x$) of a rapidly distorted flow are: the total strain $\beta = T_D \Delta U/L_D$ and the relative strain rate

$$\mathcal{T}^* = (\Delta U/L_D)T_L \qquad \text{where} \quad T_L = L_x/u_0. \tag{3.6}$$

The criterion of (3.5) implies that, if the strain rate is weak, i.e. $\mathcal{T}^* \leq 1$,

$$(\beta/\mathcal{T}^*) \ll 1, \qquad \text{or} \qquad T_D \ll T_L. \tag{3.7a}$$

But if the strain rate is strong, $\mathcal{T}^* \gg 1$, then

$$\beta/\mathcal{T}^* \quad \text{or} \quad T_D/T_L \qquad \text{are arbitrary} \tag{3.7b}$$

(see Lee *et al.* 1987). If (3.5) is satisfied the effects of random straining with large timescales are also negligible. Equation (3.5) is the essential criterion for the validity of RDT calculations of second moments of the velocity field for rapidly changing turbulent flow (RCT): it indicates that the linearisation is justifiable either if the strain rate is large enough or if the period of distortion T_D is short enough. It also shows that the neglect of nonlinear processes for the energy-containing eddies (where $\ell \sim L_x$) is better justified than for smaller-scale eddies (where $\ell \ll L_x$). For low Reynolds number turbulence this caveat is not necessary. Usually the mean straining motion selectively amplifies the turbulent vorticity $\boldsymbol{\omega}$ in one or two directions, and may even *reduce* $\boldsymbol{\omega}$ in other directions. Therefore equations (3.7*a*, *b*) are the criteria for the linear analysis to describe the growth of the components of vorticity and velocity with maximum magnitudes.

In deriving (3.5) only the nonlinear random stretching is considered which in some cases can *amplify* the anisotropy of $\boldsymbol{\omega}$ produced by the linear distortion (Lee *et al.* 1987 and Kida & Hunt 1989). However, another effect of the nonlinear terms is the random *rotation* of vortex lines by the turbulence, leading to significant transfer of vorticity into directions away from those of maximum straining. Thus the linear analysis can under or over estimate the anisotropy caused by mean strain. Therefore the criterion (3.5) for the neglect of nonlinear terms can only be applied to all vorticity and velocity components, if it is modified to allow for the reduced straining in some directions and the nonlinear rotation effect. So (3.5) becomes

$$\frac{u_0}{L_x} \left(\frac{\ell}{L_x} \right)^{-\frac{2}{3}} \ll \max \left(\frac{\Delta U}{L_D}\theta(T_D), \frac{1}{T_D} \right), \tag{3.8}$$

where $\theta(T_D) = \exp\left[(\lambda_{\min} - \lambda_{\max})\,T_D\right]$, and $\lambda_{\max}$ and $\lambda_{\min}$ are the moduli of the maximum and minimum values of principal strains of $\partial U_i/\partial x_j$. This means that the strain parameter criterion $(3.7b)$ is changed to

$$T^*\theta \gg 1. \tag{3.9}$$

So for strong enough isotropic compressive strains, where $\partial U_i/\partial x_j \propto \delta_{ij}$ and $\theta = 1$, and if $(3.7b)$ is satisfied, the nonlinear terms can be neglected for all time (Batchelor 1955). If, for any non-isotropic strain, the effective angle of rotation $\theta(T_D)$ increases with time, then the nonlinear terms eventually become significant, whatever the initial strength of the strain. (We return to this in §4.)

3.2. The statistical input and output to RDT

It is not possible in general to calculate, even using linear RDT, the changes of $\mathbf{u}$ or $\boldsymbol{\omega}$ over the domain $\mathbb{D}$ for an arbitrary input distribution $\mathbf{u}_0(\mathbf{x}, t)$ whether it is defined initially and/or over the boundary $\mathbb{B}$. But, by representing u_{0i} as a series, or integrals of orthogonal functions $\phi_{0i}^{(n)}$, where the random coefficients $S_{0i}^{(n)}$ are defined by the input distribution, it is possible to calculate the changes in $\mathbb{D}$ of $\mathbf{u}(\mathbf{x}, t)$ (and its moments) for a wide range of statistical distributions of $\mathbf{u}_0$.

In general

$$u_{0i}(\mathbf{x}, t) = \sum S_{0i}^{(n)} \phi_{0i}^{(n)}(\mathbf{x}, t), \tag{3.10}$$

where the orthogonal functions $\phi_{0i}^{(n)}(\mathbf{x}, t)$ can, in principle, be deduced from (laborious) measurements or computations of two-point moment of u_{0i} (Lumley 1965).

When the 'input' velocity field is homogeneous, Fourier integrals are used in place of the sum (3.10) and the functions $\phi_0(\mathbf{x})$ are known, so that

$$u_{0i}(\mathbf{x}, t) = \int S_{0i}(\boldsymbol{\kappa}, t) e^{i\boldsymbol{\kappa}\cdot\mathbf{x}} d\boldsymbol{\kappa}. \tag{3.11a}$$

The correlations between these random Fourier transforms S_{0i} are defined by the orthogonality relation

$$\langle S_{0i}^*(\boldsymbol{\kappa}) S_{0j}(\boldsymbol{\kappa}') \rangle = \delta(\boldsymbol{\kappa} - \boldsymbol{\kappa}') \Phi_{0ij}(\boldsymbol{\kappa}), \tag{3.11b}$$

where $\Phi_{0ij}(\boldsymbol{\kappa})$ is the energy spectrum tensor, and

$$\frac{1}{(2\pi)^3} \int R_{0ij}(\mathbf{x}, \mathbf{x} + \mathbf{r}) e^{-i\boldsymbol{\kappa}\cdot\mathbf{r}} \, d\mathbf{r} = \Phi_{0ij}(\boldsymbol{\kappa}). \tag{3.11c}$$

Changes to u_i can be expressed as the product of a non-random 'transfer function' $Q_{ij}(\boldsymbol{\kappa}; \mathbf{x}, t)$ and the original Fourier transform

$$u_i(\mathbf{x}, t) = \int_{\boldsymbol{\kappa}} Q_{ij}(\boldsymbol{\kappa}; \mathbf{x}, t)\, S_{0j}(\boldsymbol{\kappa}, t)\, d\boldsymbol{\kappa}, \qquad (3.12a)$$

where at $t = 0$, $Q_{ij} = \delta_{ij}e^{i\boldsymbol{\kappa}\cdot\mathbf{x}}$, but for $t > 0$, Q_{ij} is determined by the dynamical equations. Similar transfer functions can be defined for pressure, II_m, in terms of the initial velocity S_{0j}, and for vorticity, q_{im}, in terms of the Fourier transform of the initial vorticity. Note that Q_{ij} and q_{im} are related by

$$\epsilon_{ijk}\frac{\partial Q_{kr}}{\partial x_j} = i\, q_{in}\epsilon_{npr}k_p. \qquad (3.12b)$$

From (3.11) and (3.12) the changed two-point moments are determined by the transfer functions and the original spectrum:

$$R_{ij}(\mathbf{x}, \mathbf{x} + \mathbf{r}) = \int_{\boldsymbol{\kappa}} Q_{im}^* Q_{j\ell}\Phi_{0m\ell} d\boldsymbol{\kappa}. \qquad (3.13)$$

All the other one- and two-point second moments and spectra can be derived from (3.13) (e.g. Hunt 1973).

In most computations of rapid distortion theory, it has been assumed that the input turbulence is isotropic (and incompressible), so that $\Phi_{0ij}(\mathbf{k})$ could be expressed simply and uniquely as

$$\Phi_{0ij} = \frac{E(k)}{4\pi k^4}(k^2\delta_{ij} - \kappa_i\kappa_j), \qquad (3.14a)$$

where $k^2 = \kappa_i\kappa_j$ and $E(k)$ is the energy spectrum, whose integral is

$$\int_0^\infty E(k)dk = \frac{1}{2}u_iu_i. \qquad (3.14b)$$

To investigate how distorted homogeneous turbulence depends on its initial conditions, either the spectrum $E(k)$ or the anisotropy are varied. For axisymmetric turbulence Φ_{0ij} is defined by Batchelor (1953); for isotropic turbulence in two-dimensions, (3.14) can be applied to two dimensions only. In axisymmetric turbulence

$$\Phi_{0ij} = I_{ij}B_1(k, \boldsymbol{\kappa}.\mathbf{e}) + H_{ij}B_2(k, \boldsymbol{\kappa}.\mathbf{e}),$$

where

$$I_{ij} = \delta_{ij} - \frac{k_ik_j}{k^2}, \qquad H_{ij} = e_ie_j + \frac{(k_\ell e_\ell)^2}{k^2}\delta_{ij} - \frac{k_\ell e_\ell(e_ik_j + e_jk_i)}{k^2}. \qquad (3.15)$$

As a simple hypothesis it is usually further assumed that B_2 is zero and B_1 is a function of k only (Sreenivasan & Narasimha 1978; Maxey 1982).

Note that where the initial state of the turbulence is assumed to be homogeneous, the general representation (3.13) does not necessarily imply that the turbulence remains homogeneous while being distorted. However, in many RDT calculations (e.g. Batchelor & Proudman 1954), the turbulence is effectively homogeneous during the distortion. In that case the transfer functions Q_{in} and q_{in} can be expressed (in a suitable moving frame) as

$$\{Q_{in}(\boldsymbol{\kappa}; \mathbf{x}, t), q_{ip}(\boldsymbol{\kappa}; \mathbf{x}, t), \Pi_n(\boldsymbol{\kappa}; \mathbf{x}, t)\}$$

$$= \left\{ A_{in}(\boldsymbol{\chi}, t), a_{ip}(\boldsymbol{\chi}, t), \hat{P}_n(\boldsymbol{\chi}, t) \right\} e^{(i\boldsymbol{\chi} \cdot \mathbf{x})}, \qquad (3.16a)$$

where

$$a_{ip} = -\epsilon_{ijk}\epsilon_{nmp}k_m\chi_j A_{kn}/k^2 \qquad (3.16b)$$

and $\boldsymbol{\chi}$ is the local deformed wavenumber defined by the constraint that wave fronts are conserved, i.e.

$$d\chi_i/dt + \chi_l \partial U_l/\partial x_i = 0, \qquad (3.16c)$$

d/dt being evaluated in the moving frame, and $\boldsymbol{\chi}(t = 0) = \boldsymbol{\kappa}$. In these homogeneous distortions, a local energy spectrum tensor can be derived from the transfer function, viz.

$$\Phi_{ij}(\mathbf{x}) = A^*_{in} A_{jm}(\boldsymbol{\chi}, t)\Phi_{onm}(\boldsymbol{\kappa}). \qquad (3.17)$$

In many cases the turbulence is homogeneous in only one direction (say x_3) and in time, and then (3.16) can be generalized (following Phillips 1955 and Hunt 1973) to

$$Q_{in}(\boldsymbol{\kappa}; \mathbf{x}, t) = M_{in}(\boldsymbol{\chi}; x_1, x_2)e^{i(\boldsymbol{\chi}_3 x_3 + \omega t)}, \qquad (3.18)$$

where

$$d\chi_3/dt + \chi_3 \partial U_3/\partial x_3 = 0.$$

In this case spectra can be defined for wavenumber χ_3 or frequency ω.

3.3. Methods of solution

The essential point about RDT is that it is a method for calculating what happens to an initial velocity distribution using the linearized equations of motion under particular kinds of distortion, such as occur in the boundary-value problems classified in §2.1.2(a) and §2.1.3. In some cases, RDT provides a practical method of calculating turbulent flows at the appropriate level of moments (e.g. second order, two point) and appropriate accuracy. It is not a method of explaining how any turbulent flow arises, nor in general, a method of calculating the flow everywhere in a flow domain. However, it can be used as a diagnostic tool for studying certain aspects of the mechanics of turbulent flows. Many different kinds of distortion and initial condition have been used in a wide range of practical and fundamental studies. Different methods of solution can usefully be classified according to whether the distortions are homogeneous or inhomogeneous.

3.3.1. *Homogeneous distortion (without body forces)*

In this case the turbulent velocity and vorticity fields are homogeneous and can be represented by a three-dimensional Fourier transform throughout the distortion, as described by (3.16). This form of solution is appropriate if the rate of strain of the mean velocity field $\partial U_i / \partial x_k$ is uniform, so that the mean velocity can be expressed as $U_i = x_j \alpha_{ij}$.

(i) Using the linearized vorticity equation (3.1*b*), and substituting (3.16*a*) and (3.16*b*) leads to

$$\frac{da_{in}}{dt} = a_{jn} \frac{\partial U_i}{\partial x_j} + i\Omega_j \chi_j A_{in}, \tag{3.19}$$

which reduces to an equation for the vorticity tensor:

$$\frac{da_{in}}{dt} = \beta_{ik} a_{kn}, \qquad \text{where} \qquad \beta_{ik} = \alpha_{ik} - \epsilon_{ijk}\epsilon_{lmn}\alpha_{nm}\chi_l\chi_j/\chi^2, \tag{3.20a,b}$$

and χ_l is given by (3.16*c*). In general $\partial U_i / \partial x_k$ and $\Omega_\ell = \epsilon_{lmn}\partial U_n / \partial x_m$ are specified as functions of time. Also at $t = 0$, $a_{in} = \delta_{in}$.

For irrotational mean flows, where $\Omega_l = 0$, β_{ik} is independent of the wavenumber vector χ_l, and the change in $a_{in}(t)$ depends solely on the integral of the strain rate $\int (\partial U_i / \partial x_k)\, dt$. Also in this case the vorticity tensor $\beta_{ik} = \alpha_{ik}$, which is just the negative of the tensor given by (3.16*c*) for the rate of change of the wavenumber i.e. $\alpha_{ki} = \alpha_{ik}$. But for rotational mean flows, the change in a_{in} is dependent on the wavenumber and on the history of the changes in $\partial U_i / \partial x_k$. There is no simple relation between β_{ik} and how χ_i changes. Specific examples of sequences of irrotational and rotational strains were calculated by Townsend (1980) and Sreenivasan (1985).

Unique solutions for a_{in}, and thence A_{in} and χ_i can be expressed as

$$a_{in}(\boldsymbol{\chi}, t) = T_{ij}(t)a_{jn}(\boldsymbol{\kappa}, 0), \qquad \chi_i = \kappa_j S_{ji}$$

where the deformation tensors T_{ij}, S_{ji} can be formally expressed as integrals of the $\beta_{ik}(t)$ and $\alpha_{ij}(t)$ (e.g. Kida & Hunt 1989).

Once A_{in} and χ_i are found, the new three-dimensional spectra and cross-correlations can be derived from (3.12) and (3.16). Usually (3.20*b*) is only used for irrotational distortions, but it has been used for combinations of irrotational and rotational distortions by Kida & Hunt (1989).

(ii) The alternative approach (developed by Craya 1958, Deissler 1968 and Townsend 1976) to the calculation of the transfer function is to use the linearized momentum equations directly. For a locally homogeneous solution, the transfer function A_{in} satisfies

$$\frac{d}{dt} A_{in} = -A_{jn}\alpha_{ij} - i\chi_i \hat{p}_n. \tag{3.21a}$$

Using continuity, $\chi_j A_{jn} = 0$, and (3.16*c*) for the change of χ_j, $\hat{p}_n$ can be expressed in terms of A_{in} as

$$\hat{p}_n = i\left[\chi_i A_{jn}\alpha_{ij} - A_{in}d\chi_i/dt\right]\Big/\chi^2. \tag{3.21b}$$

Therefore $dA_{in}/dt = \mu_{ik}A_{kn}$, where

$$\mu_{ik} = -\left[\alpha_{ik} - 2\chi_i\chi_j\alpha_{jk}/\chi^2\right] \quad \text{and} \quad A_{kn}(t=0) = \delta_{kn}. \tag{3.22}$$

This solution shows how, even in this linear theory the pressure gradient generates fluctuating motions in directions perpendicular to the mean velocity $\mathbf{U}$ of the straining motion, and tends to reduce the motion in the direction of mean strain. From (3.22) A_{in} can be defined explicitly for weak distortion, i.e. $t\|\nabla\mathbf{U}\| \ll 1$, (Crow 1968) viz.

$$A_{in}(t) = (\delta_{ik} + t\mu_{ik}(t=0))\,\delta_{kn}. \tag{3.23}$$

Even for finite distortions, the equations can be integrated analytically in cases where α_{ik} is constant in time. Simple results are available for pure shear, pure rotation, and irrotational distortion (e.g. Townsend 1970; Cambon & Jacquin 1989).

3.3.2. *Inhomogeneous distortions*

Now consider the theory when the integral length scale of the initial homogeneous turbulence L_x is comparable with the length scale L_D over which the mean velocity gradients $(\nabla\mathbf{U})$ vary, or comparable with the distance n to a boundary which makes the turbulence inhomogeneous (such as a density discontinuity or a boundary with another kind of velocity field), i.e.

$$L_x > L_D \sim \|\nabla\mathbf{U}\|\Big/\|\nabla\nabla\mathbf{U}\| \quad \text{or} \quad L_x > n.$$

New solutions have to be found for the velocity, vorticity and pressure transfer functions $(Q_{in},\, q_{in},\, II_n)(\boldsymbol{\kappa}; \mathbf{x}, t)$.

So far general methods have only been found for *irrotational* flows, but for *rotational* mean flows where the turbulence is inhomogeneous, solutions have been found only for a few classes of mean flow, initial turbulence and boundary conditions. A feature of all inhomogeneous problems is that the boundary and initial conditions have a significant effect on the solution and have to be carefully specified.

Irrotational mean flow. In these problems the distortion of a weak random vorticity field by a strong irrotational mean straining flow is calculated. There may also be a significant (not necessarily weak) irrotational fluctuating velocity field. There are two methods

(a) *The vorticity method.* Since the mean vorticity is zero, i.e. $\boldsymbol{\Omega} = \nabla \wedge \mathbf{U} = 0$, the linearised vorticity equation (3.1b) reduces to

$$\partial\boldsymbol{\omega}/\partial t + (\mathbf{U}.\nabla)\boldsymbol{\omega} = (\boldsymbol{\omega}.\nabla)\mathbf{U}. \tag{3.24}$$

Then $\mathbf{u}$ is solved from $\boldsymbol{\omega} = \nabla \wedge \mathbf{u}$ and $\nabla.\mathbf{u} = 0$, subject to suitable boundary conditions.

First $\boldsymbol{\omega}$ is solved in terms of its initial or boundary vorticity (both denoted here by $\boldsymbol{\omega}_0$) and the mean velocity field, using Cauchy's theorem. Let $\mathbf{x}$ be the position of a fluid element at time t and $\mathbf{a}(\mathbf{x})$ be its position earlier at time t_0 (or $t_\mathbb{B}$ if on $\mathbb{B}$) when it is first advected by the mean flow. By considering corresponding small changes in $\mathbf{x}$ and $\mathbf{a}$, which is equivalent to considering the distortions of a line element (Batchelor 1967, Chap. 6), it follows that the solution to (3.24) is

$$\omega_i(\mathbf{x}, t) = \omega_{0j}(\mathbf{a}, t_0)\partial x_i/\partial a_j. \tag{3.25}$$

(b) *The velocity method.* In an important development in RDT, Goldstein (1978) showed that in an irrotational distortion the velocity field $\mathbf{u}$ could be computed directly without the necessity of first computing the vorticity field. In his analysis he rediscovered Weber's (1868) result that the velocity of a fluid element at $\mathbf{x}$ at time t could be expressed as the sum of an irrotational component $\nabla\phi$ and a rotational component $\mathbf{u}_R$ directly related to the velocity of the same element of $\mathbf{a}$ at time t_0, using the inverse of the same material deformation tensor as used for the change of vorticity, i.e.

$$u_i(\mathbf{x}, t) = u_{Ri} + \frac{\partial \phi}{\partial x_i}(\mathbf{x}, t) \tag{3.26a}$$

where

$$u_{Ri}(\mathbf{x}, t) = \frac{\partial a_j}{\partial x_i} u_j(\mathbf{a}, t_0) \tag{3.26b}$$

and $\phi(\mathbf{x}, t)$ is a velocity potential which is zero when $t = t_0$. To calculate ϕ it is only necessary to satisfy continuity ($\nabla.\mathbf{u} = 0$) by solving one Poisson equation:

$$\nabla^2 \phi = -\nabla.\mathbf{u}_R, \tag{3.26c}$$

subject to $(\mathbf{u}_R + \nabla\phi)$ satisfying the boundary conditions (3.25). (By comparison the method (a) requires in principle four supplementary Poisson-type equations to be solved.)

The result (3.26) can be most easily understood by considering the change of $\mathbf{u}.d\mathbf{x}$ of a line element as it moves from $\mathbf{a}$ to $\mathbf{x}$, using Kelvin's theorem (Hunt 1987). Consequently (3.26) is also valid for barotropic, compressible flows.

This method was used to calculate the distortion of turbulence in a two-dimensional contracting duct (such as an aeroengine compressor duct) where the turbulence scale L_x is comparable with the scale of the duct, so that the eddies impact on its walls, as well as being strong distorted (Goldstein & Durbin 1980).

Rotational mean flow. An essential feature of RDT solutions for homogeneous or inhomogeneous flows is that no assumptions are made about the variation with time of the transfer function Q_{in} or q_{in} in $(\boldsymbol{\kappa}; \mathbf{x}, t)$. Therefore in some parts of the flow there may be amplification and in other parts reduction of different components of the turbulence velocity. Therefore the

RDT solutions for inhomogeneous rotational straining motions are difficult to obtain analytically (for all wavenumbers) because the assumption that the time and space variations are decoupled is not made, as in the case of hydrodynamic stability theory, and because, when $\boldsymbol{\Omega} \neq 0$, the equation (3.1b) for fluctuating vorticity $\boldsymbol{\omega}$ also includes extra terms in the velocity, viz. $(\boldsymbol{\Omega}.\nabla)\mathbf{u}$ and $-(\mathbf{u}.\nabla)\boldsymbol{\Omega}$.

Interesting solutions have been obtained for inhomogeneous turbulence in various kinds of unidirectional shear flows. In the first case, consider a uniform shear flow $\mathbf{U}$ over a rigid flat boundary parallel to the mean flow, where

$$\mathbf{U} = (U_0 + \alpha x_2, 0, 0), \qquad x_2 > 0, \qquad (3.27a)$$

the turbulent vorticity $\boldsymbol{\omega}_0$ is initially homogeneous at $t = 0$, and there is no motion across the boundary, so

$$u_2 = 0 \quad \text{on} \quad x_2 = 0, \quad t > 0 \qquad (3.27b)$$

(Maxey 1978; Lee & Hunt 1989). In a uniform shear the equations yield an explicit equation for $u_2(\mathbf{x}, t)$, namely

$$\left(\frac{\partial}{\partial t} + U_1 \frac{\partial}{\partial x_1} \right) \nabla^2 u_2 = 0. \qquad (3.28)$$

Therefore a solution can be constructed in the form

$$u_2 = u_2^{(H)}(\mathbf{x}, t) + \partial \phi / \partial x_2, \qquad (3.29a)$$

where $u_2^{(H)}$ is a homogeneous solution such that

$$\nabla^2 u_2^{(H)} = \nabla^2 u_2^{(H)} (\mathbf{x}, t = 0) \quad \text{and} \quad \nabla^2 \phi = 0. \qquad (3.29b, c)$$

Thence using (3.17) and (3.29) the 'transfer function' for u_2 can be expressed as

$$Q_{2n} = A_{2n}(\boldsymbol{\chi}, t)e^{i(\boldsymbol{\kappa}.\mathbf{x})} + \frac{\partial \hat{\phi}_n}{\partial x_2}(\boldsymbol{\kappa}_{13}, y; t)e^{i(\boldsymbol{\kappa}_{13}.\mathbf{x})} \qquad (3.30a)$$

where

$$\boldsymbol{\chi} = (\kappa_1, \kappa_2 - \alpha\kappa_2, \kappa_3), \quad \boldsymbol{\kappa}_{13} = (\kappa_1, 0, \kappa_3); \qquad (3.30b)$$

$$A_{2n}(\boldsymbol{\chi}, t) = (k^2/|\boldsymbol{\chi}|^2)A_{2n}(\boldsymbol{\kappa}, t = 0) \qquad (3.30d)$$

and

$$\partial^2 \hat{\phi}_n / \partial x_2^2 - |\boldsymbol{\kappa}_{13}|^2 \hat{\phi}_n = 0,$$

subject to

$$\partial \hat{\phi}_n / \partial x_2 = A_{2n} \quad \text{on} \quad x_2 = 0. \qquad (3.30e)$$

From (3.21) – (3.23) II_n, Q_{1n} and Q_{3n} can be calculated from Q_{2n}. When $\alpha = 0$ this is the solution for rapid changes in a turbulent flow

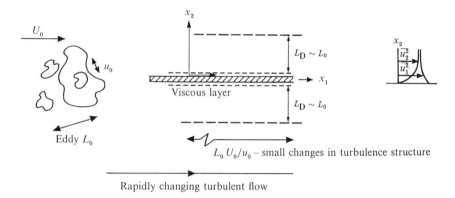

Fig. 6 A rapidly changing turbulent flow for a rigid surface introduced into a flow. $L_{\mathbb{D}}$ and L_0 are the lengthscales of the inhomogeneous layer $\mathbb{D}$, and of the incident turbulence $\mathbb{E}$.

when a rigid surface or 'wall' is introduced into the flow, and is valid over a period $0 < t \leqslant T_L$ (see Figure 6). Note that the solutions for u_1, u_2, u_3 (and their statistics) do not change over this time scale. The analysis can be generalised to allow for viscosity near the wall so that a no-slip boundary condition on the fluctuating velocity could be applied (Hunt & Graham 1978).

Other solutions of turbulence that is inhomogeneous in parallel regions (shear flows, stratified layers etc.) have been obtained by Phillips (1955), Gartshore *et al.* (1983) and Carruthers & Hunt (1986).

4. Some effects of shear on the structure of turbulence

In this section we consider in detail the solutions of RDT for turbulence in uniform shear and near boundaries, and review the relevance of these solutions to understanding slowly changing turbulent shear flows.

4.1. Uniform shear

Consider a uniform shear flow $\mathbf{U} = (\alpha x_2, 0, 0)$ rapidly distorting, over a period $t_0 < t < t_1$, a homogeneous turbulent velocity field $\mathbf{u}$ whose initial energy spectrum at $t = t_0$ is $E_0(k)$. For homogeneous distorted turbulence, the velocity transfer functions are as defined in (3.16): $Q_{in}(\boldsymbol{\kappa}; \mathbf{x}, t) = A_{in}(\boldsymbol{\chi}, t)e^{i\boldsymbol{\chi} \cdot \mathbf{x}}$, where (from (3.16c)) the changing wavenumber is

$$\boldsymbol{\chi} = (\kappa_1, \kappa_2 - \beta\kappa_1, \kappa_3), \tag{4.1}$$

and $\beta = \alpha t$. Townsend (1970) derived the analytical solutions for the components of the amplitude function, A_{in}:

$$
A_{11}\left(\boldsymbol{\chi}(\kappa,t),t\right) = \left\{ \frac{\beta \kappa_1^2}{k_{13}^2} \left(\frac{k^2 - 2\kappa_2^2 + \beta \kappa_1 \kappa_2}{\chi^2} \right) \right.
$$
$$
\left. - \frac{k^2}{k_{13}^3} \frac{\kappa_3^2}{\kappa_1} \left[\tan^{-1}\left(\frac{\kappa_2}{k_{13}} \right) - \tan^{-1}\left(\frac{\kappa_2 - \beta \kappa_1}{k_{13}} \right) \right] \right\} A_{22}(k,t_0), \qquad (4.2a)
$$

$$
A_{22}\left(\boldsymbol{\chi}(\kappa,t),t\right) = \frac{k^2}{\chi^2} A_{22}(k,t_0), \qquad (4.2b)
$$

$$
A_{33}\left(\boldsymbol{\chi}(\kappa,t),t\right) = \left\{ \frac{\beta \kappa_1 \kappa_3}{k_{13}^2} \left(\frac{k^2 - 2\kappa_2^2 + \beta \kappa_1 \kappa_2}{\chi^2} \right) \right.
$$
$$
\left. + \frac{k^2 \kappa_3}{k_{13}^3} \left[\tan^{-1}\left(\frac{\kappa_2}{k_{13}} \right) - \tan^{-1}\left(\frac{\kappa_2 - \beta \kappa_1}{k_{13}} \right) \right] \right\} A_{22}(k,t_0), \qquad (4.2c)
$$

where k is the wavenumber at $t = 0$ and $k_{13}^2 = \kappa_1^2 + \kappa_3^2$.

Note that these expressions are independent of the magnitude of the wave number (k), which is a special feature of homogeneous distortions! For the earlier history of this solution, and its generalisation to an exact solution of the Navier-Stokes equations, see Craik & Criminale (1986).

The three-dimensional energy spectrum tensor $\Phi_{ij}(\boldsymbol{\chi},t)$ of the distorted turbulence is determined by its initial value $\Phi_{0nm}(\boldsymbol{\kappa}, t = 0)$ and by $A_{in}(\boldsymbol{\chi}(\boldsymbol{\kappa},t),t)$ from (3.17). The covariances $\langle u_i u_j \rangle$ are obtained by integrating Φ_{ij} over all wavenumber space $\boldsymbol{\chi}$ or $\boldsymbol{\kappa}$. The initial spectrum $\Phi_{0nm}(\boldsymbol{\kappa})$ is characterised by the anisotropy of the variances of the components of turbulent velocity, i.e.

$$
\int_{\boldsymbol{\kappa}} \Phi_{0ij} d\boldsymbol{\kappa} \left/ \int_{\boldsymbol{\kappa}} \Phi_{0ll} d\boldsymbol{\kappa} \right.
$$

and of their distribution in wavenumber space, e.g.

$$
\int_{\boldsymbol{\kappa}} (\kappa_i \kappa_j / \kappa^2) \Phi_{0ll} \, d\boldsymbol{\kappa} \left/ \int_{\boldsymbol{\kappa}} \Phi_{0ll} \, d\boldsymbol{\kappa} \right.
$$

(c.f. Kida & Hunt 1989 and Reynolds 1989), and by the distribution of turbulent energy over different scales, i.e.

$$
E_0(k) = \frac{1}{2} \int_{|\boldsymbol{\kappa}|=k} \Phi_{0ll}(\boldsymbol{\kappa}) \, dA(\boldsymbol{\kappa}) \qquad (4.3a)
$$

(Batchelor 1953). In some cases the distribution of $\Phi_{0ij}(\boldsymbol{\kappa})$ over shells in wavenumber space is the same for all k, then

$$
\Phi_{0ij}(\boldsymbol{\kappa}) = \hat{\Phi}_{0ij}(\hat{\boldsymbol{\kappa}}) E_o(k), \qquad (4.3b)
$$

where $\hat{\boldsymbol{\kappa}} = \boldsymbol{\kappa}/k$ and

$$\int_{\hat{\boldsymbol{\kappa}}=1} \hat{\Phi}_{0ij} \, d\hat{\boldsymbol{\kappa}} = 1.$$

Since A_{in} is independent of k, the integral of (4.3) for the covariance $\langle u_i u_j \rangle$ can be written as

$$\langle u_i u_j \rangle = \int_0^\infty \left(\int_{|\boldsymbol{\kappa}|=k} A_{in}^* A_{jm} \Phi_{0nm}(\boldsymbol{\kappa}) dA(\boldsymbol{\kappa}) \right) dk. \qquad (4.4)$$

If $\Phi_{0ij}(\boldsymbol{\kappa})$ has the same form for all k, using expressions for A_{ij} normalized by k, (4.4) reduces to

$$\langle u_i u_j \rangle (t) = \langle u_{0l} u_{0l} \rangle \int_{|\boldsymbol{\kappa}|=k} A_{in}^*(\hat{\boldsymbol{\kappa}}) A_{jm}(\hat{\boldsymbol{\kappa}}) \hat{\Phi}_{0nm}(\hat{\boldsymbol{\kappa}}) \, d\hat{\boldsymbol{\kappa}}. \qquad (4.5)$$

Results for $\langle u_i u_j \rangle$ have been obtained for two cases where the turbulence is initially isotropic or initially axisymmetric about the streamwise direction x_1, and Φ_{0ij} is given by (3.14a) or (3.15) (Maxey 1982).

When the strain is finite ($\beta \gtrsim 1$), evaluating the integral (4.5) requires computation. It is found that the way that $\langle u_i u_j \rangle$ changes with small strain continues with larger strain, i.e. $\langle u_1^2 \rangle$ and $\langle u_3^2 \rangle$ increase without limit, $-\langle u_1 u_2 \rangle$ increases to a limiting value, and $\langle u_2^2 \rangle$ continues to decrease. For large strain when $\beta \gg 1$, $\langle u_1^2 \rangle \propto \beta$, $\langle u_2^2 \rangle \propto \ln\beta$ and $\langle u_2^2 \rangle \propto (\ln\beta)/\beta$; (Rogers 1991). The Reynolds shear stress is simply related to $\langle u_1^2 \rangle$ from the kinetic energy equation. Since $\langle u_i u_i \rangle \to \langle u_1^2 \rangle$, $-\langle u_1 u_2 \rangle \to \langle u_1^2 \rangle /2\beta \to$ constant as $\beta \to \infty$. The initial state of the turbulence does not effectively change the relative orders of magnitude of the moments but it does change their actual values. For isotropic turbulence $\rho_{12} \approx 0.6$, when $\beta \approx 2$, but for anisotropic axisymmetric turbulence, ρ_{12} is reduced. For $R = 2$, $\rho_{12} = 0.4$ (Maxey 1982). When $S \gg 1$, these RDT computations agree well with nonlinear results (Lee et al. 1987).

4.1.1. Spectra

To investigate how the spatial structure of the turbulence changes with shear, 'two-point' spectra can be calculated using (4.2) and (4.3a). We are particularly interested in the form of the spectra at high wavenumber to see whether there are any universal features of the small-scale turbulence in shear flow.

It is instructive to express the spectra in terms of the local wavenumber χ. In a shear flow after a finite distortion the spherical surface on which χ is constant in wavenumber space corresponds to a surface $\chi^2 = \kappa_1^2 + (\kappa_2 - \kappa_1\beta)^2 + \kappa_3^2$, which in $\boldsymbol{\kappa}$-space is a spheroid flattened in the κ_1 direction and rotated and elongated in the direction $\kappa_2 = \kappa_1\beta$. The form of the components $E_{ii}(\chi)$ of the energy spectrum $E(\chi)$ are calculated (eq. 4.8) by Hunt & Carruthers (1990).

Hunt & Carruthers (1990) derived the form of $E(\chi)$ when $\beta \gg 1$ by using asymptotic analysis of the integral (eq. 4.8 - HC) with two forms of the initial spectra

$$E_0(k) = u_0^2 L(kL)^N \exp(-k^2 L^2) \qquad (4.6a)$$

and the von Kárman energy spectrum

$$E_0(k) = u_0^2 (kL)^4 / \left[1 + (kL)^2 \right]^p , \qquad (4.6b)$$

where N is an integer ($N \geq 0$) and $p > 1$. They found that whatever the initial spectrum, provided it decreases faster than k^{-2}, (which is characteristic of a spectrum of vortex sheets,) for large enough strain, over an increasing range of wavenumber, the energy spectrum tends to the limiting form of

$$E(\chi) \propto \hat{\beta} \chi^{-2}. \qquad (4.7)$$

where $\hat{\beta} = \beta \int_0^\infty k E_0(k) \, dk$ and β is a constant. This results holds for broader classes of spectra than those specified in (4.6). However if the initial spectrum decreases slower than k^{-2} as k increases, in high Reynolds number turbulence ($Re_\lambda > 10^4$) the effect of shear is to maintain a $-5/3$ spectrum, which has been well established in many field experiments (e.g. see Monin & Yaglom 1971; Wyngaard & Cote 1972).

Computations of $E_{ii}(\chi)$ for the streamwise and vertical velocity components are displayed in fig. 7 (for the simple spectrum $E(k) \propto \exp\left(-k^2 L^2\right)$. These show a tendency to the χ^{-2} spectrum, for the streamwise component and χ^{-4} for the vertical component when the initial spectrum decays faster than k^{-2}. (By the action of nonlinear processes the spectra of all components tend to have the same form as the most amplified—as shown by the DNS results of Rogallo 1981)

The similarity of k^{-2} spectra is associated with the existence of vortex sheets (on a scale much less than χ^{-1}), rather than being characteristic of any cascade process, which is found at much higher Reynolds numbers. In the former case, the 'vortex sheets' are highly localized gradients $\partial u_1 / \partial x_2, \partial u_1 / \partial x_3$, which surround long, narrow regions or 'streaks', where the streamwise fluctuations are positive or negative. These phenomena have been seen in DNS of individual flow realisations (Lee et al. 1987).

The results above imply that the essential mechanism responsible for the formation of the streaks in most turbulent shear flows is contained in the *linear* theory rather than being the effect of nonlinear interactions. The algebraic form of the spectra is consistent with the existence of discontinuities in velocity of velocity gradients on the scale of k^{-1}.

The spectra for the small scales of turbulent flows appear to be highly dependent on the Reynolds number (e.g. Re_λ) of the turbulence and the on mean velocity distribution. For fully developed turbulence generated by grids in wind tunnels, or obtained by DNS, without mean shear (where $Re_\lambda \lesssim 300$) the spectra decrease rapidly with k, typically $E(k) \propto e^{-k^2 L^2}$

(a)

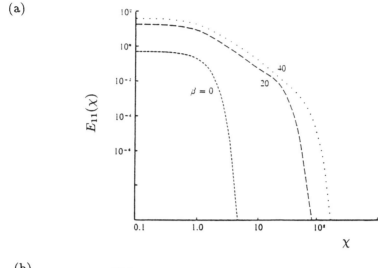

(b)

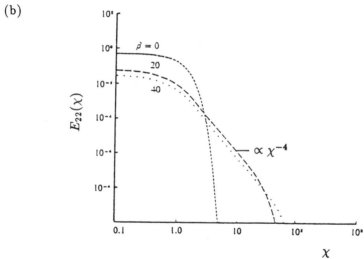

Fig. 7 RDT calculation of $E_{ii}(\chi)$ for increasing shear rate β. The initial spectrum is $E_0(k) = u_0^2 L e^{-k^2 L^2}$. (a) $E_{11}(\chi)$; (b) $E_{22}(\chi)$.

at the lowest values of Re_λ (Champagne *et al.* 1970; Rogallo 1981), and $E(k) \propto k^{-n_H}$, where $n_H \geqslant 2$, at the higher values of Re_λ. However, in shear flows, even at these ranges of Re_λ, it is quite usual to find that $E_0(k)$ decays algebraically, i.e. $E_0(k) \propto k^{-n}$ when $-5/3 \leqslant n \leqslant 2$ – over a significant range. Rogallo (1981) found $n \to 2$ in his computations of turbulence in a homogeneous shear flow which extended to $\beta = 18$ and Rogers & Moin (1987) found that $n \approx 3$ for $\beta = 8$. In Champagne *et al.*'s (1970) wind-tunnel measurements they found that $n = 2$, even for quite small strain.

4.2. Turbulence near a boundary

The idealised RCT analysis in §3.3.2 of turbulence in a uniform flow over a boundary (obtained by putting $\alpha = 0$) showed how the profiles of one point moments (e.g., the variance) of the normal component u_2 of turbulence are affected by the boundary but do not change with time. There is a simple physical reason, namely, that the normal velocities of the eddies are 'blocked' by the surface. It is found that, in high Reynolds number flow,

$$\langle u_2^2(y) \rangle \simeq C_2 \varepsilon^{\frac{2}{3}} y^{\frac{2}{3}} \quad \text{as} \quad y/L \to 0, \tag{4.8}$$

where L is the integral lengthscale away from the boundary and C_2 can be obtained in terms of the Kolmogorov constants α_k, and the moment of u_2 between two points y and y_1 $(y \leqslant y_1)$ of the normal velocity component, is given by

$$\hat{r}_{22}(y, y_1) = \langle u_2(y)u_2(y_1) \rangle / \langle u_2^2(y_1) \rangle \simeq y/y_1. \tag{4.9}$$

For the same reasons as explained in §2.2, these linear 'statistical eigensolution' which do not vary with time, can be expected to be approximately valid in SCT, for the same flow. (4.8) has been found to compare well with measurements in the convective turbulent boundary layer, below the liquid surface of a turbulent channel flow, and at a rigid surface moving with the free stream – in all these flows ε is constant as $y \to 0$. It is not valid at the rigid surface below a boundary layer where $\varepsilon \propto y^{-1}$ (e.g. Hunt 1984). (4.9) has been found to agree with field measurements of convective boundary layer, and neutral boundary layer (even though $\varepsilon \propto y^{-1}$) (see Hunt *et al.* 1989).

Furthermore, the concept of (4.9) has been found to apply in shear boundary layers, even though both the shear and the blocking affect the velocity of the large eddies. The measurements of v in the atmospheric boundary layer by Kaimal & Gaynor on the 300m at the Boulder Atmospheric Observatory, (using the same methods described in the paper by Hunt, Kaimal & Gaynor 1988) fit a curve $\hat{r}_{22} = f(y/y_1)$ which is about 15% below the simple expression of (4.9) (see figure 8). Data was taken when there was a strong shear in the atmpsphere; in one case these was neutral stratification and in the other case stable stratification (with the Monin-Obukhov length of 27m).

4.3. Use of Kinematic Simulation to demonstrate structure

In the following section, we show that in purely kinematic simulations which use random Fourier modes constrained to have a given energy spectrum (see Fung 1990 and Fung *et al.* 1990), similar 'streaks' were observed. By following the work of Wray & Hunt (1989), we objectively characterise these and other structures in an ensemble of realisations of the flow field,

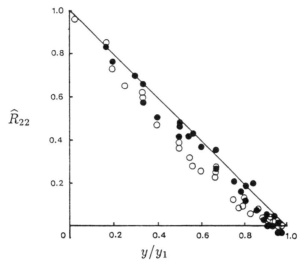

Fig. 8 Cross correlation of u_2 at height y and y_1 normalised by $\langle u_2^2 \rangle$ at y_1.
Symbols, data from atmospheric measurements at Boulder Atmospheric
Observatory; ———, theoretical predictions of Hunt (1984): $\hat{R}_{22} \simeq$
y/y_1.

and show how the shape and distribution of different regions change when
shear is applied.

4.3.1. *Methods of simulation*

The numerical method of the Kinematic simulation of turbulence (KS) with
mean shear distortion is essentially the same as in the case of homogeneous,
isotropic turbulence simulation i.e. the technique of KS is to superpose
linearly space-time Fourier modes with random coefficients and to allow
the large scales to advect the small scales. The simulations are started
from an initially-random velocity field with a prescribed Eulerian energy
spectrum. The distortion of the velocity field by the mean shear is included
by varying the wave-vectors $\boldsymbol{\kappa}$ and the Fourier coefficients $\mathbf{a}_n(\boldsymbol{\kappa})$ and $\mathbf{b}_n(\boldsymbol{\kappa})$
with time according to the transfer function (4.2) (see Fung 1990).

4.3.2. *Detailed flow structures*

In this section we investigate the detailed structure of the computed flow
field in uniform shear. This will be done by examining contour plots of
instantaneous velocity and vorticity fields. We also see how the structures
changes with different initial spectra.

    In order to examine the effect of low shear rate on the turbulence
structure, contours of streamwise turbulent velocity u from a homogeneous
shear flow simulation with small $\alpha = 20.0$ and at $\beta = 9.4$ are plotted on

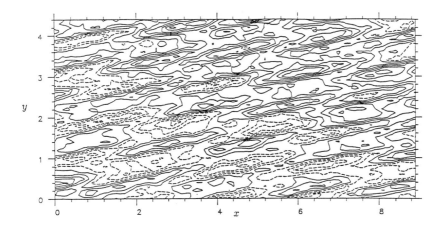

Fig. 9 Contours of the streamwise turbulent velocity u for a homogeneous turbulent field with von Kármán energy spectrum at low shear rate: $\alpha = 20.0$
and $\beta = 9.4$. Positive values are contoured by solid lines and negative
values are contoured by dashed lines.

a horizontal (x, y)-plane in figure 9 which indicates that at low shear rate,
structures are not highly elongated in the streamwise direction.

In figures 10a and b, similar u-contour plots in a homogeneous turbulent shear flow field at high shear rate ($\alpha = 43.2$) are shown at different
times $\beta = 4.7$ and 9.4 respectively. At the earlier time ($\beta = 4.7$), figure 10a
shows structures somewhat elongated in the streamwise direction. The contour plot at a later time ($\beta = 9.4$ in figure 10b) clearly shows the existence
of highly elongated high- and low-speed streak structures alternating in the
spanwise direction similar to the streaks observed in the near-wall region
of turbulent boundary layers (Kline *et al.* 1967). In contrast, the normal
and spanwise velocity components do not exhibit streaky structures. This
means that, as shear is prolonged, the lateral extent of the streaks is mainly
controlled by the effect of shear, and the presence of a solid boundary is not
necessary, also that the essential mechanism responsible for the formation
of the streaks is contained in the linear theory (Lee & Hunt 1989).

Figure 11a shows the contour plot of spanwise vorticity fluctuation, ω_z
in the same (x, z)-plane as in figure 10b.

Streamwise streak pattern of y-vorticity can also be seen in figure 11b.
Conditions are identical to those of figure 11a, and careful comparison with
that figure shows that the vorticity spots reside in the relatively quiescent
corridors between the high velocity streaks.

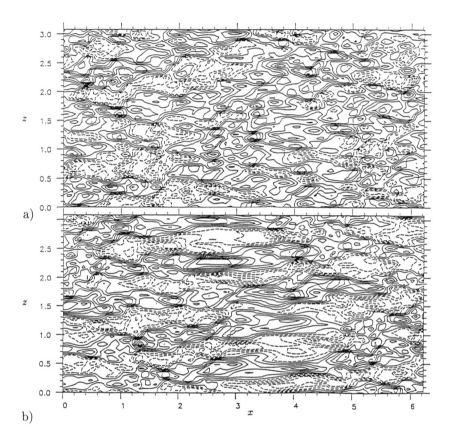

Fig. 10 Contours of the streamwise turbulent velocity u for a homogeneous turbulent field with von Kárman energy spectrum at high shear rate: $\alpha = 43.2$
and (a) $\beta = 4.7$; (b) $\beta = 9.4$.

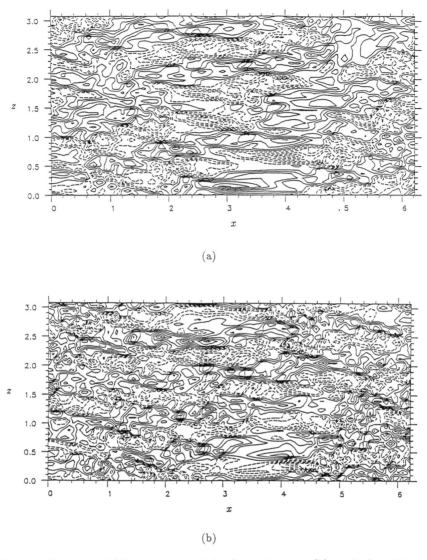

(a)

(b)

Fig. 11 Contours of (a) spanwise vorticity fluctuation ω_y; (b) vertical vorticity fluctuation ω_z in the (x, y)-plane.

4.3.3. *Boundary effects*

The use of kinematic simulation with Rapid Distortion Theory has also been used to look at the flow structures near boundaries and density interfaces, in zero mean shear. RDT has been used in a succession of papers to calculate the 'transfer coefficients' Φ_{ij} in a series of increasingly complex flows. In Hunt & Graham (1978) and Hunt (1984) a rigid boundary was considered whilst in Carruthers and Hunt (1986, 1988), turbulent flow is bounded by one or two stably stratified layers. In these cases both propagating and trapped internal gravity waves are excited by the turbulence, which has defined statistical properties; away from the interface. The reader is referred to the papers for details. Here we shall merely demonstrate some of the flow structures.

Figure 12a shows a cross section through the interface in the plane perpendicular to the interface for the two layer model; The parameter N is the buoyancy frequency of the stably stratified layer . The figure shows that strong wave motion generated in the stably stratified layer and large convective structures in the turbulence. Figure 12b shows contours of vertical velocity in a vertical cross section in the case where the turbulence is bounded by an inversion layer (thickness h and buoyancy frequency N_2) with a less stably stratified (buoyancy N_3) layer beyond. Resonant waves are excited in the very stable layer but there is little excitation of waves in the deep stable layer. In both simulations the vertical velocity is continuous at the interface bordering the turbulence but the horizontal velocity is discontinuous; locally this can result in very strong shears.

5. Discussion and tentative conclusions

In this paper we have reviewed some of the aspects of turbulence structure and some developments in the techniques, the applications, and the interpretation of RDT. We have shown how it provides some insights into the structure of turbulent shear flow.

5.1. RDT 'Statistical eigensolutions'

There are certain features of turbulent flow structure in shear flow as calculated by RDT, in which moments of certain components of the turbulent velocity reach a steady state, or change very slowly, even when the turbulence is being rapidly distorted. These are 'eigensolutions', in the sense that, if the initial turbulence was specified to have these forms, the particular statistical features of the turbulence would change little under the action of the distortion (provided it was rapid). In the first case of inhomogeneous turbulence, near a rigid surface, $x_2 = 0$, with or without mean shear, these features are the moments such as $\langle u_2(x_2')u_2(x_2)\rangle / \langle u_2^2(x_2)\rangle$,

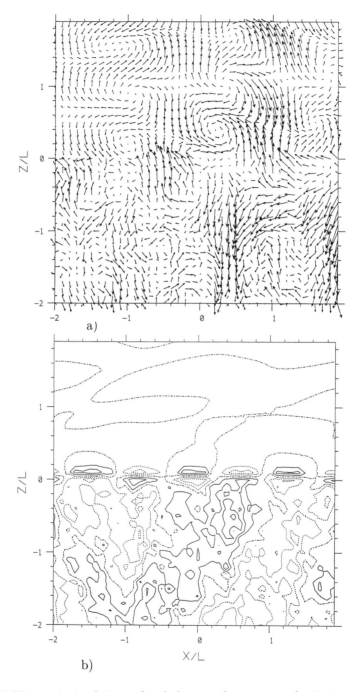

Fig. 12 Kinematic simulations of turbulence and waves near density interfaces.
(a) Two-layer model: vertical cross-section of velocity vectors through
the interface, $NL/u' = 6.0$; (b) Three-layer model: vertical cross-section
of contours of vertical velocity, $N_2L/u' = 50.0$, $N_3L/u' = 4.0$ and $h/L = 0.1$.

$\langle u_2(x_2')u_2(x_2)\rangle \big/ (\langle u_1^2\rangle \langle u_2^2\rangle)^{\frac{1}{2}}$, and other components. These correlations depend weakly on the initial anisotropy and the form of the energy spectra.

In the second case, of locally homogeneous turbulence in a uniform shear these features include the shear-stress cross correlation coefficient

$$\rho_{12} = -\langle u_1(x_1)u_2(x_2)\rangle \big/ [\langle u_1^2\rangle \langle u_2^2\rangle]^{\frac{1}{2}},$$

which changes very slowly ($\propto \ln^{-\frac{1}{2}} \beta$, for $\beta \gg 1$) and the structure function, which, for a wide class of flows, becomes proportional to the spacing, $|\mathbf{r}|$, when it is much less than the largest scales and much greater than the scales controlled by viscosity, i.e.

$$\left\langle [u_i(\mathbf{x}) - u_i(\mathbf{x}+\mathbf{r})]^2 \right\rangle = B|\mathbf{r}|. \tag{5.1}$$

(This is equivalent to the energy spectra $E(\chi)$ becoming proportional to χ^{-2} over a range of χ, when $\beta \gg 1$). The proportionality factor B ($\sim \beta u_0^2 L_0$ with dimensions LT^{-2}) increases with time, with the shear dU_1/dx_2, and with the initial kinetic energy of the turbulence u_0^2, but inversely with the initial length scale L_0 (defined by $\int_0^\infty k^{-1}E(k)dk/u_0^2$). In this case the shear-stress cross-correlation coefficient depends on the anisotropy of the initial spectrum, but not the form of the spectrum, whereas the *form* of the structure function and the energy spectra (over a given range of $|\mathbf{r}|$ and χ) tends to become independent of the initial turbulence, provided that in the initial spectra $E_0(k) = o(k^{-2})$ when $k \gg L^{-1}$.

5.2. Extrapolation of results to slowly changing turbulence

The nonlinear processes in a turbulent flow can only be estimated and modelled approximately; but it is clear that they affect the energy and anisotropy of the turbulence on a time scale L/u_0. So, if the turbulence is distorted significantly on this time scale by a linear process (i.e. $T^* \gg 1$), the effect of the nonlinear terms is approximately equivalent to a continual change in the initial conditions of RDT calculations. Therefore if certain results of the RDT calculation (the 'statistical eigensolutions') are not only changing slowly with time, but are approximately independent of the anisotropy and energy spectrum of the initial turbulence, the form of these RDT solutions also approximately describes turbulent flows that persist over many timescales, i.e. slowly changing turbulent flows, such as shear flows and flows bounded by a rigid surface (Hunt 1984) or a density interface (Carruthers & Hunt 1986). The numerical values of the coefficients, such as ρ_{12}, of these 'eigensolutions', depend on the initial anisotropy, and consequently different coefficients can be expected in different turbulent flows with different initial and boundary conditions.

This is essentially the argument for using the results of RDT to provide new insights and practical models for many kinds of slowly changing turbulent flows SCT.

More complete models of SCT require modelling and understanding of the effects of the nonlinear processes, one aspect of which is the random mean distortion of small-scale turbulence by large scales, and the transfer of energy between the scales.

5.3. Characteristic structure of the turbulence

Linear rapid distortion theory has shown how a velocity field of random Fourier components undergoing mean shear produces many of the characteristic structures in the flow that are observed in experiments and DNS at moderate Reynolds number. In the remaining part of this section, following the work by Hunt *et al.* (1988) and Wray and Hunt (1990), we objectively characterise the flow structure in terms of the local values of the deformation tensor $u_{i,j}$, specifically the second invariant $II = \dfrac{\partial u_i}{\partial x_j} \dfrac{\partial u_j}{\partial x_i}$ of the velocity field and the pressure p. The flow is divided into eddy regions (E), corresponding to regions of high swirl where $II < -II_0$, $p < -p_0$, convergence regions (C), where $II > II_0$, $p > p_0$, and streaming regions (S) where $-II_0 \leqslant II \leqslant II_0$ and $u^2 \geqslant u_0^2$. Here II_0, p_0 and u_0 are the r.m.s. values of II, p and u values. By using these criteria, we show how the shape and distribution of the (E), (S) and (C) regions change when shear is applied.

Figure 13a shows the streaming regions of an initially isotropic velocity field distributed on the surfaces of a cuboid. They seem to be connected, their shapes are somewhat elongated with no preferred directions and they are uniformly distributed. However, after shear is applied at $\beta = 5.0$ (Fig. 13b), the distributions of streaming regions seem to be rather discrete and much more elongated in the direction of shear.

Figures 14a and 15a show the eddy and convergence regions of an initially isotropic velocity field. It can be seen that both regions are distributed uniformly in space with some of the eddy regions connected and elongated, whilst the convergence regions are of round shapes. After shear is applied at $\beta = 5.0$, figure 14b and 15b show that both regions are elongated in the direction of shear. In order to find out the angle of inclination after distortion, we assume that individual eddy or convergence regions have elliptical shapes, and ellipses have been fitted to these regions, according to the following criteria: (i) equal position of centroid of area; (ii) equal area; (iii) equal 2nd moments of area. The results are shown in figure 16 and the average angle of inclination to the direction of the flow after an average of 10 realisations (of volume $(2L)^3$) is $9°$ for streaming regions, $12°$ for eddy regions and $13°$ for convergence regions. Obviously, these angles of inclination will be different for different values of $t(DU/dy)$. This is surprising since vorticity is at $45°$ to the flow direction (Rogers and Moin 1987). The angle of the principle axes of the Reynolds stress tensor is defined as

$$\alpha_s = \frac{1}{2}\tan^{-1}\left[\frac{2\langle u_1 u_2\rangle}{\langle u_1^2\rangle - \langle u_2^2\rangle}\right].$$

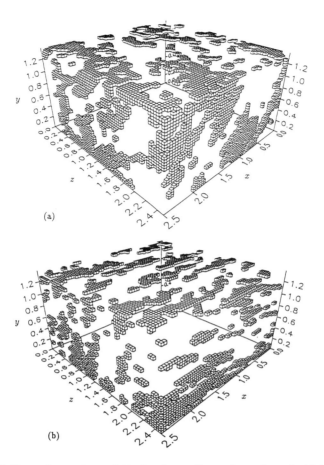

Fig. 13 Streaming regions plotted only on the surface of a cuboid (a) initially homogeneous, isotropic turbulence with $\beta = 0.0$; (b) after shear is applied with $\beta = 5.0$.

Rogers and Moin (1987) found that it increases as Reynolds number decreases and typically $20° \lesssim \alpha_s \lesssim 30°$.

5.4. Turbulence structure

In §2 we reviewed boundary and initial conditions on different turbulent flows, and showed that there are many turbulent flow where the time that fluid elements spend in the flow domain is of the order of or less than the turbulence time scale (which may itself be varying). Consequently in the turbulence structure of such flows, there are certain features that are *specific* to these flows. The sensitivity depends on whether they are RCT ($T_D \ll T_L$) or SCT flows where $T_D \sim T_L$, and on the nature of the flows (c.f. strain flows to shear flows). In addition large eddies with significant energy extend over the width of any inhomogeneity that are present.

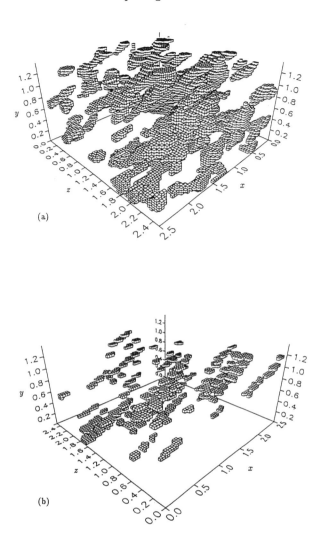

Fig. 14 Eddy regions plotted only in the volume of a cuboid.

On the other hand we showed in §4 that in certain turbulent flows or certain regions of a wide class of turbulent flows (e.g. near surfaces or in well-developed shear layers), there are general feature of the turbulence structure of the large scale motion, some aspects of which are largely determined by local linear mechanics.

Our conclusion is that there are classes of flows where there are common features of turbulence structure and limited sensitivity to initial conditions. Also, in these and other classes of flows the intrinsic nonlinear 'scrambling' processes are too slow for the turbulence to develop a form

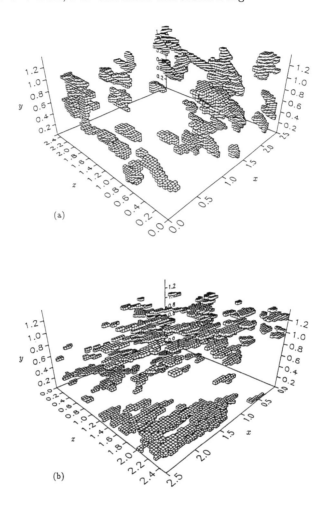

Fig. 15 Convergence regions plotted in the volume of a cuboid.

completely independent of the initial conditions. An important point that
RDT has shown is that certain *linear* processes (especially shear) are effective in changing the turbulence into a general form that is insensitive to
initial conditions, and also that linear distortions can change the spectra so
as to enhance greatly the nonlinear processes giving the turbulence a more
general structure (Hunt & Vassilicos 1991).

    There are reasons why general models of turbulence (e.g. Lumley 1978,
Launder 1975) are effective for certain classes of flow (especially shear flow
near boundary), but can be quite poor in the turbulent flows which are
sensitive to their initial and boundary conditions.

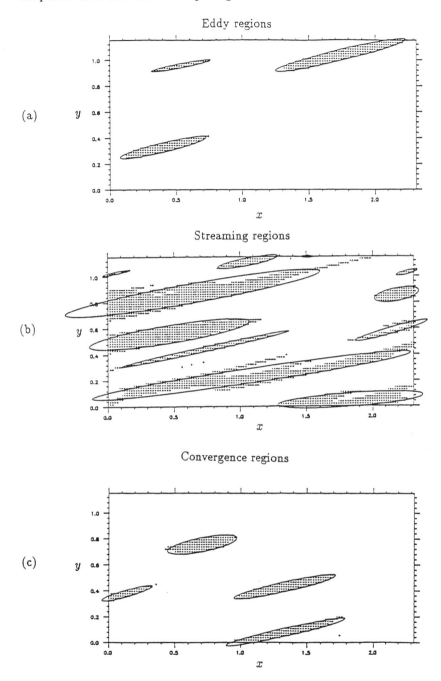

Fig. 16 Elliptics have been fitted to (a) eddy regions; (b) streaming regions and
(c) convergence regions.

Acknowledgements

J.C.R.H. is grateful to the organisation of the Newport conference for being invited to give this paper. D.J.C. acknowledges financial support from NERC and SERC. J.C.H.F. acknowledges financial support from Peterhouse and the use of computing resources at UKAEA Harwell. We are grateful to R.J. Perkins for his help in developing the method of Kinematic Simulation. We also gratefully acknowledge many conversations with many colleagues, especially M. Lee, P. Moin and A. Wray (C.T.R. NASA Ames), F. Hussain (Houston), C. Cambon (Lyon), M. Lesieur (Grenoble).

References

Adrian, R.J & Moin, P. 1988 Stochastic estimation of organized turbulent structure: homogeneous shear flow. *J. Fluid Mech.* **190**, 531–559.

Antonia, R.A. & Bisset, D.K. 1990 Spanwise structure in the wall region of a turbulent boundary layer. *J. Fluid Mech.* **210**, 437–458.

Batchelor, G.K. 1953 *The Theory of Homogeneous Turbulence.* Cambridge University Press, 197pp.

Batchelor, G.K. 1955 The effective pressure, exerted by a gas in turbulent motion. In: *Vistas in Astronomy* (ed. A. Beer), vol. 1, pp. 290–295, Pergamon.

Batchelor, G.K. 1967 *An Introduction to Fluid Dynamics.* Cambridge University Press, 615pp.

Batchelor, G.K. & Proudman, I. 1954 The effects of rapid distortion of a fluid in turbulent motion. *Q.J. Mech. Appl. Math.* **7**, 83.

Bavilagau, P.M. & Lykads, P.S. 1978 Turbulence memory in self-presering wakes. *J. Fluid Mech.* **89**, 589–606.

Britter, R.E., Hunt, J.C.R. & Richards, K.J. 1981 Analysis and wind-tunnel studies of speed-up, roughness effects and turbulence over a two-dimensional hill. *Q.J. Roy. Met. Soc.* **107**, 91–110.

Cambon, C. & Jacquin, L. 1989 Spectral approach to non-isotropic turbulence subjected to rotation. *J. Fluid Mech.* **202**, 295–318.

Carruthers, D.J., Fung, J.C.H., Hunt, J.C.R. & Perkins R.J. 1989 The emergence of characteristic eddy motion in turbulent shear flows. *Proc. Organized Structures and Turbulence in Fluid Mechanics* (ed. M. Lesieur & O. Metais), Kluwer Academic Publishers.

Carruthers, D.J. & Hunt, J.C.R. 1986 Velocity fluctuations near an interface between a turbulent region and a stably stratified layer. *J. Fluid Mech.* **165**, 475–501.

Carruthers, D.J. & Hunt, J.C.R. 1988 Turbulence, waves, and entrainment near density inversion layers. Proc. I.M.A. Conf. on "Stably Stratified Flow and Dense Gas Dispersion" (Ed. J.S. Puttock), Clarendon Press, pp. 77–96.

Champagne, F.H., Harris, V.G. & Corrsin, S. 1970 Experiments on nearly homogeneous turbulent shear flow. *J. Fluid Mech.*   **41**, 81–139.

Comte-Bellot, G. & Corrsin, S. 1966 The use of a contraction to improve the isotropy of grid-generated turbulence. *J. Fluid Mech.* **25**, 657–682.

Craik, A.D.D. & Criminale, W.O. 1986 Evolution of wavelike disturbances in shear flow: a class of exact solutions of the Navier-Stokes equations. *Proc. R. Soc. Lond.* **A406**, 13–26.

Craya, A. 1958 Contribution à l'analyse de la turbulence associée à des vitesses moyennes. *P.S.T. Ministère de l'Air* **345.**,

Crow, S.C. 1968 Turbulent Rayleigh shear flow. *J. Fluid Mech.*   **32**, 113–120.

Davidson, P.A., Hunt, J.C.R. & Moros, A. 1988 Turbulent recirculating flows in liquid metal magnetohydrodynamics. *Prog. in Astronaut. Aeronaut.* **111**, 400–420.

Deissler, R.G. 1968 Effects of combined two-dimensional shear and normal strain on weak locally homogeneous turbulence and heat transfer. *J. Math. Phys.* **47**, 320.

Domaradzki, J.A. & Rogallo, R.S. 1990 Local energy transfer and nonlocal interactions in homogeneous isotropic turbulence. *Phys. Fluids* **A2**, 413–426.

Dritschel, D. 1989 Contour dynamics and contour surgery: numerical algorithms for extended, high-resolution modelling of vortex dynamics in two-dimensional, inviscid, incompressible flows. *Comput. Phys. Rep.* **10**, 77–146.

Farge, M. & Holschneider, M. 1990 Wavelet analysis of coherent structures in two-dimensional flows. In *Proc. IUTAM Symp. On Topological Fluid Mechanics*, pp. 765–776 (Ed. H.K. Moffatt & A. Tsinober), Cambridge University Press.

Ferré, J.A., Mumford, J.C., Savill, A.M. & Giralt, F 1990 Three-dimensional large-eddy motions and fine-scale activity in a plane turbulent wake. *J. Fluid Mech.* **210**, 371–414.

Favre, A., Guitton, H., Lichnerowicz, A.&Wolff, E. 1988 *De la causalité à la finalite – a propos de la turbulence,* Maloine.

Fung, J.C.H. 1990 Kinematic Simulation of turbulent flow and particle motions. Ph.D dissertation University of Cambridge.

Fung, J.C.H., Hunt, J.C.R., Malik,N.A. and Perkins,R.J. 1991 Kinematic simulation of homogeneous turbulent flows generated by unsteady random Fourier modes. Submitted to *J. Fluid Mech.*

Gartshore, I.S., Durbin, P.A. & Hunt, J.C.R. 1983 The production of turbulent stress in a shear flow by irrotational fluctuations. *J. Fluid Mech.* **137**, 307–329.

Gaster, M., Kit, E. & Wygnanski, I. 1985 Large-scale structures in a forced turbulent mixing layer. *J. Fluid Mech.* **150**, 23–39.

Gilbert, A. 1988 Spiral structures and spectra in two-dimensional turbulence. *J. Fluid Mech.*   **193**, 475–497.

Goldstein, M.E. 1978 Unsteady vortical and entropic distortion of potential flow round arbitrary obstacles. *J. Fluid Mech.* **89**, 433–468.

Goldstein, M.E. & Durbin, P.A. 1980 The effect of finite turbulence spatial scale on the amplification of turbulence by a contracting stream. *J. Fluid Mech.* **98**, 473–508.

Hayakawa, M. & Hussain, F. 1989 Three dimensionality of organized structures in a plane turbulent wake. *J. Fluid Mech.* **206**, 375–404.

Ho, C.M. & Huerre, P. 1984 Perturbed free shear layers. *Ann. Rev. Fluid Mech.* **16**, 365–424.

Hunt, J.C.R. 1973 A theory of turbulent flow round two-dimensional bluff bodes. *J. Fluid Mech.* **61**, 625–706.

Hunt, J.C.R. 1978 A review of the theory of rapidly distorted turbulent flow and its applications. *Proc. of XIII Biennial Fluid Dynamics symp., Kortowo, Poland, Fluid Dyn. Trans.* **9**, 121–152.

Hunt, J.C.R. 1984 Turbulence structure in thermal convection and shear-free boundary layers. *J. Fluid Mech.* **138**, 161–184.

Hunt, J.C.R. 1987 Vorticity and vortex dynamics in complex turbulence flow. *Trans. Can. Soc. Mech. Eng.* **11**, 21–35.

Hunt, J.C.R. 1988 Studying turbulence using direct numerical simulation: 1987 Center for Turbulence Research NASA Ames-Stanford Summer Programme. *J. Fluid Mech.* **190**, 375–392.

Hunt, J.C.R. 1990 Review of *De la causalité à la finalite – a propos de la turbulence*, Maloine, by Favre, A., Guitton, H., Lichnerowicz, A. & Wolff, E., 1988. Euro. J. of Mech. B/Fluids.

Hunt, J.C.R. & Graham, J.M.R. 1978 Free .ream turbulence near plane boundaries. *J. Fluid Mech.* **84**, 209–235.

Hunt, J.C.R., Kaimal, J.C. & Gaynor, E. 1988 Eddy structure in the convective boundary layer – new measurements and new concepts. *Q. J. Roy. Met. Soc.* **114**, 827–858.

Hunt, J.C.R., Moin, P., Lee, M., Moser, R.D., Spalart, P. & Mansour, N.N. 1989 Cross correlation and length scales in turbulent flow near surfaces. In *Proc. of the Second European Turbulence Conference.* Advances in Turbulence 2 (ed. H.H. Fernholz & H.E. Fiedler), pp. 128–134, Springer-Verlag.

Hunt, J.C.R., Stretch, D.D. & Britter R.E. 1988 Length scales in stably stratified turbulent flows and their use in turbulence models. In *Proc. IMA Conf. on Stably Stratified Flow and Dense Gas Dispersion* (ed. J.S. Puttock), Chester, England, April 1986, pp. 285–321, Clarendon.

Hunt, J.C.R. & Carruthers, D.J. 1990 Rapid distortion theory and the 'problem' of turbulence. *J. Fluid Mech.* **212**, 497–532.

Hunt, J.C.R.& Vassilicos, J.C. 1991 Kolmogrov's contributions to the physical and geometrical understanding of turbulent flows and recent developments. *Proc. Roy. Soc. A* , July 1991.

Hussain, A.K.M.F. 1986 Coherent structures and turbulence. *J. Fluid Mech.* **173**, 303–356.

Husain, H.S. & Hussain, F. 1990 Subharmonic resonance in a shear layer. In *Proc. of the Second European Turbulence Conference.* Advances in Turbulence 2 (ed. H.H. Fernholz & H.E. Fiedler), pp. 96–101, Springer-Verlag.

Jeandel, D., Brison, J.F. & Mathieu, J. 1978 Modelling methods in physical and spectral space. *Phys. Fluids* **21**, 169–182.

Kida, S. & Hunt, J.C.R. 1989 Interaction between turbulence of different scales over short times. *J. Fluid Mech.* **201**, 411–445.

Klebanoff, P.S. & Sargent, L.H. 1962   The three-dimensional nature of boundary layer. *J. Fluid Mech.* **12**, 1–34.

Kellogg, R.M. & Corrsin, S. 1980 Evolution of a spectrally local disturbance in grid-generated, nearly isotropic turbulence. *J. Fluid Mech.* **96**, 641–669.

Kline, S.J. 1981 Universal or zonal modelling – the road ahead. In *Proc. of Complex Turbulent flows.* Stanford University, California. vol. II, pp. 991–998.

Kline, S.J., Reynolds, W.C., Schraub, F.A. & Runstadler, P.W. 1967 The structure of turbulent boundary layers. *J. Fluid Mech.* **30**, 741–773

Kolmogorov, A.N. 1941 The local structure of turbulence in incompressible viscous fluid for very large Reynolds numbers. *Acad. Sci., USSR* **30**, 301–305.

Komori, S., Ueda, H., Ogino, F. & Mizushina, T. 1983 Turbulence structure in stably stratified open channel flow. *J. Fluid Mech.* **130**, 13–26.

Landahl, M.T. 1990 On sublayer streaks. To appear in J. Fluid Mech.

Launder, B.E., Reece, G.J. & Rodi, W. 1975 Progress in the development of a Reynolds stress turbulence closure. *J. Fluid Mech.* **68**, 537–566.

Launder, B.E.&Spalding, D.B. 1972   *Mathematical Models of turbulence.* Academic.

Lee, M.J. & Hunt, J.C.R. 1988 The structure of sheared turbulence near a boundary. *Report, Center for Turbulence Research, Stanford, No. CTR-S88,* pp. 221–242.

Lee, M.J. & Hunt, J.C.R. 1989   The structure of shear turbulence near a plane boundary. In *Proc. of the Seventh Sypm. on Turbulent Shear Flows.* Stanford University. pp. 8.1.1–8.1.6.

Lee, M.J. Kim, J. & Moin, P. 1987 Turbulent structure at high shear rate. In: *Sixth Symposium on Turbulent Shear Flows, Toulouse, France* (ed. F. Durst *et al.*), pp. 22.6.1–22.6.6.

Lesieur, M. 1987 *Turbulence in Fluids.* Martinus Nijhoff.

Lesieur, M & Metais, O. 1989 *Proc. Pole European Pilote de Turbulence (PEPIT), ERCOFTAC Summer School, Lyon July 1989.*

Leslie, D.C. 1973 *Developments in the Theory of Turbulence.* Clarendon.

Liu, J.T.C. 1989 Contributions to the understanding of large-scale coherent structures in developing free turbulent shear flows. *Adv. in Appl. Mech.* **26**, 535–540.

Lumley, J. 1965 The structure of inhomogeneous turbulent flows. *Proc. Int. Coll. on Radio Wave Propagation.* (ed. A.M. Yaglom & V.I. Takasky). *Dokl. Akad. Nauk. SSSR.* 166–178.

Lumley, J.L. 1978 Computational modelling of turbulent flows. *Adv. in Appl. Mech.* **18**, 126–176.

Malkus, W.V.R. 1956 Outline of a theory of turbulent shear flow. *J. Fluid Mech.* **1**, 521–539.

Mason, P.J. & Sykes R.I. 1982 A two-dimensional numerical study of horizontal roll vortices in an inversion capped planetary boundary layer. *Q. J. Roy. Met. Soc.* **108**, 801–823.

Maxey, M.R. 1978 Aspects of Unsteady Turbulent Shear Flow. Ph.D. Dissertation, University of Cambridge.

Maxey, M.R. 1982 Distortion of turbulence in flows with parallel streamlines. *J. Fluid Mech.* **124**, 261–282.

Melander M.V. & Hussain, F. 1990 Cut-and connect of antoparaellel vortex tubes. In *Proc. IUTAM Symp. On Topological Fluid Mechanics.* Cambridge University Press.

Moffatt, H.K. 1967 On the suppression of turbulence by a uniform magnetic field. *J. Fluid Mech.* **28**, 571–592.

Moffatt, H.K. 1984 Simple topological aspects of turbulent vorticity dynamics. In: *Turbulence and Chaotic Phenomena in Fluids* (ed. T. Tatsumi), pp. 223–230, Elsevier.

Monin, A.S.& Yaglom, A.M. 1971 *Statistical Theory of Turbulence,* vol. II. MIT Press.

Mumford, J.C. 1982 The structure of the large eddies in fully developed turbulent shear flows. Part 1. The plane jet. *J. Fluid Mech.* **118**, 241–268.

Murakami, S. & Mochida, A. 1988 3D numerical simulation of airflow around a cubic model by means of a $k - \epsilon$ model. *J. Wind. Eng. Indust. Aerodyn.* **31**, No. 2.

Naish, A.& Smith, F.T. 1988 The turbulent boundary layer and wake of an aligned flat plate. *J. Eng. Sci. Maths.* **22**, 15–42.

Phillips, O.M. 1955 The irrotational motion outside a free turbulent boundary. *Proc. Camb. Phi. Soc.* **51**, 220-229.

Reynolds, W.C. 1989 Effects of rotation on homogeneous turbulence. In *Proc. of Australasian Conf. on Fluid Mechanics.*

Rogallo, R.S. 1981 Numerical experiments in homogeneous turbulence. *NASA Tech. Memo.* 81315.

Rogers, M.M. 1991 The structure of a passive scalar field with a uniform mean scalar gradient in rapidly sheared homogeneous turbulent flow. *Phys. Fluids* **A3**, 144-154.

Rogers, M.M. & Moin, P. 1987 The structure of the vorticity field in homogeneous turbulent flows. *J. Fluid Mech.* **176**, 33–66.

Sabot, J. & Comte-Bellot 1976 Intermittency of coherent structures in the core region of fully develop turbulent pipe flow. *J. Fluid Mech.* **74**, 767–796.

Savill, A.M. 1987 Recent developments in rapid-distorion theory. *Ann. Rev. Fluid Mech.* **19**, 531–570.

Sreenivasan, K.R. 1985 The effect of a contraction on a homogeneous turbulent shear flow. *J. Fluid Mech.* **154**, 187–213.

Sreenivasan, K.R. & Narasimha 1978 Rapid distortion of axisymmetric turbulence. *J. Fluid Mech.* **84**, 497–516.

Sulem, P.L., She, Z.S., Scholl, H. & Frisch, U. 1989 Generation of large-scale structures in three dimensional flow lacking partity-invariance. *J. Fluid Mech.* **205**, 341–358.

Tennekes, H.&Lumley, J.L. 1971 *A First Course in Turbulence.* MIT Press.

Townsend, A.A. 1970 Entrainment and the structure of turbulent flow. *J. Fluid Mech.* **41**, 13–46.

Townsend, A.A. 1976 *Structure of turbulent Shear Flow.* Cambridge University Press.

Townsend, A.A. 1980 The response of sheared turbulence to additional distortion. *J. Fluid Mech.* **98**, 171–191.

van Haren, L. 1991 1991 ERCOFTAC Summer School in Rutherford Appleton Laboratory, U.K.

Weber, W. 1868 Über eine Transformation der hydrodynamischen Gleichungen. *J. Reine Angew. Math.* **68**, 286.

Wong, H.Y.W 1985 Shear-free turbulence and secondary flow near angled and curved surfaces. Ph.D dissertation University of Cambridge.

Wray, A. & Hunt, J.C.R. 1990 Algorithms for classification of turbulent structures. In *Proc. IUTAM Symp. On Topological Fluid Mechanics* , pp. 95–104 (Ed. H.K. Moffatt & A. Tsinober), Cambridge University Press.

Wyngaard, J.C. & Cote, O.R. 1972 Modelling the buoyancy driven mixed layer. *J. Atmos. Sci.* **33**, 1974–1988.

Yakhot, V. & Orszag, S.A. 1986 Renormalization group analysis of turbulence. I. Basic theory. *J. Sci. Comput.* **1**, 3.

Zhou, M.D. 1989 A new modelling approach to complex turbulent shear flow. In *Proc. of the Second European Turbulence Conference.* Advances in Turbulence 2 (ed. H.H. Fernholz & H.E. Fiedler), pp. 146–150, Springer-Verlag.

3

Order and Disorder in Turbulent Flows

John L. Lumley

1. Introduction

Examination of pictures of turbulent flows reveals a wide range of order and apparent disorder, seeming to depend on the geometry, Reynolds number and initial conditions. The ordered component may arise as the remnant of an instability of the laminar flow that first gave rise to the turbulence, or as an instability of the turbulent profile and transport. Sometimes it is necessary to take the ordered component explicitly into account in a description of the flow (if the ordered component has arisen from a source different from that of the smaller scale, apparently disordered, component), and sometimes it is not necessary, if everything has arisen from the same source, both components are in equilibrium with each other, and there is only one scaling law for both. The disordered component, on the evidence of exact numerical simulations, is deterministic. By analogy with temporal chaos in mechanical systems with small numbers of degrees of freedom, it is tempting to identify it as resulting from a strange attractor, although there is no real evidence for this. The question seems to be of primarily philosophical interest. What evidence there is suggests that at reasonable Reynolds numbers the dynamical structure would in any event be too complex to be computable, nor are we interested in the details. It seems that a statistical approach to the structure in the phase space would make most sense, since we are interested presumably in global statements, bounds, and the like. At least for a single homogeneous scalar in one dimension, the representation of a disordered function as a superposition of ordered entities occurring with disordered phase and strength is well-known. Things become more complicated when the orientation also becomes disordered. In inhomogeneous situations, an attempt to identify order in chaos leads to empirical eigenfunctions. These can be shown to give the fastest-converging representation of a disordered field. In situations close to the critical value of the parameter controlling the flow (e.g. — close to the wall, where the Reynolds number is low), the complexity of the flow is low, and a large fraction of the energy is in the ordered component. In these situations, severely truncated dynamical models can be constructed using these eigenfunctions, as we shall describe below, and the models display many of the salient features of the modeled flow. It may be possible to use these models in a scheme to control the turbulent boundary layer, reducing the drag, or increasing the mixing. Recent work indicates that the boundary layer resulting from polymer addition, and the boundary layer due to this type of control, should be identical. The work to be described below is contained largely in [1, 2, 3].

2. The Proper Orthogonal Decomposition

Lumley [4] proposed a method of identification of coherent structures in a random turbulent flow. This uses what Love [5] called the Proper Orthogonal Decomposition, and which is often called the Karhunen-Love expansion. An advantage of the method is its objectivity and lack of bias. Given a realization of an inhomogeneous, energy integrable velocity field, it consists of projecting the random field on a candidate structure, and selecting the structure which maximizes the projection in quadratic mean. In other words, we are interested in the structure which is the best correlated with the random, energy-integrable field. More precisely, given an ensemble of realizations of the field, the purpose is to find the structure which is the best correlated with all the elements of the ensemble. Thus we want to maximize a statistical measure of the magnitude of the projection, which can be given by the mean square of its absolute value. The calculus of variations reduces this problem of maximization to a Fredholm integral equation of the first kind whose symmetric kernel is the autocorrelation matrix. The properties of this integral equation are given by Hilbert Schmidt theory. There is a denumerable set of eigenfunctions (structures). The eigenfunctions form a complete orthogonal set, which means that the random field can be reconstructed. The coefficients are uncorrelated and their mean square values are the eigenvalues themselves. The Kernel can be expanded in a uniformly and absolutely convergent series of the eigenfunctions and the turbulent kinetic energy is the sum of the eigenvalues. Thus every structure makes an independent contribution to the kinetic energy and Reynolds stress.

The most significant point of the decomposition is perhaps the fact that the convergence of the representation is optimally fast since the coefficients of the expansion have been maximized in a mean square sense. The mean square of the first coefficient is as large as possible, the second is the largest in the remainder of the series once the first term has been subtracted, etc.

We have described here the simplest case, that of a completely inhomogeneous, square-integrable, field. If the random field is homogeneous in one or more directions, the spectrum of the eigenvalues becomes continuous, and the eigenfunctions become Fourier modes, so that the proper orthogonal decomposition reduces to the harmonic orthogonal decomposition in those directions. See Lumley [4, 6, 7] for more details.

2.1 APPLICATION OF THE PROPER ORTHOGONAL DECOMPOSITION TO THE SHEAR FLOW OF THE WALL REGION

The flow of interest here is three dimensional, approximately homogeneous in the streamwise direction (x_1) and spanwise direction (x_3), approximately stationary in time (t), inhomogeneous and of integrable energy in the normal direction (x_2).

We want a three dimensional decomposition which can be substituted in the Navier-Stokes equations in order to recover the phase information carried by the coefficients. We have to decide which variable we want to keep. Time is a good candidate since we are particularly interested in the temporal dynamics of the structures. The idea is to measure the two velocities at the same time and determine $\langle u_i(x_1, x_2, x_3, t)u_j(x_1', x_2', x_3', t)\rangle = R_{ij}$. Since the flow is quasistationary, R_{ij} does not depend on time and nor do the eigenvalues and eigenfunctions. The information in time is carried by the coefficients $a(n)$ which are still "stochastic," but now evolve under the constraint of the equations of motion. Thus the decomposition becomes

$$(1) \qquad u_i(x_1, x_2, x_3, t) = \sum_n \int a_{k_1 k_3}^{(n)}(t)e^{2\pi i(k_1 x_1 + k_3 x_3)}\phi_{i k_1 k_3}^{(n)}(x_2)dk_1 dk_3,$$

$$(2) \qquad \int \Phi_{ij}(x_2, x_2')\phi^{(n)}(x_2')dx_2' = \lambda^{(n)}\phi_i^{(n)}(x_2),$$

and we have to solve equation (2.2) for each pair of wave numbers (k_1, k_3). Φ_{ij} now denotes the Fourier transform of the autocorrelation tensor in the x_1, x_3 directions.

Our second change to the decomposition is a transformation of the Fourier integral into a Fourier series, assuming that the flow is periodic in the x_1 and x_3 directions. The periods $L1, L3$ are determined by the first nonzero wave numbers chosen. Finally, each component of the velocity field can be expanded as the triple sum

$$(3) \qquad u_i(x_1, x_2, x_3, t) = \frac{1}{\sqrt{L_1 L_3}}\sum_{k_1 k_3 n} e^{2\pi i(k_1 x_1 + k_3 x_3)}a_{k_1 k_3}^{(n)}(t)\phi_{i k_1 k_3}^{(n)}(x_2).$$

In this case, a "structure" is defined by:

$$\frac{1}{\sqrt{L_1 L_3}}\sum_{k_1 k_3} e^{2\pi i(k_1 x_1 + k_3 x_3)}a_{k_1 k_3}^{(n)}(t)\phi_{i k_1 k_3}^{(n)}(x_2).$$

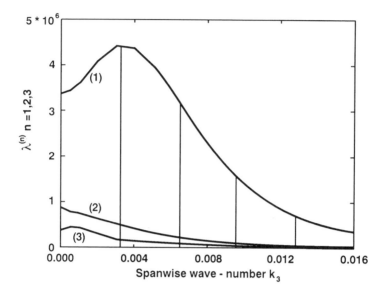

FIGURE 1. Convergence of the proper orthogonal decomposition in the near-wall region ($x_2^+ = 40$) of a pipe flow according to the experimental data. Turbulent kinetic energy in the first three eigenmodes. $\lambda^{(n)}(n = 1, 2, 3)$ function of the spanwise wavenumber (from [8]).

and the entire velocity field is recovered by the sum of all the structures (over n).

3. Experimental Results

The candidate flow we are investigating is the wall region (which reaches $x_2^+ = 40$; x_2^+ is the distance from the wall normalized by kinematic viscosity and friction velocity) of a pipe flow with almost pure glycerine (98%) as the working fluid [8]. The Reynolds number based on the centerline mean velocity and the diameter of the pipe is 8750. The corresponding Reynolds number based on the shear velocity ut is 531. From this data the auto-correlation tensor at zero time lag ($t - t' = 0$ between the two velocities), $R_{ij}(x_1 - x_1', x_2, x_2', x_3 - x_3')t - t' = 0$, was obtained and the spatial eigenfunctions were extracted by numerical solution of the eigenvalue problem. The results show that approximately 60% of the total kinetic energy is contained in the first eigenmode (figure 1.) and that the first three eigenmodes capture essentially the entire flow field as far as these statistics are concerned.

4. The Dynamical Equations

We decompose the velocity—or the pressure—into the mean (defined using a spatial average) and fluctuation in the usual way. We substitute this decomposition into the Navier-Stokes equations. Taking the spatial average of these equations we obtain, in the quasi-stationary case, an approximate relation between the divergence of the Reynolds stress and the mean pressure and velocity.

$$(4) \qquad \langle u_i, j u_j \rangle = -\frac{1}{\rho} P_{,i} + \nu U_{,jj} \delta_{i1}.$$

(where $u_{i,j}$ indicates the derivative with respect to x_j of u_i, and similarly for the other terms; repeated indices are summed). This is substituted in the Navier-Stokes equations, giving an equation for the fluctuating velocity. Equation (4) may be solved to give the mean velocity U in terms of the Reynolds stress $\langle u_1 u_2 \rangle$ in a channel flow in a manner which gives some feedback to the system of equations as the fluctuation varies. We will see that this feedback is necessarily stabilizing for the first structure (according to the experimental results) and increases as the Reynolds stress gets stronger. In other words, this term controls the intensity of the rolls, by reducing the mean velocity gradient as the rolls intensify, thus weakening the source of energy.

The expansion of the Fourier transform $\hat{u}_i$ of the fluctuating velocity u_i, is achieved by use of the complete set of eigenfunctions $f(m)$'s in an infinite sum:

$$(5) \qquad \hat{u}_i(k_1, k_3, x_2, t) = \sqrt{L_1 L_3} \sum_{m=1}^{\infty} a_{k_1 k_3}^{(m)}(t) \phi_{i k_1 k_3}^{(m)}(x_2).$$

After taking the Fourier transform of the Navier Stokes equations and introducing the truncated expansion, we apply Galerkin projection by multiplying the equations by each successive eigenfunction in turn, and integrating over the domain.

Finally we obtain a set of ordinary differential equations of the form:

$$(6) \qquad A \frac{da_{\mathbf{k}}^{(n)}}{dt} = B a_{\mathbf{k}}^{(n)} + \text{N.L.}$$

where A and B are matrices. Here A is the identity matrix (since the complete set of eigenfunctions is orthogonal) and N.L. are nonlinear terms. The non linear terms are of two sorts: quadratic and cubic. The quadratic terms come from the non linear fluctuation- fluctuation interactions and represent energy transfer between the different eigenmodes and Fourier modes. Their signs vary. The role of the Reynolds stresses $\langle u_i u_j \rangle$ on these terms should be mentioned. They vanish for all wave number pairs except for

$(k_1, k_3) = (0, 0)$ for which they exactly cancel the quadratic term. Therefore they prevent this mode from having any kind of quadratic interactions with other Fourier modes. Since the cubic terms are zero too, the (0,0) mode just decays by action of viscosity and does not participate in the dynamics of the system.

The cubic terms come from the mean velocity - fluctuation interaction corresponding to the Reynolds stress $\langle u_1 u_2 \rangle$ in the mean velocity equation (the other part of this equation leads to a linear term). Since the streamwise and normal components of the first eigenfunction have opposite signs, they make a positive contribution to the turbulence production and hence provide negative cubic terms which are thus stabilizing. We remark that this is not necessarily the case for higher-order eigenfunctions.

By use of the continuity equation and the boundary conditions (vanishing of the normal component at the wall, and at infinity) it can be seen by integration by parts that the pressure term would disappear if the domain of integration covered the entire flow volume. Since this is not the case (rather the domain is limited to $X_2^+ = 40$, where X_2^+ indicates the value of x_2^+ at the upper edge of the integration domain), there remains the value of the pressure term at X_2, which represents an external perturbation coming from the outer flow.

5. Energy Transfer Model

The exact form of the equations obtained from the decomposition, truncated at some cut-off point (k_{1c}, k_{3c}, n_c), does not account for the energy transfer between the resolved (included) modes and the unresolved smaller scales. The influence of the missing scales will be parameterized by a simple generalization of the Heisenberg spectral model in homogeneous turbulence. Such a model is fairly crude, but we feel that its details will have little influence on the behavior of the energy-containing scales, just as the details of a sub-grid scale model have relatively little influence on the behavior of the resolved scales in a large eddy simulation. This is a sort of St. Venant's principle, admittedly unproved here, but amply demonstrated experimentally by the universal nature of the energy containing scales in turbulence in diverse media having different fine structures and dissipation mechanisms (see [9] for a fuller discussion). The only important parameter is the amount of energy absorbed.

We will refer to α_1 as a Heisenberg parameter. We will adjust α_1 upward and downward to simulate greater and smaller energy loss to the unresolved modes, corresponding to the presence of a greater or smaller intensity of smaller scale turbulence in the neighborhood of the wall. This might correspond, for example, to the environment just before or just after a bursting event, which produces a large burst of small scale turbulence, which is then diffused to the outer part of the layer.

A term representing the energy fluctuation in the unresolved field due to the resolved field appears in the equation for the resolved field. This term could be combined with the pressure term and would not have any dynamical effect if the integration domain covered the entire flow volume. In our case, it needs to be computed since, like the pressure term, it leads to a term evaluated at X_2. We assume that the deviation (on the resolved scale) in the kinetic energy of the unresolved scales is proportional to the rate of loss of energy by the resolved scales to the unresolved scales. This pseudo-pressure term gives some quadratic feed-back. Because this approximation involves a further assumption, and to give ourselves greater flexibility, we call this parameter α_2, although in all work presented in this paper, we have set $\alpha_1 = \alpha_2$.

Thus the Heisenberg model introduces two parameters in the system of equations, one, α_1, in the linear term, the other one, α_2, in the quadratic term. The equations therefore have the following form:

$$(7) \qquad da_{k_1 k_3}^{(n)}/dt = L + (\nu + \alpha_1 \nu_T)L' + Q + \alpha_2 Q' + C,$$

where L and L' represent the linear terms, Q the direct quadratic terms, Q' the quadratic pseudo-pressure term and C the cubic terms arising from the Reynolds stress.

6. Numerical Simulations and Further Analysis of the Model ODEs

Numerical integrations of 3, 4, 5 and 6 mode models have been carried out, but we shall only report in detail on the 6 mode (5 active mode) simulations here.

We summarize here the behavior of the system. For $\alpha > 1.61$ a unique circle of globally attracting stable fixed points exists. For $1.37 < \alpha < 1.61$, an S^1-symmetric family of globally attracting double homoclinic cycles Γ exists, connecting pairs of saddle points which are p out of phase with respect to their second (x_2, y_2) components. The existence of the cycles Γ implies that, after a relatively brief and possibly chaotic transient, almost all solutions enter a tubular neighborhood of Γ and thereafter follow it more and more closely. As they approach Γ, the duration of the "laminar" phase of behavior (in which r_2 and r_4 remain non-zero and almost constant and r_1, r_3, r_5 grow exponentially in an oscillatory fashion) increases while the bursts (in which r_1, r_3 and r_5 collapse) remain short. In an ideal, unperturbed system, the laminar duration would grow without bound, but small numerical perturbations, such as truncation errors, presumably prevent this occurring in our numerical simulations. More significantly, the pressure perturbation will limit the growth of the laminar periods. Thus there is an effective maximum duration of events, which is reduced as α is decreased from the critical value $\alpha_b \sim 1.61$.

The existence of some of these solutions—notably the traveling waves and quasiperiodic motions—follow from rather general group theoretic considerations, cf., papers on $O(2)$ symmetry in [10, 11].

For the present study the intermittent behavior exhibited by the six mode model for α between 1.35 and 1.61 appears to be of greatest interest, since it is reminiscent in several ways of the physical instability, sweep and ejection event observed in boundary layer experiments.

The existence of the fixed points requires explanation. The exact solution of the Navier Stokes equations for x_1—invariant streamwise rolls has no fixed points. The equation for the streamwise vorticity has no steady solutions—there is no source of energy for the streamwise vorticity, and it must decay [12]. Initially the rolls have their planes of circulation tipped backward, so that the net vorticity vector is in the direction of the maximum positive strain rate; in this configuration the eddy can extract energy from the mean flow. As the eddy evolves, however, the planes of circulation are progressively tipped forward, until the net vorticity lies in the streamwise direction. At that point, the eddy must decay. Hence, typically, the eddy will at first grow in energy, pass through a maximum, and then decay [13]. Another way of seeing the same thing is to consider the Reynolds stress which, normalized by the energy, goes to zero as the direction of the vorticity approaches the streamwise. In our case, however, the ratio of the streamwise and normal components of the eddy velocity is fixed by the structure of the first eigenfunction. Hence, the Reynolds stress normalized by the energy is a fixed quantity, and cannot evolve. The eddy is thus supplied with energy continuously, and this is why the fixed points exist. In the real boundary layer, these eddies are born, go through their life cycle, and decay, first in one location and then in another. Statistically, the process is stationary, and the eddy captured by the Proper Orthogonal Decomposition Theorem is always in mid-life. In effect, the use of this theorem is a closure assumption, that provides a continual energy source for the eddy.

7. Implications for the Flow in the Wall Region

We describe here the reconstructions of the velocity field from the expansion, using the computed values of x_i and y_i. Probably the most interesting sets of solutions are those exhibiting intermittency, obtained for $1.3 \le \alpha \le 1.61$. In the flow field, the rapid event which follows the slowly growing oscillation and the repetition of the process reminds one of the bursting events experimentally observed [14, 15]. For that reason we call it a "burst." We will analyse its effect on the streamwise vortices.

In Figure 2 we show the time histories of the modal coefficients for $\alpha = 1.45$. A description of the motion of the eddies during a burst is given in Figure 3 for $\alpha = 1.4$ by plotting u_2 and u_3 at 14 different times during one

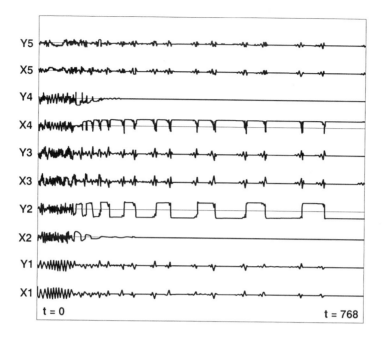

FIGURE 2. Time histories of the real (x_i) and imaginary (y_i) parts of the coefficients of the Heisenberg parameter of $\alpha = 1.45$.

of the transitions shown in figure 2. Before and after the event, two pairs of streamwise vortices are present in the periodic box. However, pictures 1 and 14 are shifted in the spanwise direction by $x_3^+ = 333/4$, corresponding to the phase shift $\Delta\theta_2 = \pi, \Delta\theta_4 = 2\pi$. Moreover it is possible to adjust the value of the Heisenburg parameter $(\alpha \sim 1.5)$ so that the bursting period is 100 wall units as experimentally observed [14]. It is found that, in this case, the "burst" lasts 10 wall units which is also the right order of magnitude. During one of these events there is a sudden increase in Reynolds stress, as observed. An event consists of a sudden intensification and sharpening of the updraft between eddies (5, 6 & 7, fig.3), followed by a drawing apart of the eddies, and the establishment of a gentle downdraft between them (9, 10 & 11, fig. 3): these are similar respectively to the ejection and sweep events that are observed.

In the chaotic state the behavior is less regular and isolated vortices sometimes emerge. This is consistent with the flow visualization experiments of Smith and Schwarz [16] who observed a significant number of solitary vortices among the predominant vortex pairs. These patterns are also very rich in dynamics. We intend to study the regime further in future work.

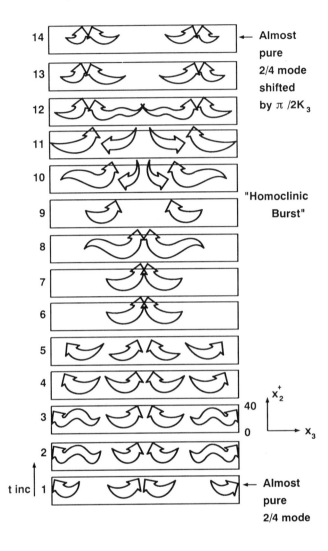

FIGURE 3. Intermittent solution (corresponding to a Heisenberg parameter $\alpha =$ 1.4) during a burst, times equally spaced from the bottom. Each snapshot is a cross section of the flow (normal to the streamwise direction) from the wall (bottom) to $x_2^+ = 40$ (top), of width $\Delta x_3^+ = 333$.

8. Physical Interpretation

We do not have present in our modal population a mechanism to represent the production of higher wavenumber energy when an intense updraft is formed, presumably as a result of a secondary instability. Thus, although our eddies are capable of exhibiting the basic bursting and ejection process, the labor is in vain—there is no sequel, no production of intense higher wavenumber turbulence. A contribution is made only to the low wavenumber part of the streamwise fluctuating velocity and the Reynolds stress. We could easily simulate the production of this high wavenumber turbulence, although we do not expect its inclusion to change the dynamics of our system qualitatively. However, we have held the value of our Heisenberg parameter constant, whereas its value should rise and fall with the level of this intense higher wavenumber turbulence in the vicinity of the wall. The transport effectiveness of the intense higher wavenumber turbulence would be expected to damp the system, suppressing the interesting dynamics until the higher levels are either blown downstream or lifted and diffused to the outer edge of the boundary layer.

Initially we did not exercise the pressure term. Recall that the pressure term appeared due to the finite domain of integration. It represents the interaction of the part of the eddy that we have resolved with the part above the domain of integration, which is unresolved. The order of magnitude that we estimated for this term was small, and for that reason we at first neglected it. It has, however, an important effect, while not changing the qualitative nature of the solution.

The term has the form of a random function of time, with a small amplitude. This slightly perturbs the solution trajectory constantly; away from the fixed points this has little effect, but when the solution trajectory is very close to these points, the perturbation has the effect of throwing the solution away from the fixed point, so that it need not wait long to spiral outward. This results in a thorough randomization of the transition time from one solution to the other, while having little effect on the structure of the solution during a burst. While in the absence of the pressure term (and round-off error), the interburst time tends to lengthen as the solution trajectory is attracted closer and closer to the heteroclinic cycle, with the pressure term, the mean time stabilizes.

Probably the most significant finding of this work is the suggestion of the etiology of the bursting phenomenon. That is, presuming that the abrupt transitions from one fixed point to the other can be identified with a burst, these bursts appear to be produced autonomously by the wall region, but to be triggered by pressure signals from the outer layer. Whether the bursting period scales with inner or outer variables has been a controversy in the turbulence literature for a number of years. The matter has been obscured by the fact that the experimental evidence has been measured in boundary layers with fairly low Reynolds numbers lying in a narrow range, so that

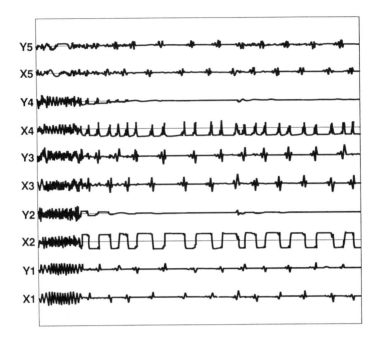

FIGURE 4. Similar to figure 2, but with the pressure term operative. Note that the inter-burst period is randomized, and on the average, stabilized.

it is not really possible to distinguish between the two types of scaling. The turbulent polymer drag reduction literature is particularly instructive, however, since the sizes of the large eddies, and the bursting period, all change scale with the introduction of the polymer [17, 18]. The present work indicates clearly that the wall region is capable of producing bursts autonomously, but the timing is determined by trigger signals fromthe outer layer. This suggests that events during a burst should scale unambiguously with wall variables. Time between bursts will have a more complex scaling, since it is dependent on the first occurrence of a *large* enough pressure signal *long* enough after a previous burst; "long enough" is determined by wall variables, but the pressure signal should scale with outer variables.

9. Further Consequences

We are, of course, concerned about the robustness of our findings. We have tried eigenfunctions generated from exact numerical simulations of channel flow, by Moser and Moin at the Center for Turbulence Research (Stanford/NASA Ames). These eigenfunctions are superficially similar to

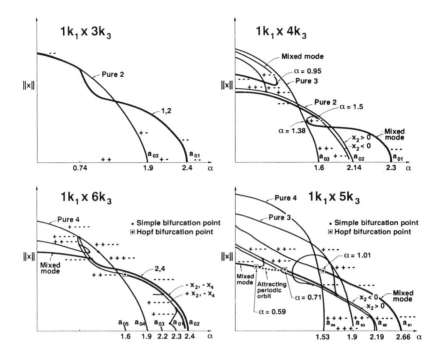

FIGURE 5. Bifurcation diagrams for models with varying numbers of cross-stream moders (from [2]). Note in the six-mode model that modes 3 and 5 have been suppressed for clarity. Note the similarity of the basic structure.

those from Herzog's data, but result in changes of the order of 20% in the values of the coefficients in the equations. The bifurcation diagram is similar, but the fixed points are replaced by limit cycles. Physically, this means that the eddies are wiggling from side to side instead of sitting still. This makes no essential difference, and is even more realistic physically. The intermittent behavior remains.

In her thesis, Stone [2] investigated models with various numbers of cross-stream modes: 3, 4, 5 and 6. She found that the bifurcation diagrams had a backbone common to all of these systems, and were all structurally similar. In particular, the intermittent behavior was common to all. This is illustrated in figure 5.

In addition, Stone [2] found that a small change in the value of the coefficients of the third order terms could change the heteroclinic cycles from attracting to repelling. This is illustrated in Figure 6 (lower), where one can see that the system begins on a traveling wave, but is gradually attracted to the heteroclinic cycle. In Figure 6 (upper) we see the opposite

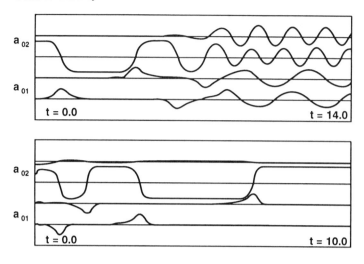

FIGURE 6. A model system with two streamwise modes (one active) and three cross-stream modes (two active). We show only the 01 and 02 modes (the others are unexcited). In the upper figure one of the third order coefficients is -2.69, and the heteroclinic cycle is repelling; in the lower, it is -3.00, and it is attracting [2].

—the system starts on the heteroclinic cycle, but is repelled by it, and ends on a traveling wave. (We show only the values of the first two transverse Fourier modes—the others are quiescent).

This would be a dynamical systems curiosity, if we could not relate it to the physics. However, if we consider the case $\partial_t U_1 = \kappa U_1$ (an exponential increase or decrease of the mean velocity) we find that this results in a change in the real part of the cubic terms for $k_1 = 0$. When κ changes sign (from acceleration to deceleration) the addition to the real part of the cubic term changes sign. The absolute value of the change to the cubic term can only be determined by a case-specific analysis of the empirical eigenfunctions, which has not yet been done. However, it is clear that this phenomenon must be related to the destabilization and stabilization known to be induced by deceleration and acceleration of the flow (as by an adverse or favorable pressure gradient). Although we have discussed here the effect of temporal acceleration and deceleration, the same qualitative effect is obtained from a spatial acceleration and deceleration. Making the heteroclinic cycle more attractive would increase the time between bursts, stabilizing the flow, and vice versa.

Stone [2] also predicted and measured histograms of the bursting period (Fig. 7a). These look reasonably similar to measurements of the same by Kline et al. [14], (Fig. 7b).

Bloch and Marsden [3] have also shown that it is possible to stabilize this

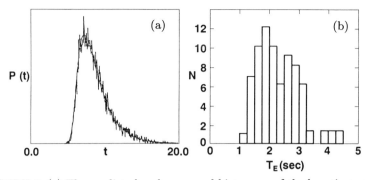

FIGURE 7. (a) The predicted and measured histogram of the bursting period of our model [2]. (b) The measured histogram of the bursting period in the turbulent boundary layer, from [14].

system by feedback, in the absence of noise. That is, if an eigenfunction is fed back with the proper phase, the system can be held in the vicinity of a fixed point for all time. In the presence of noise, however, (such as the pressure perturbation from the outer layer) the system cannot be stabilized completely; however, it can be held in a neighborhood of the fixed point for a longer time. When the system finally wandersso far from the fixed point as to make it uneconomical to recapture it, it is allowed to leave. The same procedure is carried out at the other fixed point. The effect is to increase the mean time between bursts, and hence to reduce the drag. Of course, the system can be made to work the other way, also, kicking the system away from the fixed point whenever it comes too close, resulting in a decrease in the mean time between bursts, and an increase in drag. This would be useful in avoiding separation, for instance.

In drag reduction by polymer additives, one of the accepted mechanisms [17, 18] is the stabilization of the large eddies in the turbulent part of the flow, allowing the eddies to grow bigger and farther apart, as observed. This stabilization is caused by the extensional viscosity associated with the polymers. Aubry *et al.* [19] tried stretching the eddy structure in the wall region, producing drag reduction, and found the bifurcation diagrams morphologically unchanged, except that the bifurcations occurred for larger and larger values of the Heisenberg parameter. This suggests that the motions giving rise to the bifurcations are more and more unstable, the more the region is stretched, requiring a larger and larger value of the Heisenberg parameter to stabilize them. Now, the Heisenberg parameter represents the loss of energy to the unresolved modes. However, the crudeness of the model is such, that it cannot distinguish between loss to the unresolved modes and loss to any other dissipation mechanism, such as viscosity or extensional viscosity. As far as the large scales are concerned, all losses are the same. Hence the findings of Aubry *et al.* [19] are completely consistent with the

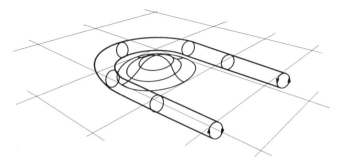

FIGURE 8. Schematic of a piezoelectrically produced welt on the surface, over-turning the vorticity in the boundary layer and producing a negative eigenfunction.

idea of the larger eddies being less stable, and able to grow to this larger, less stable size due to the stabilizing effect of the polymer.

Finally, Bloch and Marsden [3] have shown that, within the assumptions of the scendario above, polymer drag reduction is equivalent to control of the wall region. That, is they showed that an increase of the Heisenberg parameter was equivalent to control by feeding back eigenfunctions, and would lead to a reduction in the bursting rate, and hence to a decrease in the drag. According to the scenario above, this would lead to a stabilization, and result in a growth of the eigenfunctions. This has the important consequence that a controlled boundary layer would be very similar to a polymer drag-reduced boundary layer. From experience with the polymer-drag-reduced boundary layer, we know that it would be a robust layer, still turbulent though with a reduced bursting rate, relatively insensitive to roughness and external disturbance. This is important from the standpoint of applications. Other drag reduction schemes connected with stabilization of the laminar layer are not robust in this sense, and are very sensitive to external disturbances and surface roughness.

From a practical point of view, the feed back could be implemented by an array of hot film sensors to detect the presence, location and strength of an eigenfunction. Piezoelectrically raised welts could produce a negative eigenfunction (see figure 8) by overturning the vorticity in the boundary layer.

References

1. Aubry, N., Holmes, P., Lumley, J. L. and Stone, E. 1988. The dynamics of coherent structures in the wall region of a turbulent boundary layer. J. Fluid Mech. 192: 115-173.

2. Stone, E. 1989. A Study of Low Dimensional Models for the Wall Region

of a Turbulent Boundary Layer. Ph. D. Thesis. Ithaca, NY: Cornell University.

3. Bloch, A. M. and Marsden, J. E. 1989. Controlling Homoclinic Orbits. Theoretical and Computational Fluid Dynamics. In Press.

4. Lumley, J.L. 1967. The structure of inhomogeneous turbulent flows. In Atmospheric Turbulence and Radio Wave Propagation, A.M. Yaglom and V.I. Tatarski:, eds.: 166-178. Moscow: Nauka.

5. Love, M. 1955. Probability Theory . New York: Van Nostrand

6. Lumley, J.L. 1970. Stochastic tools in turbulence. Academic Press, New York.

7. Lumley, J.L. 1981. Coherent structures in turbulence. Transition and Turbulence, edited by R.E. Meyer, Academic Press, New York: 215-242.

8. Herzog, S. 1986. The large scale structure in the near-wall region of turbulent pipe flow. Ph.D. thesis, Cornell University.

9. Tennekes, H. and Lumley, J.L. 1972. A first course in turbulence. Cambridge, MA: M.I.T. Press.

10. Golubitsky, M. & Guckenheimer, J. 1986. (eds). Multiparameter Bifurcation Theory. A.M.S. Contemporary Mathematics Series, No. 56. American Mathematical Society, Providence, R.I.

11. Armbruster, D., Guckenheimer, J. and Holmes, P. 1987. Heteroclinic cycles and modulated traveling waves in systems with O(2) symmetry. Physica D (to appear).

12. Moffat, H. K. 1989 Fixed points of turbulent dynamical systems and suppression of non-linearity. In Whither Turbulence, ed. J. L. Lumley. Heidelberg: Springer. In press.

13. Lumley, J. L. 1971. Some Comments on the energy method. In Developments in Mechanics 6, eds. L. H. N. Lee and A. H. Szewczyk, pp. 63-88. Notre Dame IN: Notre Dame Press.

14. Kline, S.J., Reynolds, W.C., Schraub, F.A. and Rundstadler, P.W. 1967. The structure of turbulent boundary layers. J. Fluid Mech. 30(4): 741-773.

15. Corino, E. R., and Brodkey, R.S. 1969. A visual investigation of the wall region in turbulent flow. J. Fluid Mech. 37(1):1-30.

16. Smith, C.R. and Schwarz, S.P. 1983. Observation of streamwise rotation in the near-wall region of a turbulent boundary layer. Phys. Fluids 26(3) 641-652.

17. Kubo, I. and Lumley, J. L. 1980. A study to assess the potential for using long chain polymers dissolved in water to study turbulence. Annual Report, NASA-Ames Grant No. NSG-2382. Ithaca, NY: Cornell.

18. Lumley, J. L. and Kubo, I. 1984. Turbulent drag reduction by polymer additives: a survey. In The Influence of Polymer Additives on Velocity and Temperature Fields. IUTAM Symposium Essen 1984. Ed. B. Gampert. pp. 3-21. Berlin/Heidelberg: Springer.

19. Aubry, N., Lumley and Holmes, P. , J. L. 1989 The effect of drag reduction on the wall region. Submitted to Theoretical and Computational Fluid Dynamics.

Text Copyright Springer Verlag. Will appear simultaneously as "Order and Disorder in Turbulent Flows" in the proceedings of the 1989 Newport Conference on Turbulence, and as "Low Dimensional Models of the Wall Region of a Turbulent Boundary Layer, and the Possibility of Control" in the proceedings of the Tenth Australasian Fluid Mechanics Conference. Supported in part by: the U.S. Air Force Office of Scientific Research, The U.S. Office of Naval Research (Mechanics Branch and Physical Ocenaography Program), The U.S. National Science Foundation (programs in Applied Mathematics, Fluid Mechanics, Meteorology and Mechanics, Structures & Materials) and the NASA Langley Research Center.

4

The Turbulent Fluid as a Dynamical System

David Ruelle

Très
Très
Très
Très
Très

(Henri Michaux -
L'infini turbulent)

Abstract

This paper reviews the applications of the theory of differentiable dynamical systems to the understanding of chaos and turbulence in hydrodynamics. It is argued that this point of view will remain useful in the still elusive strongly turbulent regime.

1. Introduction: setting up a dynamical system

A dynamical system is just a pompous name for a time evolution. More precisely — as we shall see in a minute — it is an autonomous time evolution. If the functions involved in the description of the time evolution are differentiable, we have a *differentiable dynamical system*. As it happens, differentiable dynamical systems have many remarkable properties, which H. Poincaré started to unravel at the end of the last century. *Differentiable dynamics* has prospered ever since, and holds now a vast body of highly nontrivial results.

The problem of turbulence is to try to understand certain fluid motions with complicated space–time structure called turbulent flows. Understanding turbulence is a *physical* problem, and its mathematical translation and analysis has fascinated a number of great scientists of this century. In fact, the problem is fraught with many difficulties, and remains largely open to-day.

A natural way to start a study of turbulence is by writing a time evolution equation for the fluid motion:

$$\frac{d\xi}{dt} = F(\xi, t) \tag{1.1}$$

Here ξ denotes the velocity, pressure and temperature of the fluid as a function of space position and time, so that ξ is an element of some infinite–dimensional functional space. The function F contains the effect of friction, external forces and boundaries. Suppose that the time evolution (1.1) is autonomous, i.e., that $F(\xi, t)$ does not explicitly depend on t. One can then rewrite (1.1) as

$$\frac{d\xi}{dt} = F(\xi) \tag{1.2}$$

Suppose furthermore that we have an existence and uniqueness theorem for solutions of (1.2) in a suitable functional space. The solution $\xi(t)$ at time $t \geq 0$ can then be expressed in terms of the initial condition $\xi(0)$ by the formula

$$\xi(t) = f^t \xi(0)$$

where f^t is a nonlinear *time evolution operator* satisfying the semigroup properties

$$
\begin{aligned}
f^0 &= \text{identity} \\
f^s f^t &= f^{s+t}
\end{aligned}
$$

These semigroup properties (or group properties if t is allowed to be negative) express that we have a dynamical system. Similarly, one considers discrete time dynamical systems where t is an integer (possibly restricted to be positive). Differentiable dynamics corresponds to the situation where $(\xi, t) \mapsto f^t \xi$ is differentiable (where f^1 is differentiable if the time is discrete).

The restriction to autonomous system is a nuisance, which can be avoided as follows. Suppose that (1.1) can be rewritten as

$$\frac{d\xi}{dt} = G(\xi, g^t \eta(0)) \tag{1.3}$$

where (g^t) satisfies the semigroup properties ($g^0 = \text{identity}$, $g^s g^t = g^{s+t}$). If we have existence and uniqueness of solutions of (1.3) we can write

$$(\xi(t), \eta(t)) = f^t(\xi(0), \eta(0))$$

where f^t again satisfies the semigroup properties. We are thus again in the framework of dynamical systems.

If we turn now to the time evolution equations for fluids, we have to face the problem that establishing existence and uniqueness of solutions is not at all an easy formality. *The problem becomes harder as the space dimension increases.* (Besides the physical $d = 3$ one likes to discuss $d = 2$ because it is easy, and also because it is relevant to atmospheric circulation. Note however that so–called two–dimensional turbulence is quite different from three–dimensional turbulence). *The problem becomes easier if one takes the fluid as incompressible.* (This is a good approximation — in fact surprisingly good experimentally — when the fluid velocities are small compared with the speed of sound. Of course sound waves are ignored, and shock waves do not occur in this approximation. The equation of state of the fluid is taken to be density $=$ constant, so that once can ignore the distribution of temperature in the fluid. Note that in the discussion of convection one often uses the Boussinesq approximation in which temperature differences cause motion through buoyancy effects, but density differences are otherwise ignored). *The problem becomes harder if one takes the fluid as inviscid.* (Viscosity tends to slow the fluid down so that one may hope that the time evolution is asymptotic to a compact set in a suitable functional space. For an inviscid fluid no such hope is possible, even if there is no driving force). *The problem becomes easier for a fluid enclosed in a bounded domain* $\Omega \subset \mathbf{R}^d$. (This is again to try to ensure that the time evolution is asymptotic to a compact set in a suitable functional space).

In view of the above remarks we shall mostly restrict our attention to the case of a *viscous incompressible* fluid enclosed in a *bounded domain* $\Omega \subset \mathbf{R}^d$ with $d = 2$ or 3. Because the fluid is viscous, the time evolution operator f^t is defined, if at all, only for $t \geq 0$ (reversing the time direction amounts to taking a negative viscosity, which produces catastrophic time evolutions). The standard time evolution equation for a viscous incompressible fluid is the *Navier–Stokes equation:*

$$\frac{\partial}{\partial t}v_i = -\sum_{j=1}^{d} v_j \frac{\partial}{\partial x_j}v_i + \nu \Delta v_i - \frac{\partial}{\partial x_i}p + g_i \qquad (1.4)$$

completed by the incompressibility condition.

$$\sum_{i=1}^{d} \frac{\partial}{\partial x_i}v_i = 0 \qquad (1.5)$$

and by the boundary condition

$$v_i \mid \partial\Omega = v_i^{\partial} \qquad (1.6)$$

In (1.4), (v_i) is the velocity field, which is a function of the position $(x_i) \in \Omega$ and of the time t, ν is the kinematic viscosity, p is the pressure (divided by the density) and (g_i) the force per unit of mass. It is assumed that the velocity field v_i^{∂} imposed on the boundary $\partial\Omega$ of Ω extends to a divergence-free field v_i^{∂} on Ω. Note that one can impose external forces on the fluid

either through the term (g_i) of (1.4) or through the boundary condition
(1.6).

Let us briefly discuss the physical meaning of the Navier-Stokes equation (1.4). It expresses the acceleration $(\partial_i v_i + \sum v_j \partial_j v_i)$ in terms of three forces. There is a viscosity term $(\nu \Delta v_i)$ corresponding to self-friction, then the gradient of the pressure term, and finally the external force term (g_i). The viscosity term is obtained by expressing the self-friction forces in terms of the rate of deformation of the fluid *in the linear approximation*; this term can therefore only be trusted for small velocity gradients. Suppose in particular that a solution of the Navier-Stokes equation develops a singularity (infinite velocity gradient), then it is clear that the term $(\nu \Delta v_i)$ is no longer physically correct. It is thus tempting to consider instead of (1.4) an equation with a more general nonlinear viscosity term, but the simplicity of (1.4) makes it well worth investigating in detail before going to something more complicated.

The next thing to do is to convert (1.4), (1.5), (1.6) into a single equation of the form (1.1). One trick involved in doing this consists in replacing the velocity field (v_i) by $(\tilde{v}_i) = (v_i - v_i^\partial)$. The boundary condition (1.6) is then implemented by saying that $(\tilde{v}_i)$ belongs to a suitable space of functions $\Omega \mapsto \mathbf{R}^d$ vanishing on $\partial \Omega$. Notice also that the Hilbert space $L^2(\Omega, dx) \otimes \mathbf{R}^d$ of square integrable vector fields $\Omega \mapsto \mathbf{R}^d$ is the orthogonal sum of a space of divergence-free vector fields, and a space of gradients. If we project (1.4) orthogonally on the space of divergence-free vector fields, the pressure term disappears, and we are left with an equation of the form (1.1) for $\xi = (v_i)$. After this has been solved, p can be recovered from (1.4) if desired. This is only a rough sketch of how one converts (1.4), (1.5), (1.6) to the form (1.1), in particular one has to be sure that the nonlinear terms of (1.4) make sense. For details we refer the reader to the literature on the Navier-Stokes equation*.

We suppose now that the boundary $\partial \Omega$ of the bounded region Ω is sufficiently nice, as well as the data (g_i) and (v_i^∂), and we ask the big question: can we choose a space M of velocity fields such that the Navier-Stokes time evolution is well defined in M? In other words, is there a good existence and uniqueness theorem in M for the "initial value problem"? The answer is positive if $d = 2$, i.e., for two dimensional fluids, and unknown in the more realistic case $d = 3$.

In three dimensions it is at least conceivable that an initially smooth solution later develops singularities. Leray has shown how one can define *weak solutions*, for which singularities are allowed but, as soon as a singularity occurs, uniqueness is no longer guaranteed. (Singularities are points at

---

*An excellent review is given by Foias and Temam [1], and a discussion from the point of view relevant here in Ruelle [2]. Standard monographs on the Navier Stokes equation include Ladyzhenskaya [3], Lions [4], Temam [5], Girault and Raviart [6].

which the velocity gradient diverges). Scheffer, following Leray, has shown how to obtain bounds on the size of the set of singularities of a weak solution of the Navier-Stokes equation. The best result in this direction is due to Caffarelli, Kohn and Nirenberg [7] who showed that the set of singularities in 4–dimensional space–time has zero one-dimensional measure (this means that a curve of singularities is already not possible). The existence of singularities has been neither proved nor disproved, but recent work by Scheffer [8] indicates that they may well be present after all.

Weak solutions with singularities were called by Leray turbulent solutions, and this has caused some confusion in the literature as to what is turbulence. One would not at present want to define turbulence as the presence of singularities in solutions of the Navier–Stokes equation; it is however conceivable that such singularities exist and are relevant to the study of strong turbulence. In other regimes (weak turbulence) it is believed that no singularities occur; the Navier–Stokes time evolution then defines a well-behaved differentiable dynamical system (in some infinite dimensional space).

What, then, is turbulence? A little bit of light on that rather complex question has come from studying the *onset of turbulence*, or *weak turbulence* which is an aspect of what is now called *chaos*. As we shall see, the fact that we are dealing with a differentiable dynamical system is essential to the understanding of weak turbulence, but the precise choice of the evolution equation is not. In fact, the spatial structure of the flow is not very important in weak turbulence. For strong turbulence, by contrast, spatial structure is essential, as is the specific choice of the evolution equation.

Our theoretical understanding of strong turbulence is rather limited. In fact, studies of strong turbulence are centered on a hope: that in the limit of very strong turbulence some simple scaling laws are verified, this is called *universality*. There is actually a good approximate theory of turbulence which is universal: the theory of Kolmogorov. This theory is based on simple physical assumptions (energy cascade in the inertial range, homogeneity and isotropy) and its conclusions are dictated by purely dimensional arguments. The problem is that real turbulence is not homogeneous (one says that it is *intermittent*). Efforts to produce a scaling theory of intermittent turbulence have failed thus far. In another jargon, we have no "universal" theory of fully "developed turbulence" which correctly describes intermittency, and it appears possible that no such theory exists. A dynamical systems approach brings no miraculous solution to this problem, but we shall see that some interesting conclusions are obtained by combining the ideas of differentiable dynamics with specific estimates about the Navier–Stokes equation.

In what follows we shall review three areas where differentiable dynamics has contributed to our understanding of hydrodynamic turbulence. We begin with the *onset of turbulence*, or onset of chaos, which is now fairly well understood in terms of the *geometric theory of differentiable dynami-*

cal systems. Somewhat beyond the onset of turbulence, the time evolution becomes too complicated for a geometric description to be practical, but the methods of the *ergodic theory* of differentiable dynamical systems apply to this regime of *weak turbulence.* Finally, the methods of ergodic theory should apply beyond the weakly turbulent regime, we combine them with Navier-Stokes estimates to obtain results relevant to *strong turbulence.*

2. Geometric differentiable dynamics: the onset of turbulence

Let us rewrite the time evolution equation (1.2) in the form

$$\frac{d\xi}{dt} = F_\mu(\xi) \tag{2.1}$$

where the *bifurcation parameter* μ is at our disposal. In the description of fluid motion, μ usually measures the external action exerted on the fluid; typically, μ is a Reynolds number, or Rayleigh number, etc. When $\mu = 0$, there is no external action, and the fluid goes (when $t \to \infty$) to a state of rest ξ_0. Experimentally, there appears to be a wide range of values of μ for which (2.1) gives rise to various kinds of complicated motion, but where no singularities of the velocity gradient of the fluid occurs. The dynamical systems description is then adequate. Of course, the time evolution takes place in an infinite dimensional space (a Hilbert space) but one can show (we shall discuss this in Section 4) that every solution $\xi(t) = f^t\xi(0)$ of (2.1) is asymptotic to a compact set with finite (Hausdorff) dimension. Therefore we do not expect that the study of (2.1) for a fluid will be essentially different from the corresponding study for a finite dimensional dynamical system.

For large time t, the solutions of evolution equations of the type (2.1) are usually seen to be asymptotic to one or more sets called *attractors.* It is the asymptotic motion on these attractors which interests us. For $\mu = 0$, we have a fixed point attractor ξ_0 corresponding to rest (that ξ_0 is an attractor reflects the fact that rest is a stable equilibrium state). For small μ, one can show (implicit function theorem) that (2.1) still has an attracting fixed point ξ_μ; this fixed point corresponds to a stable steady state. When μ is further increased, more interesting attractors usually appear. There is a great variety of those, but a common development is that the attracting fixed point ξ_μ is replaced by an attracting periodic orbit γ_μ through a process called *Hopf bifurcation.* The asymptotic behavior of solutions of (2.1) is then periodic. Further increase of μ produces further *bifurcations,* i.e., qualitative changes of the global structure of solutions of (2.1). Among the new attractors produced, we shall discuss *quasi periodic attractors* and *strange attractors.*

A quasi periodic attractor is an attracting torus T^k embedded in ξ-space (the phase space of our dynamical system) such that one can parametrize T^k by k angle variables $\varphi_1, \ldots, \varphi_k$ varying linearly with time:

$$(\varphi_1(t), \ldots, \varphi_k(t)) = f^t(\varphi_1(0), \ldots, \varphi_k(0)) = (\omega_1 t + \varphi_1(0), \ldots, \omega_k t + \varphi_k(0))$$
$$(2.2)$$

The φ_i's are taken modulo 1 or modulo 2π. We also assume that the frequencies $\omega_1, \ldots, \omega_k$ are not rationally related. The *quasiperiodic motion* described by (2.2) results, if k is sufficiently large, in complicated oscillations. It was proposed by E. Hopf [9] (the man of the Hopf bifurcation) and independently by Landau [10] that these complicated quasiperiodic oscillations constitute turbulence.

If we make a small change $\delta\varphi(0)$ in the initial condition $\varphi(0) = (\varphi_1(0), \ldots, \varphi_k(0))$ for the dynamical system (2.2), there results a change $\delta\varphi(t)$ at time t, and the length $\mid \delta\varphi(t) \mid$ is equal to the length $\mid \delta\varphi(0) \mid$. Going back to the ξ variables we see that $\mid \delta\xi(t) \mid \leq$ const $\mid \delta\xi(0) \mid$. For a quasiperiodic motion, then, a small change of $\xi(0)$ produces a uniformly small change of ξ at any later time. By contrast, there are many dynamical systems which exhibit *sensitive dependence on initial condition*, i.e., for which $\mid \delta\xi(t) \mid$ grows exponentially. ($\delta\xi$ is to be taken here as an infinitesimal or tangent vector; the distance $\text{dist}(\xi(t), \xi^*(t))$ cannot grow indefinitely if ξ, ξ^* belong to a bounded attractor).

Sensitive dependence on initial condition was discovered by Hadamard at the end of the 19th century in the geodesic flow on surfaces of negative curvature. The physical and philosophical implications of this discovery were already clear at the beginning of this century to such people as Duhem and Poincaré. In fact, Poincaré [11] in his philosophical writings already points out (among other examples, like the hard sphere gas) that sensitive dependence on initial condition in atmospheric dynamics introduces fundamental limitations in weather forecasting. These are amazingly modern ideas but, due to the limitations of physics and mathematics at Poincaré's time, they remained at the philosophical level and, in fact, quickly fell into oblivion. Much later, and independently, Lorenz [12] hit again upon the fact that sensitive dependence on initial condition limits weather predictability, and he worked out in detail a specific (if somewhat unrealistic) example of turbulent fluid convection. Lorenz' study makes essential use of a digital computer (one of the tools which Poincaré did not have at his disposal).

Later (and again independently of the earlier ideas of Poincaré and Lorenz), Ruelle and Takens [13] made the general proposal that turbulence is not described by quasi periodic attractors as proposed by Landau and Hopf, but by more general *strange attractors*, for which the prototypes are the strange Axiom A attractors of Smale (which exhibit sensitive dependence on initial condition by definition). In retrospect, this paper had two important features. On one hand it made a general statement about turbulence (which was received rather coldly at the time). On the other it attracted attention on the onset of turbulence, and claimed that the Landau-Hopf quasiperiodic scenario did not apply. This is an experimentally verifiable assertion, and the experiments (Ahlers [14], Gollub and Swinney [15], and many more afterwards) showed that turbulence is not quasiperiodic, but *chaotic* as one would now say.

There followed a period of extensive investigations — theoretical and experimental — of pathways to turbulence, i.e., of scenarios by which an increase of the bifurcation parameter μ in (2.1) leads from a stable steady state to a strange attractor. There are infinitely many pathways to chaos or turbulence, of which three have been particularly studied: the *intermittent* scenario of Manneville and Pomeau [16], the *period doubling cascade* of Feigenbaum [17], [18] and the *quasiperiodic scenario* of Ruelle and Takens [13].

The saga of the elucidation of the mechanisms of the onset of turbulence cannot be told here *, but a technical point deserves mention because it will occur again later. This is the problem of reconstructing an attractor and its dynamics from the experimental measurement of (usually) a single scalar variable $u(t)$ as a function of time. In other words how does one get a signal $\xi^*(t) \in \mathbf{R}^m$ from the signal $u(t) \in \mathbf{R}$? One idea is to use a vector of time derivatives: $\xi^* = (u, \dot{u}, \ddot{u}, \ldots)$. Since the time derivatives are in fact estimated by finite differences, once can just as well use $\xi^*(t) = (u(t), u(t+\tau), u(t+2\tau), \ldots)$ where τ is a *time delay*; this is the *time delay method* which has been largely used in the analysis of time series$(u(t))$ **

Let us now discuss the physical conclusions of the study of the onset of turbulence. It is very striking that the Navier–Stokes equation, or the fact that we are dealing with a fluid system, seem to play no role here. The interpretation of experimental data is in terms of generic bifurcations of differentiable dynamical systems, and the same bifurcations are observed in the analysis of experimental time series from viscous fluids or other dissipative physical systems, or in the analysis of numerical time series obtained by computer integration of rather arbitrary differential equations. This lack of specificity in the onset of fluid turbulence may be understood by the fact that spatial structure plays no role here: the fluid oscillates as a whole, and its spatial structure remains coherent, even though the temporal structure has become complicated and chaotic.

At this point we must raise a problem of terminology. Traditionally, fluid motion is either *laminar* or *turbulent*, neither term being precisely defined. What we have said above suggest that we *define* fluid motion as laminar

---

*We refer the reader to Cvitanović [19] and Hao Bai–Lin [20] for collections of reprints of the original papers, which deal also with the ergodic theory of chaos to be discussed in the next section. See also Eckmann [21], Bergé, Pomeau and Vidal [22].

**The early advocacy of the time delay method by the present author is recorded in footnote 8 of Packard, Crutchfield, Farmer and Shaw [23].

when there is no sensitive dependence on initial condition, and as turbulent when there is sensitive dependence on initial condition. This has the advantage of precision, and gives a satisfactory definition of laminarity. One may however argue that a fluid with chaotic time evolution but without complicated spatial structure is not really *turbulent*, and should only be described as *chaotic*. What real turbulence is remains then to be precisely defined. Whatever the definition, we now know that *sensitive dependence on initial condition is present in turbulence*, and this is fundamental progress in our understanding of fluid motions. In particular, this proves that no reasonable theory of turbulence could be based on the quasiperiodic proposal of Landau and Hopf.

Let us summarize. As above we denote by μ a parameter (like the Reynolds or Rayleigh number) which describes the state of excitation of the fluid. Then for μ sufficiently small the dynamics of the viscous fluid is undistinguishable from that of a generic low dimensional differentiable dynamical system; the fluid does not exhibit the complicated spatial structure characteristic of strong turbulence *. The time behavior is *laminar* if there is no sensitive dependence on initial condition, *turbulent* or *chaotic* when there is. The transition between the two regimes (*onset of turbulence* or *transition to chaos*) has been much studied and is reasonably well understood in terms of smooth dynamics and without any reference to Navier-Stokes.

3. Ergodic theory of chaos: weak turbulence

The study of the onset of chaos has been done largely by visualizing the geometry of low dimensional attractors. This becomes rapidly impossible when the dimension increases. It remains however feasible to determine important characteristics of the motion by referring to the *ergodic* theory of differentiable dynamical systems.

The basis of the ergodic approach is the reasonable assumption that there is a probability measure ρ describing the average fraction of time spent by the physical system under consideration in various parts of phase space. The measure ρ is *invariant under time evolution*, and we shall assume that it is ergodic, i.e., it cannot be decomposed into different invariant measures. A dynamical system usually has many invariant measures, and it is not obvious which ones are physically relevant (they are probably

---

*We must briefly mention a situation which deviates from the above description and which is encountered in studies of convection in flat boxes (containers of "large aspect ratio"). One has then almost invariance under horizontal translations and the transition to chaos produces complicated oscillations described as *phase turbulence*. This is however quite different from strong 3–dimensional turbulence.

determined by stability under small stochastic pertubations, as proposed by Kolmogorov). Once however ρ is given one can associate with it various physically interesting (and measurable) qualities.

First we define *characteristic exponents* (or *Lyapunov exponents*). If an infinitesimal vector φ is added to ξ in the evolution equation (1.2) we obtain

$$\frac{d\varphi}{dt} = (D_\xi F)\varphi \qquad (3.1)$$

where the linear operator $D_\xi F$ is the derivative of F with respect to ξ. We may consider φ as a tangent vector rather than an infinitesimal vector, and it satisfies the *linearized equation* (3.1). The limit

$$\lambda = \lim_{t\to\infty} \frac{1}{t} \log \| \varphi(t) \|$$

is a characteristic exponent. It depends on the ergodic measure ρ associated with the orbit ξ. In fact there are several characteristic exponents $\lambda_1 \geq \lambda_2 \geq \dots$ corresponding to different choices of the initial $\varphi(0)$ *. Note that the condition $\lambda_1 > 0$. i.e., the positivity of the largest characteristic exponent, precisely expresses sensitivity to initial condition.

Another quantity associated with the ergodic measure ρ is the *entropy* h which measures the amount of information produced by the system per unit of time. It may seem strange that a deterministic system is said to produce information but, because of sensitivity to initial condition, it may happen that an imperceptible change in initial condition produces a perceptible change in the state of the system at time t: time evolution has thus given us some information on the system which we did not know. One can prove the inequality

$$h \leq \sum \quad \text{positive characteristic exponents} \qquad (3.2)$$

In fact, it appears that for the ergodic measures ρ which are physically relevant, (3.2) is usually replaced by *equality*.

Still another important quantity is the *information dimension* $\dim_H \rho$, which is the minimum of the Hausdorff dimension of a set S such that $\rho(S) = 1$. This information dimension turns out to be a more readily measurable quantity than the dimension of the attractor carrying ρ. In fact, let $B_r(x)$ be the ball of radius r centered at x, and suppose that the limit

$$\lim_{r\to 0} \frac{\log \rho(B_r(x))}{\log r} = \alpha \qquad (3.3)$$

exists for ρ–almost all x, then this limit is $\dim_H \rho$ as noted by L.–S. Young [25]. This gives a possibility of obtaining a dimension from experimental

---

*For a general description of the ergodic theory of chaos, with references to the original publications see the review by Eckmann and Ruelle [24].

data and in fact an algorithm related to (3.3) to measure the *correlation dimension* has been proposed by Grassberger and Procaccia [26] and widely used. Note also that the information dimension is related by an inequality to the characteristic exponents *.

Given a differentiable dynamical system, and an ergodic measure ρ, one can thus define various ergodic quantities. Besides those mentioned above there is the *dimension spectrum* associated with the "multifractal structure" of attractors **, and also the position of resonances † in the power spectrum. These ergodic quantities are in principle measurable, and in fact many estimates of dimension have been made — using the Grassberger–Procaccia algorithm — on mildly excited fluids (see for instance [31]). Characteristic exponents are also accessible (see [32]), and entropy, which has been somewhat neglected, could also be estimated.

The concepts and methods mentioned above apply to fluids excited a bit more than at the onset of chaos, and allow the study of what has been called *weak turbulence*. In this regime, which was completely beyond earlier methods of analysis, one can now determine correlation dimensions around 3 or 4, and estimate one (or possibly two) positive characteristic exponents. The systems under consideration still do not exhibit the spatial features of strong turbulence and behave, as far as one can ascertain, as generic moderately excited dynamical systems. Perhaps one physically interesting remark is that systems with apparently quite complicated dynamics still have relatively low correlation dimension.

4. Ergodic theory of the Navier–Stokes equation

It is conceivable that a 3-dimensional strongly turbulent fluid does not behave as a deterministic dynamical system. This would indeed occur if the uniqueness of solutions of the Navier–Stokes equation fails because of singularities, and if the actual evolution of the fluid is determined by microscopic (thermal) fluctuations. The discussion in this section will thus be conditional on the absence of singularities, which means that it will at least apply to the 2-dimensional situation.

Another problem is whether ergodic theory can be useful in situations where the actual interest is about spatial structures which have no expression in a pure dynamical systems approach *. In addition note that, in

---

*An identity which was conjectured by Frederikson, Kaplan, Yorke and Yorke [27] was turned into a rigorous inequality by Ledrappier [28]. The domain of applicability of the conjecture in [27] is still unknown.

**There are many references, for which we refer to Collet, Lebowitz and Porzio [29].

†See Ruelle [30].

*I am indebted to R.H. Kraichnan for emphasizing this point to me.

principle, ergodic considerations involve longer and longer times as the dimension of the attractor representing the system becomes larger: for strong turbulence these times are unreasonably large. As we shall see, the ergodic approach to strong turbulence indeed has severe limitations, but yields nevertheless useful physical information. The saving grace is that the detailed discussion naturally leads to a spatial localization which takes care of the problems raised above.

The approach to the ergodic theory of the Navier–Stokes equation outlined below follows Ruelle [33], [34], who introduced the use of *Schrödinger estimates* into the study of the Navier–Stokes equation. These Schrödinger estimates are due to Lieb and Thirring [35], and Lieb [36] (in fact the results in [36] were proved by Lieb at the request of the present author). Other studies of bounds on the dimension of Navier–Stokes attractors have been undertaken by Babin and Vishik [37] (this paper contains lower bounds on the dimension of attracting sets) and by Constantin, Foias, Temam and coworkers (first without, then with use of Schrödinger estimates). The memoir of Constantin, Foias and Temam [38] gives a general presentation of their approach, which appeals less explicitly to ergodic theory, but is not otherwise fundamentally different from that of [33], [34].

Let thus the Navier–Stokes time evolution take place in a space V (Hilbert space of velocity fields with the Dirichlet norm). We assume that there is a nonempty open set $M \subset V$ such that the time evolution $f^t \xi$ is well defined for $\xi \in M, t > 0$ (no singularities) and that there is some $T_0 \geq 0$ such that $f^t M$ is relatively compact in M for $> T_0$. (If $d = 2$ one can take $M = \{\xi : \| \xi \| < R\}$ for sufficiently large R; if $d = 3$ we are making a nontrivial assumption). The attracting set

$$\Lambda = \bigcap_{t > T_0} f^t M \qquad (4.1)$$

is then compact; if $\xi \in \Lambda, t > T_0$, the operator, $D_\xi f^t$ (derivative) is compact and injective, and Λ has finite Hausdorff dimension (Mallet–Paret [39]).

To estimate the characteristic exponents for the Navier–Stokes time evolution we have to study the linearized equation (3.1), which takes the form

$$\frac{\partial \varphi_i}{\partial t} = -\sum_{j=1}^{d} v_j \frac{\partial}{\partial x_j} \varphi_i - \sum_{j=1}^{d} \varphi_j \frac{\partial}{\partial x_j} v_i + \nu \Delta \varphi_i - \frac{\partial}{\partial x_i} \tilde{p} \qquad (4.2)$$

The right–hand side is to be understood as a linear operator $\mathcal{K}$ acting on elements (φ_i) of the space of divergence free velocity fields vanishing on $\partial \Omega$ (this space is identified as the Hilbert space of divergence free vector fields with the Dirichlet norm). The rate of exponential growth of $\| \varphi(t) \|$ is determined by the hermitean part of $\mathcal{K}$, which has the simple form PH, where

$$(H\varphi)_j = \nu \Delta \varphi_j - \frac{1}{2} \sum_i \left(\frac{\partial v_i}{\partial x_j} + \frac{\partial v_j}{\partial x_i} \right) \varphi_i \qquad (4.3)$$

and P is the projection on divergence–free fields.

Notice that the operator H is, *up to sign*, essentially the Schrödinger operator. The usual Schrödinger operator acts on scalar function, here we have a vector function (with values in $\mathbf{R}^d$) and the "potential" is a traceless $d \times d$ matrix, but these are minor differences. Note that the matrix "potential" is bounded by $w(x)$, where

$$w^2(x) = \frac{1}{4} \sum_{i,j} \left(\frac{\partial v_i}{\partial x_i} + \frac{\partial v_j}{\partial x_i} \right)^2 = \frac{1}{2\nu} \varepsilon(x) \tag{4.4}$$

and ε is the Navier-Stokes rate of energy dissipation per unit volume. It must be stressed that the simple formula (4.4) is a lucky accident, and due to the simple form of the Navier-Stokes equation.

The instantaneous rate of growth of $\| \varphi(t) \|$ is determined by H, and this gives control on average rates of growth, which are by definition characteristic exponents. Notice that the large eigenvalues of H correspond (because of the change of sign) to the small eigenvalues of the quantum mechanical Schrödinger operator (i.e. to bound state energies). Since the bound states of the quantum problem are localized near low values of the potential, we see that rapid growth of $\| \varphi(t) \|$ is localized in regions of strong dissipation. That sensitive dependence on initial condition is linked to strong dissipation is of course physically reasonable (even if the Landau-Hopf theory did not allow for this), but here we have a precise description relating sensitive dependence on initial condition to dissipation in a local manner. In particular if there are two subregions of Ω separated by a zone with little dissipation, there should be a dynamical factorization which could be investigated experimentally.

To proceed, estimates on the spectrum of the Schrödinger operator are needed, which cannot be discussed in detail here (see [35], [36]). Let us just say that these estimates are related to the classical approximation of quantum mechanics. From the Schrödinger estimates one obtains estimates on the characteristic exponents, and from there also on the entropy, and on the dimension of the attracting set Λ of (4.1). In fact these estimates extend to the case where the fluid is subjected to time dependent forces.

The bounds on the entropy and on the information dimension of the ergodic measure ρ are of the form

$$h(\rho) \leq A_d' \nu^{1/2-3d/4} \langle \int_\Omega \varepsilon^{1/2+d/4} dx \rangle_\rho \tag{4.5}$$

$$\frac{\dim_H \rho}{|\Omega|} \leq A_d'' \nu^{-3d/4} \left(\langle \int \varepsilon^{1/2+d/4} \frac{dx}{|\Omega|} \rangle_\rho \right)^{\frac{d}{d+2}} \tag{4.6}$$

and numerical estimates for A_d', A_d'' are known. (As already indicated, these bounds are very general, and extend to the case of time dependent forces, and to the replacement of $\dim_H \rho$ by $\dim_H \Lambda$). For $d = 2$, the exponent of ε is 1, and therefore the estimates involve the average dissipation, for which there are a priori estimates in terms of the forces (explicit estimates can be found in references given above, and also in Eckmann and Ruelle [40], and Lafon [41]). For $d = 3$, the exponent of ε is $5/4$ and, as one would expect, it is not possible to obtain bounds in terms of the forces applied to the system.

In spite of the satisfaction which one may have in writing rigorous inequalities originating from a nontrivial *linear* theory, it must be said that the great difficulty which remains is to understand the *nonlinear* objects of turbulence: vortices in 2 dimensions and the elusive eddies in 3 dimensions.

Conclusion

The problem of turbulence is not just to find more accurate formulae for various physical quantities associated with a turbulent fluid, but also to obtain a conceptually satisfactory theory based on first principles. The ideas of differentiable dynamics have been particularly helpful in this respect: even if the problem of strong turbulence remains rather open, it is now clear that its understanding involves certain features, like sensitive dependence on initial condition, which were not part of the classical theories. The successes of the dynamical systems approach have been mainly in the study of the onset of chaos and of weak turbulence. The fact that dissipation in turbulent fluids is localized (intermittency), and the fact that positive characteristic exponents are tied to dissipation suggest that the dynamical viewpoint will also be important for the elucidation of strong turbulence.

Références

[1] C. Foias and R. Temam. Some analytic and geometric properties of the solutions of the evolution Navier-Stokes equations. J. Math. pures et appl. **58**, 339-368 (1979).

[2] D. Ruelle. Differentiable dynamical systems and the problem of turbulence. Bull. Amer. Math. Soc. **5**, 29-42 (1981).

[3] O.A. Ladyzhenskaya. *The mathematical theory of viscous incompressible flow.* 2nd ed., Nauka, Moscow, 1970; 2nd English ed., Gordon Breach, New York, 1969.

[4] J.-L. Lions. *Quelques méthodes de résolution des problèmes aux limites non-linéaires*, Dunod, Paris, 1969.

[5] R. Temam. *Navier-Stokes equations.* Revised ed., North Holland, Amsterdam, 1979.

[6] V. Girault and P.-A. Raviart. *Finite element methods for Navier-Stokes equations*. Springer, Berlin, 1986.

[7] L. Caffarelli, R. Kohn and L. Nirenberg. Partial regularity of suitable weak solutions of the Navier-Stokes equations. Commun. pure appl. Math. **35**, 771-831 (1982).

[8] V. Scheffer. A self-focussing solution to the Navier-Stokes equations with a speed-reducing external force. pp 1110-1112 in Proceedings of the International Congress of Mathematicians 1986 Amer. Math. Soc., Providence R.I., 1987.

[9] E. Hopf. A mathematical example displaying the features of turbulence. Commun. pure appl. Math. **1**, 303-322 (1948).

[10] L.D. Landau. On the problem of turbulence. Dokl. Akad. Nauk SSSR **44** N° 8, 339-342 (1944).

[11] H. Poincaré. *Science et méthode*. Ernest Flammarion, Paris, 1908.

[12] E.N. Lorenz. Deterministic nonperiodic flow. J. Atmos. Sci. **20**, 130-141 (1963).

[13] D. Ruelle and F. Takens. On the nature of turbulence. Commun. Math. Phys. **20**, 167-192 (1971).

[14] G. Ahlers. Low temperature studies of the Rayleigh-Bénard instability and turbulence. Phys. Rev. Lett. **33**, 1185-1188 (1974).

[15] J.P. Gollub and H.L. Swinney. Onset of turbulence in a rotating fluid. Phys. Rev. Lett. **35**, 927-930 (1975).

[16] Y. Pomeau and P. Manneville. Intermittent transition to turbulence in dissipative dynamical systems. Commun. Math. Phys. **74**, 189-197 (1980).

[17] M.F. Feigenbaum. Quantitative universality for a class of nonlinear transformations. J. Statist. Phys. **19**, 25-52 (1987).

[18] M.J. Feigenbaum. The universal metric properties of nonlinear transformation. J. Statis. Phys. **21**, 669-706 (1979).

[19] P. Cvitanović. *Universality in Chaos*. Adam Hilger, Bristol, 1984.

[20] Hao Bai-Lin. *Chaos*. World Scientific, Singapore, 1984.

[21] J.-P. Eckmann. Roads to turbulence in dissipative dynamical systems. Rev. Mod. Phys. **53**, 643-654 (1981).

[22] P. Bergé, Y. Pomeau and Chr. Vidal. *Order within Chaos*. J. Wiley, New York, 1987.

[23] N.H. Packard, J.P. Crutchfield, J.D. Farmer and R.S. Shaw. Geometry from a time series. Phys. Rev. Letters **45**, 712-716 (1980).

[24] J.-P. Eckmann and D. Ruelle. Ergodic theory of chaos and strange attractors. Rev. Mod. Phys. **57**, 617-656 (1985).

[25] L.-S. Young. Dimension, entropy and Lyapunov exponents. Ergod. Th. and Dynam. Syst. **2**, 109-124 (1982).

[26] P. Grassberger and I. Procaccia. Measuring the strangeness of strange attractors. Physica 9D, 189-208 (1983).

[27] P. Frederickson, J.L. Kaplan, E.D. Yorke and J.A. Yorke. The Lyapunov dimension of strange attractors. J. Diff. Equ. **49**, 185-207 (1983).

[28] F. Ledrappier. Some relations between dimension and Lyapunov exponents. Commun. Math. Phys. **81**, 229-238 (1981).

[29] P. Collet, J.L. Lebowitz and A. Porzio. The dimension spectrum of some dynamical systems. J. Statist. Physics **47**, 609-644 (1987).

[30] D. Ruelle. Resonances of chaotic dynamical systems. Phys. Rev. Letters **56**, 405-407 (1986).

[31] B. Malraison, P. Atten, P. Bergé and M. Dubois. Dimension of strange attractors: an experimental determination for the chaotic regime of two convective systems. J. Physique-Lettres **44**, L-897 - L-902 (1983).

[32] J.-P. Eckmann, S. Oliffson Kamphorst, D. Ruelle and S. Ciliberto. Lyapunov exponents from time series. Phys. Rev. A **34**, 4971-4979 (1986).

[33] D. Ruelle. Large volume limit of the distribution of characteristic exponents in turbulence. Commun. Math. Phys. **87**, 287-302 (1982).

[34] D. Ruelle. Characteristic exponents for a viscous fluid subjected to time dependent forces. Commun. Math. Phys. **93**, 285-300 (1984).

[35] E. Lieb and W. Thirring. Inequalities for the moments of the eigenvalues of the Schrödinger equation and their relation to Sobolev inequalities; pp 269-303 in *Essays in honor of Valentine Bargman* (edited by E. Lieb, B. Simon, and A.S. Wightman) Princeton University Press, Princeton, NJ, 1976.

[36] E. Lieb. On characteristic exponents in turbulence. Commun. Math. Phys. **92**, 473-480 (1984).

[37] A.V. Babin and M.I. Vishik. Attractors for partial differential equations of evolution and estimation of their dimension. Usp. Mat. Nauk **38** N°4 (232), 133-182 (1983).

[38] P. Constantin, C. Foias and R. Temam. *Attractors representing turbulent flows.* Amer. Math. Soc. Memoirs N°314. Providence, R.I., 1985.

[39] J. Mallet-Paret. Negatively invariant sets of compact maps and an extension of a theorem of Cartwright. J. Diff. Eq. **22**, 331-348 (1976).

[40] J.-P. Eckmann and D. Ruelle. Two-dimensional Poiseuille flow. Physica Scripta. **T9**, 153-154 (1985).

[41] A. Lafon. Borne sur la dimension de Hausdorff de l'attracteur pour les équations de Navier-Stokes à deux dimensions. C.R. Acad. Sc. Paris **298**, Sér. I, 453-456 (1984).

5

Empirical Eigenfunctions and Low Dimensional Systems

Lawrence Sirovich

1. Introduction

In the course of this lecture I hope to cover a wide and diverse range of topics which are relevant to the study of near chaotic, chaotic and turbulent flows. A thread which joins these topics is the Karhunen-Loève (K-L) procedure for the generation of the *empirical eigenfunctions*. Lumley introduced this procedure into turbulence theory[1] and suggested that it might be used to unambiguously extract coherent structures in a turbulent flow. This idea will be one of the topics to be touched on later.

Lumley's pioneering contribution appeared in 1967 (see also references 2 and 3) and although a number of applications of the method have been made, its potential in turbulence has been far from realized. The short explanation for its lack of use is that the procedure has an enormous appetite for data. Only in recent times have we been able to generate numerical and experimental databases of sufficient size to satisfy the needs of the procedure.

2. Empirical Eigenfunctions

The Karhunen-Loève procedure according to one recent book[4] can be traced back to the nineteenth century and in particular to Beltrami and Jordan. Its presentation is textbook material[5,6] and only a sketch of the derivation, in the framework of turbulence, will be given.

Imagine a time stationary flow, $\mathbf{v}(\mathbf{x}, t)$, where $\mathbf{v}$ represents the state variables, say

$$(2.1) \qquad \mathbf{v} = (u, v, w),$$

for incompressible flow. (In later discussion $\mathbf{v}$ will also include the temperature.) After some initial transient period, the flow may be supposed to lie in the attracting set of the flow, on which the system trajectory moves, in general in a chaotic fashion. We will assume, as usual, that the trajectory is ergodic on the attractor, along with all that goes with such a statement.

The trajectory is sampled on a sufficiently coarse, but uniform spaced set of times (ideally at uncorrelated times),

$$(2.2) \qquad \mathbf{v}^{(n)} = \mathbf{v}(\mathbf{x}, t_n).$$

This defines the ensemble of states, which is denoted by

$$\{\mathbf{v}^{(n)}(\mathbf{x})\}.$$

It is advisable to take

$$(2.3) \qquad < \mathbf{v}^{(n)}(\mathbf{x}) >= 0,$$

which may be done without loss of generality. (This has a number of obvious advantages. One less obvious benefit follows from the belief that the mean flow is a relatively strong function of the Reynolds number, whereas the fluctuations and the eigenfunctions derived below are relatively weak functions of Re.) A functional geometry is defined by an inner product and we take this to be,

$$(2.4) \qquad (\mathbf{f}, \mathbf{g}) = \int_V \sum_{k=1}^{D} f_k(\mathbf{x}) g_k(\mathbf{x}) d\mathbf{x},$$

where D is the dimension of $\mathbf{v}$, and V is the domain of the flow.

To introduce the K-L procedure we first consider representing the flow $\mathbf{v}(\mathbf{x}, t)$ in terms of an admissible set of functions,

$$\{\boldsymbol{\phi}^{(n)}(\mathbf{x})\}.$$

(Loosely speaking, by admissible, we mean a complete orthonormal set which satisfies the boundary and other relevant conditions of the problems.)

Thus if we write

$$(2.5) \qquad a_n = (\boldsymbol{\phi}^{(n)}, \mathbf{v})$$

we can write

$$(2.6) \qquad \mathbf{v} \approx \mathbf{v}_N = \sum_{n=1}^{N} a_n \boldsymbol{\phi}^{(n)}(\mathbf{x}),$$

with an error

$$(2.7) \qquad \epsilon_N =\| \mathbf{v} - \mathbf{v}_N \|,$$

where the norm is based on the inner product, (2.4). Of all admissible systems *the empirical eigenfunctions have the property that* $< \epsilon_N >$ *is minimal for all* N. Another interesting property is

$$(2.8) \qquad < a_n a_m >= \lambda_n \delta_{nm},$$

i.e., the modes corresponding to the empirical eigenfunctions are statistically orthogonal or decorrelated. *On the long term, the modes based on the empirical eigenfunctions, do not talk to one another.* Each of these properties can be used as a basis for showing that the empirical eigenfunctions are generated by

$$(2.9) \qquad \int_V K_{ij}(\mathbf{x}, \mathbf{x}') \phi_j^{(n)}(\mathbf{x}') d\mathbf{x} = \lambda_n \phi_i^{(n)}(\mathbf{x}),$$

where

$$(2.10) \qquad \mathbf{K} = < \mathbf{v}(\mathbf{x}) \otimes \mathbf{v}(\mathbf{x}') >,$$

is the two point correlation function and $\otimes$ indicates the outer product. Before leaving these preliminary remarks we mention two other connections in which the empirical eigenfunctions can be developed.

Minimal Representational Entropy

If we denote the total mean energy of the flow by

$$(2.11) \qquad e = < (\mathbf{v}, \mathbf{v}) >$$

then

$$(2.12) \qquad p_k = \frac{\lambda_k}{e}$$

may be interpreted as being the probability that the flow is in the kth state. In general, given any orthonormal basis set, $\{\psi^k\}$, we can compute a probability

$$(2.13) \qquad q_k = < (\psi^{(k)}, \mathbf{v})^2 > /e$$

that the system lies along the $\psi^{(k)}$ coordinate. From the optimal property of the empirical eigenfunctions it follows that

$$(2.14) \qquad \sum_{k=1}^{M} p_k \geq \sum_{k=1}^{M} q_k$$

for all M, and from this that

$$(2.15) \qquad \mathcal{S}_p = -\sum p_k \ln p_k \leq -\sum q_k \ln q_k = \mathcal{S}_q.$$

In words, the representational entropy (as defined by the summations in (2.15)) is minimal if the basis set is the empirical eigenfunctions. The implication of this remark from the point of view of information theory is that the empirical eigenfunctions ideally compress the information content of the flow.

Connection with Linear Theory

In a certain sense (and with some fine print) the K-L procedure can be regarded as generalizing the usual eigentheory of linear operators. If

$$(2.16) \qquad \frac{\partial \mathbf{v}}{\partial t} = \mathbf{D}(\mathbf{v})$$

represents the non-linear Navier-Stokes equations the K-L procedure generates eigenfunctions of $< \mathbf{v} \otimes \mathbf{v} >$. If instead $\mathbf{D}$ is some linear operator then under certain circumstances we can show that the eigenfunctions of $< \mathbf{v} \otimes \mathbf{v} >$ are also eigenfunctions of $\mathbf{D}$! For example, if a given database comes from the motion of a string stretched between two fixed points, then K-L procedure generates the appropriate trigonometric functions as the empirical eigenfunctions. Examples of this and an exposition of these ideas is presented in reference 7.

In addition to being interesting in its own right this last feature, has a direct bearing on the empirical eigenfunctions for the N-S equations. For incompressible flow continuity is a linear relation. A consequence of this is that the *empirical eigenfunctions themselves are incompressible* flows.

3. Data Compression

An examination of the K-L procedure indicates that it depends in an essential way on the choice of the inner product. For example another kernel for the generation of empirical eigenfunctions based on the vorticity, $\boldsymbol{\omega}$, is given by,

$$(3.1) \qquad \boldsymbol{\kappa} =< \boldsymbol{\omega} \otimes \boldsymbol{\omega} >,$$

which might be referred to as the enstrophy kernel. The eigenvalues (which are energies) associated with $\boldsymbol{\kappa}$ are weighted more towards the higher wavenumbers than those of $\mathbf{K}$, (2.10). Other possible criteria for choosing a kernel will be mentioned in Section 6.

In later sections which deal with applications, we will see that the empirical eigenfunctions give us a handle on how to view and analyze the vast amounts of data now being generated numerically and experimentally. In this section the focus is on a related but different application of the empirical eigenfunctions. Specifically, on the somewhat mundane topic of data storage and compression, and how we might avoid the present practice, which is forced upon us by overhead considerations, of discarding vast amounts of data. To illustrate this point, consider the Kim, Moin and Moser[8] simulation of channel flow (at $Re = 3300$). This simulation calculation, which is one of the most muscular thus far performed in computational fluid mechanics follows three flow quantities at $\approx 4 \times 10^6$ grid points. The entire calculation, corresponding to roughly fifty eddy turnover times, took

250 CPU hours on a Cray XMP. Roughly 100 data dumps were taken. Since this involved $O(10^9)$ words or $O(10^2)$ tapes, the overhead problem is seen to be considerable. Except for detailed record at selected points, information at intermediate times is thus lost. Thus for example we cannot generate any continuous record for purpose of viewing the flow.

An alternate strategy for packing the data is furnished by the empirical eigenfunctions. In the above channel flow example, it is our estimate that if a one to two percent error is tolerable, then an $O(10^2)$ storage reduction takes place if the data is packed with the use of the empirical eigenfunctions. It is of importance to note that the tolerance error is based on the energy of the flow (this is a result of the inner product, (2.4)). An obvious shortcoming to this approach is that the dissipation spectrum is less well accounted for. To remedy this one can in addition compress the data by packing it according to the vorticity eigenfunctions discussed in relation to (3.1).

One last remark concerns the question of how long a calculation is necessary before one has accurate forms for the empirical eigenfunctions. The information in this respect is somewhat anecdotal but as will be seen does not introduce any essential problems. The principal eigenfunctions carry information on the large scale motions, and it is therefore no surprise that they are reasonably well resolved after a few turnover times. As the index of the eigenfunctions is increased the scale sizes of the eigenfunctions diminish and the length of calculation needed increases for resolution of the eigenfunction. On the other hand a complete set (though not fully resolved) of empirical eigenfunctions are determined after only a few turnover times. These can be used to compress the data in the early stages of the calculation. As the calculation proceeds in time, the empirical eigenfunctions can be recalculated at convenient stages. The already recorded data can then be *recompressed* according to the new empirical eigenfunctions. This process can be repeated a number of times, until at the end of the calculation, well resolved empirical eigenfunctions are obtained and all past data recompressed according to these. (It is believed that this only introduces relatively small errors in the early records.) Since the eigenfunctions calculations involve only diagonalization of matrices, and the determination of **K** just add-multiplies, all of this is computationally cheap.

4. Dimension Considerations

Rayleigh-Bénard (R-B) convection will be used as more or less of a hobby horse problem in order to illustrate a number of ideas. Before going further, it is important to take note of the fact that R-B convection is a *closed* flow, since for some of the remarks that will be made there is a profound difference between closed and open flows. Some further comments along these lines will be made in later in this section.

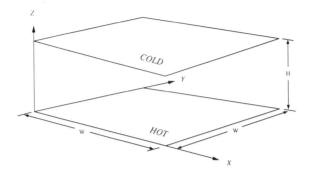

FIGURE 1. Rayleigh-Bénard geometry.

R-B convection is governed by the Boussinesq equations[9,10]

$$(4.1) \quad \begin{cases} \nabla \cdot \mathbf{u} = 0 \\\\ \dfrac{d\mathbf{u}}{dt} + \nabla p = RaPr\, \mathbf{e}_z T + Pr\nabla^2 \mathbf{u} \\\\ \dfrac{dT}{dt} = w + \nabla^2 T. \end{cases}$$

A sketch of the geometry is shown in Figure 1.

In a numerical simulation we must be able to resolve the smallest relevant scale. For the R-B problem (and $Pr = O(1)$) this is the height of the thermal sublayer which we denote by δ. On dimensional grounds this must be of the same order as the Kolmogorov scale, η. [11,12] Thus if H is the height of the cell then we need H/δ degrees of freedom in the vertical direction (i.e. we need at least one grid spacing in this *linear* thermal sublayer) and for a roughly cubical cell we need $(H/\delta)^3$ degrees of freedom. Since the Nusselt number is

$$(4.1) \qquad\qquad Nu = H/2\delta$$

(which can be taken as defining δ), the number of degrees of freedom, D, is given by

$$(4.2) \qquad\qquad D = O(Nu^3).$$

If we use the *free fall* relation[13,14]

$$(4.3) \qquad\qquad Nu = Ra^{1/3}$$

(but see Heslot et al[15] and Castaing et al[16]) then

$$(4.4) \qquad\qquad D = O(Ra),$$

which is a well-known result. This derivation is essentially the same as that used by Landau to arrive at the well known $O(Re^{9/4})$ [17,18] estimate of the number of degrees of freedom in a flow.

If the R-B cell is not a cube but say the square planform shown in Figure 1, then we obtain

$$(4.5) \qquad D = HW^2/\delta^3 = \left(\frac{W}{H}\right)^2 \cdot O(Ra)$$

From this it is clear that dimension is an extensive quantity and if $(W/H) \uparrow \infty$, the number of degrees of freedom become infinite!

We can explore the impact of this last observation from a slightly different viewpoint. Imagine simulating the flow with the use of sinusoids in the horizontal directions, i.e. the flow is W-periodic in horizontal directions. This is sketched in Figure 2(a), where the largest scale mode, a pair of counter rotating rolls, is indicated. A consequence of this is that the flow is now correlated at a distance W. A possible form for the spatial correlation function appears as a heavily drawn curve. Physically this is clearly wrong, but if we try to remedy this by increasing the horizontal dimension we only defer the unphysical correlation to a larger length scale. For example in Figure 2(b) this is indicated by doubling the cell width to $2W$. The correlation function as indicated now shows correlation at $2W$. But for a vigorous flow we can estimate that the correlation scale is bounded by $O(H)$ provided $H < W$ and the introduction of a *false* correlation at horizontal distances larger than the scale of H should not have an affect on the results of the calculation. On the other hand certain *reasonable* modes which are large scale are excluded by forcing periodicity. Thus, for example in Figure 2(a), the mode of six rolls indicated as dotted curves is possible for $2W$ but its counterpart of three rolls is not possible for width W. Thus, although, the estimate of the number of degrees of freedom is extensive, practical considerations would imply that the relevant number of degrees of freedom is more accurately given by (4.4).

Another issue has as its base the purpose of the simulation. If the purpose of the calculation is to explore a W-periodic geometry in its own right, then the above deliberations are not relevant. One only has to guarantee that the smallest scale, the Kolmogorov scale, be well resolved in order for the simulation to be physically meaningful. On the other hand if the goal is to describe the physical flow which occurs between infinite parallel planes then the above considerations are very relevant. The core question is: How large must the planform of the cell be in order for the simulation to closely resemble the flow in the infinite domain? One criterion is that the energy in the *lowest mode*, indicated by a dashed curve in Figure 2(a), have significantly less energy than its second harmonic. Another, somewhat more precise, criterion is that the spatial correlation fall close enough to zero in the computational domain. The situation is far from clear cut. For example,

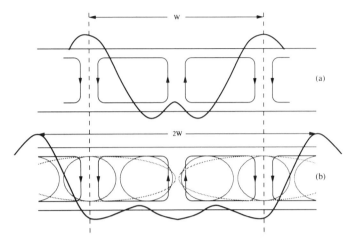

FIGURE 2. Two dimensional schematic of R-B flow. (a) Streamlines of the lowest harmonic for width W. Possible form of the spatial correlation is indicated by the heavy curve. (b) Result of doubling the width. First, second and third harmonics indicated by dashed, continuous and dotted curves, respectively.

one has to be concerned with the mode indicated by dots in Figure 2(b) not appearing in Figure 2(a). Also by imposing a square geometry prevents a mode from turning slowly in time as it would in an infinite, or circular, geometry. These and other issues merit further study and are important in the wider context of all fluid simulations.

LYAPUNOV DIMENSION

The issue of the dimensionality of the space needed to depict a turbulent flow is more directly confronted by measuring the Hausfdorf dimension, D_H, or *capacity* of the attractor.[19,20] A computational approach to this measurement is given by the Lyapunov dimension, D_L, which has been shown to give a sharp upper bound on D_H in cases of practical importance.[18] Briefly stated, the Lyapunov dimension is obtained by estimating the local *volume* of the attractor and averaging this by assaying the entire attractor.[21] This procedure is impractical for experimental data,[22] and since it is computational intensive it becomes prohibitive when D_L becomes large. For open systems it seems likely that D_L is never small. Keefe[23] has calculated the Lyapunov dimension for channel flow, and found that D_L is $O(750)$ for the Reynolds number only 15% above its critical value. Since this is very likely a subcritical situation, an instability *explosively* entrains a large number of modes.

In Figure 3 we show D_L versus Ra which was obtained in a parametric

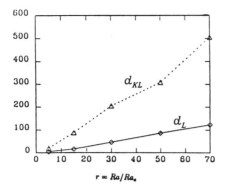

FIGURE 3. Dimension estimates for R-B convection. Continuous line gives Lyapunov dimension and the dashed curve the K-L dimension.

study of the R-B problem.[24,25] Only at lower values of Ra are the values of D_L thoroughly calculated. At the higher values of Ra, D_L was computed by assuming (1) that the largest Lyapunov exponent is measured by the correlation time, and (2) that Lyapunov exponents fill in uniformly independently of Ra. Both these assumptions have been tested only for $Ra/Ra_c < 100$, and therefore must be regarded as conjectural. The result of this is that

$$(4.6) \qquad\qquad D_L = O(Ra^{2/3})$$

which is sharper than the $O(Ra)$ of (4.4). This sharpening of the estimate may be due to a correlation or relationship among scales, something which is not accounted for in the bookkeeping that goes into calculating the Landau formula, (4.4).

KARHUNEN-LOÉVE DIMENSION

As is well-known the calculation of Lyapunov dimension does not provide a set of *fitting functions* for the attractor. Since the empirical eigenfunctions are optimal in the sense described in the previous section we can attempt to use these to *parametrize* the attractor. As a preliminary step we propose that as $Ra \uparrow \infty$ we might expect to enter a scaling regime, so that the K-L dimension, D_{KL}, and D_L scale with one another. (We define D_{KL} in more precise terms in a moment.) In Figure 4 we plot the K-L energy, contained in D_L.[25] Specifically the ordinate is

$$(4.7) \qquad\qquad \mathcal{E}_L = \sum_{k=1}^{D_L} \lambda_k / \mathcal{E}$$

where $\mathcal{E}$ is the total energy. It is remarkable (if indeed the scaling region has been reached) that scaling appears at these low values of Ra.

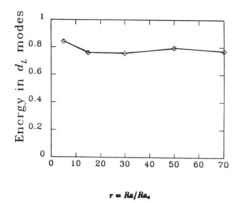

FIGURE 4. Relation between D_L and D_{KL} (see text).

For the definition of K-L dimension we take D_{KL} to be the number of modes at which ninety percent of the energy is achieved and no neglected mode has energy greater than one percent of the principal mode. [Thus, a neglected amplitude, on average, is less than one-tenth that of the principal amplitude.] This, admittedly is arbitrary, and is only intended to give a nominal criterion. D_{KL} is plotted in Figure 3, and as can be seen is more than twice D_L.

5. Rayleigh-Bénard Problem[11,12,24,25,26,27,28]

The results described in the previous section are based on a range of simulations of R-B convection. These will be described in brief including one large scale calculation which lies in the hard turbulence regime as defined by the Chicago group.[15,16] Comparison with their experiments will also be made.

For each of the flows discussed in this section the geometry is that shown in Figure 1. The planform is a square and the aspect ratio $2\sqrt{2}$. The vertical boundary conditions are,

$$(5.1) \qquad T = \mathbf{u} \cdot \mathbf{e}_z = \frac{\partial}{\partial n}\mathbf{u} \wedge \mathbf{e}_z = 0, \ z = 0, H,$$

where T is the departure from the conduction profile. Note that the slip boundary condition is being applied to the velocity. In the horizontal directions, W-periodic boundary conditions are imposed. A range of Ra numbers were considered with $Ra = 70 \cdot Ra_c$, the most thoroughly investigated case, thus far.[26,27,28,29] In Figures 5 we show the average temperature fluctuation $< T >$ and the rms vertical component of velocity, $< w^2 >^{1/2}$. The

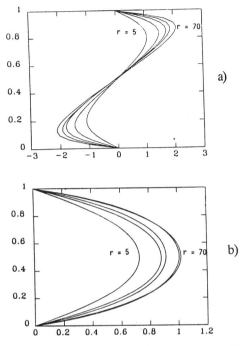

FIGURE 5. Parametric variation in mean flow quantities. (a) Mean temperature
(b) Rms variation of the vertical velocity.

scaling used for these are the characteristic velocity in the boundary layer

$$(5.2) \qquad u_c = \sqrt{2g\alpha\Delta T\delta},$$

and the corresponding temperature,

$$(5.3) \qquad \Theta_c = \kappa\Delta T/(2\delta u_c).$$

These appear to be appropriate even at very large Ra.

Since both horizontal directions are translationally invariant, $\mathbf{K} = \mathbf{K}(x - x', y - y', z, z')$, and all empirical eigenfunctions have the form

$$(5.4) \qquad \phi_{mn}^{(q)} = \Phi_{mn}^{(q)}(z)e^{2\pi i(mx+ny)/W}.$$

Here q represents the vertical quantum number. The mean energy in a
mode (5.4) (which is the eigenvalue that appears in (2.9)) will also carry
three indicies, $\lambda_{mn}^{(q)}$. In Figure 6 we show the principal empirical eigenfunction $\phi_{01}^{(1)}$. This is four-fold degenerate, as a result of the symmetries of the
flow geometry. The mechanical motion corresponding to this eigenfunction
is two-dimensional and is indicated by the streamlines shown in the Figure 6. Isotherms are indicated by dashed curves. Other members of the

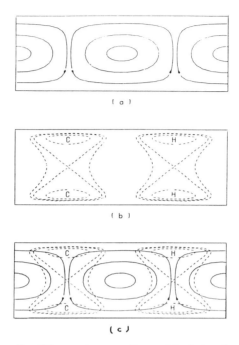

FIGURE 6. Schematic of the most energetic empirical eigenfunction. (a) Streamlines (b) Isotherms (c) Composite picture.

invariant subspace are gotten by translation in the x-direction by a quarter wavelength and rotation about the z-axis by $\pi/2$. The streamlines corresponding to $\phi_{11}^{(1)}$ are shown in Figure 7. This is also a four-fold degenerate case.

Table 1 shows the average energy in each of the modes as the ratio

$$(5.5) \qquad\qquad r = Ra/Ra_c$$

is varied. The principal eigenfunction which is a rolling motion, and loosely speaking is the *wind* in the flow, goes from $\approx 65\%$ at $r = 5$ to 40% at $r = 70$. For $r = 10^4$ we find 25% of the energy in this mode.[11,12] It is not clear whether this share of the energy will decrease to zero or asymptote, as $Ra \uparrow \infty$.

In Figure 8 we show the time course of the *energies* of the four principal modes, i.e. if

$$(5.6) \qquad\qquad a_{mn}^{(q)} = (\phi_{mn}^{(q)}, \mathbf{v}),$$

then $|\,a_{mn}^{(q)}\,|^2$ is the quantity that is plotted. The horizontal line is $\lambda_{mn}^{(q)}$, the mean energy. Studying the first of these plots, we observe that the crossing

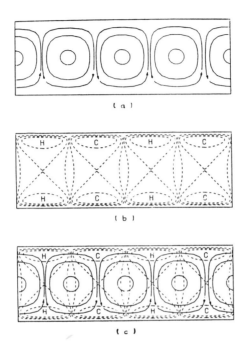

(a)

(b)

(c)

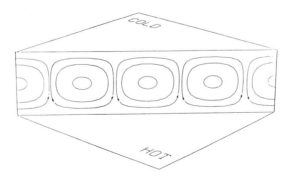

FIGURE 7. Second most energetic mode. 7(a-c) Same description as Figure 6. 7(d) Orientation of the fluid motion.

TABLE 1. Eigenvalues as a function of (k_1, k_2, q) for the indicated values of r. The modes for $k_1 = k_2 = 0$ can be either purely mechanical (degeneracy two) or purely thermal (degeneracy one).

	$r = 5$			15			30			50			70		
	mode	energy	deg	mode	energy	deg	mode	energy	deg	mode	energy	deg	mode	energy	deg
1	(0, 1, 1)	64.9	4	(0, 1, 1)	51.8	4	(0, 1, 1)	43.2	4	(0, 1, 1)	42.4	4	(0, 1, 1)	39.2	4
2	(1, 1, 1)	19.3	4	(1, 1, 1)	14.3	4	(1, 1, 1)	12.6	4	(1, 1, 1)	11.8	4	(1, 1, 1)	8.76	4
3	(0, 1, 2)	4.40	4	(0, 1, 2)	4.93	4	(0, 1, 2)	5.72	4	(0, 1, 2)	5.61	4	(0, 1, 2)	4.44	4
4	(1, 2, 1)	2.61	8	(0, 0, 1)	1.06	2	(0, 0, 1)	1.23	2	(0, 0, 1)	1.27	2	(0, 0, 1)	1.49	2
5	(0, 2, 1)	1.05	4	(1, 2, 1)	3.95	8	(1, 2, 1)	3.91	8	(1, 2, 1)	3.57	8	(0, 2, 1)	2.32	4
6	(0, 1, 3)	0.092	4	(0, 2, 1)	1.94	4	(0, 2, 1)	1.87	4	(0, 2, 1)	1.63	4	(1, 2, 1)	3.29	8
7	(0, 0, 1)	0.021	1	(0, 0, 3)	0.45	1	(0, 2, 2)	1.77	4	(1, 1, 2)	1.54	4	(0, 1, 3)	1.64	4
8	(1, 1, 2)	0.054	4	(0, 1, 3)	1.56	4	(0, 1, 3)	1.37	4	(0, 1, 3)	1.13	4	(1, 1, 2)	1.47	4
9	(1, 1, 3)	0.049	4	(0, 1, 4)	1.26	4	(0, 1, 4)	1.30	4	(0, 1, 4)	1.09	4	(0, 0, 3)	0.30	1
10	(0, 3, 1)	0.046	4	(0, 2, 2)	1.23	4	(1, 1, 2)	1.24	4	(0, 2, 2)	1.04	4	(0, 1, 4)	1.10	4

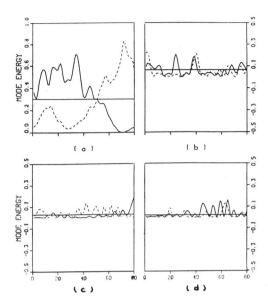

FIGURE 8. Time histories of the amplitudes of the first four empirical eigenfunctions for $r = 70$. (a) $| a_{01}^1 |$ and $| a_{10}^{(1)} |$, (b) $| a_{11}^{(1)} |$ and $| a_{1-1}^{(1)} |$, (c) $| a_{01}^{(2)} |$ and $| a_{01}^{(2)} |$ and (d)$| a_{10}^{(2)} |$ $| a_{00}^{(1)} |$.

of modes indicate a $\pi/2$ rotation of the rolls. As we go deeper into the stack of modes, their time courses become more highly oscillatory.

It is also of interest to examine the change in the empirical eigenfunctions as the Ra number is varied. Specifically, we consider a sequence of values of $r = Ra/Ra_c$, which is the essential bifurcation parameter for R-B convection. Since the eigenfunctions are factorable, (5.4), we can restrict attention to variation in the vertical direction. In Figure 9 we show the logarithm of the spatial power spectrum versus wavenumber for the temperature portion of the principle eigenfunction for the five indicated values of r. Each eigenfunction is an admixture of sinusoids and we might expect that the appropriate scaling should be based on the sublayer thickness, δ, and not the spacing, H. This implies that we should plot power versus k/Nu, or in view of the observed fall-off, (Nu log power) versus k. This leads to a plot with near universal form. Only at the lowest values of r is there a significant departure from the universal form.[24]

HARD TURBULENCE

We close this section with the description of a recent calculation at a high Rayleigh number.[11,12] A principal purpose of this calculation is to make

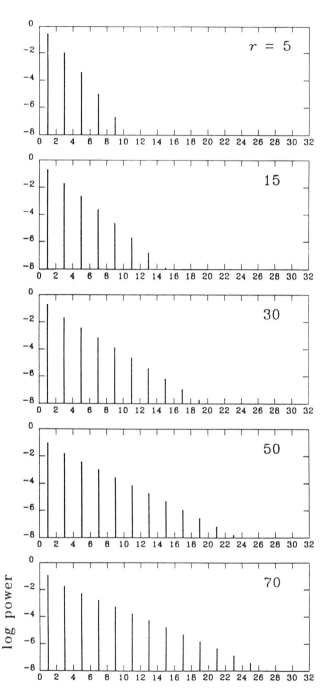

FIGURE 9. Spatial power spectrum for the temperature component of the most energetic mode as a function of r.

comparison with the results of the Chicago group,[15,16,30] who found a new flow regime at high Rayleigh numbers, which they refer to as *hard* turbulence. The experiments of the Chicago group were carried out in circular cylinders of unit aspect ratio, heat flux boundary conditions at top and bottom, insulating side walls and a no slip velocity condition everywhere. In our simulation we maintained the same conditions mentioned above, (5.1). Thus the two cases, as relates to geometry and boundary conditions are vastly different. To compare the two cases we assume that the relevant active parameter is the ratio, (5.5). In these terms the Chicago group found that hard turbulence first appears at[15]

$$(5.7) \qquad\qquad r_h = 6.9 \times 10^3.$$

In the calculations reported in references 11 and 12, $r \approx 10^4$, which therefore lies within the hard turbulence regime.

A number of flow features change as $r = Ra/Ra_c$ passes through r_h. (Turbulence below r_h has been termed soft turbulence by the Chicago group.) For example, the Nusselt number was found to vary as[30]

$$(5.8) \qquad\qquad Nu \propto r^{1/3}$$

for soft turbulence, but as

$$(5.9) \qquad\qquad Nu \propto r^{.282}$$

in the hard turbulence regime.[15,16] Therefore heat transfer is less efficient in the hard turbulence regime!

Another feature of the hard turbulence regime is the finding that temperature fluctuations fall on an exponential probability distribution, instead of the Gaussian distribution found in the soft turbulence regime. Thus, indicating a higher degree of intermittency for $r > r_h$. Probability distributions at three elevations are shown in Figure 10.[11,12] At the mid-station, the probability is

$$(5.10) \qquad\qquad p(\theta) \propto \exp(-1.25\theta/\Delta_c),$$

where Δ_c is the boxed averaged rms temperature fluctuation and θ represents the temperature fluctuation. The comparison with the Chicago group is very good – it is estimated that they find 1.2 versus the 1.25 exponential coefficient in (5.10). The remaining two p.d.f.s in Figure 10 show that the distributions become skewed toward negative fluctuations as the lower boundary layer is approached and entered.

It is important to note again that the geometry in the Chicago experiment was substantially different than for the above simulation. To summarize these difference; the aspect ratio in their apparatus was 1 (versus $2\sqrt{2}$); their cell was a vertical circular cylinder; (versus rectangular); their velocity boundary conditions were no-slip; (versus slip and periodic); they

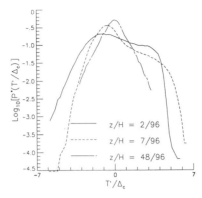

FIGURE 10. Probability distribution functions for the temperature fluctuations at the indicated elevations.

had adiabatic side walls (versus periodic). In addition they found a sustained circulating *wind*, presumably this symmetry breaking was due to the presence of measuring devices. Since the wind is of importance in their explanation of the phenomena [15,16] some further discussion is warranted.

Although no sustained wind appeared in our simulations, it is our belief that the circulating roller motion, shown in Figure 6, plays the role of the wind in our simulation. Since 25% of energy lies in this mode, roughly half the velocity amplitude comes from this mode. It was also estimated that the time scale of the rolling mode is an order of magnitude larger than the time scale for plume formation.[11,12] Thus, relatively speaking, a sustained wind was present during the time for plume formation and release, in our case.

Other Problems

The above discussion made use of the fact that there was only one inhomogeneous direction in the problem. Since this reduces the determination of the empirical eigenfunctions to a one-dimensional integral equation this is an important point. This feature was crucial to a number of other studies in which the empirical eigenfunctions were derived.[31-34] However, it has been shown recently that the presence of homogeneous directions is not an essential feature for the successful calculation of the empirical eigenfunctions and that fully three dimensional inhomogeneous turbulence can be developed in the empirical eigenfunctions.[35] In particular, a fully three dimensional Rayleigh-Bènard problem was simulated and the corresponding three-dimensional empirical eigenfunctions determined.[36,37] The *method of snapshots* developed in[35] was also applied to two-dimensional laser sheet experimental data obtained from a turbulent jet.[38] Furthermore, the determination of the empirical eigenfunctions for fully three dimensional flows is well within the reach of the computational power available today. Given the advances being made in the use of optical methods it is likely that experimental databases for complicated geometries will be available in the near future.

6. Dynamical Systems and Approximate Inertial Manifolds

The empirical eigenfunction, whether determined from computation or experiment may be viewed as ideal *fitting functions* to be used in a Galerkin procedure. In particular we denote the empirical eigenfunctions by $\{\phi^{(n)}\}$ and write

$$(6.1) \qquad \mathbf{v}_N = \sum_{n=1}^{N} a_n(t)\phi^{(n)}(\mathbf{x}) = P_N\mathbf{v}$$

where

$$(6.2) \qquad a_n(t) = (\phi^{(n)}, \mathbf{v}),$$

and P_N is the projection operator. If the governing equations are formally written as

$$(6.3) \qquad \frac{d\mathbf{v}}{dt} = \mathbf{F}(\mathbf{v}),$$

the Galerkin procedure is given by

$$(6.4) \qquad \left(\phi^{(k)}, \frac{d\mathbf{v}_N}{dt}\right) = \left(\phi^{(k)}, \mathbf{F}(\mathbf{v}_N)\right), k = 1, \ldots N.$$

This procedure has been applied to the Ginzburg-Landau (G-L) equation (Refs. 39–41), and with remarkable success. Relatively few ordinary equations are able to faithfully describe the complex sequence of transitions to and from chaos found in the case of the G-L equation. In as much this material has been recently reviewed in,[42] we do not survey it here. A similar approach was adopted for the Rayleigh-Bènard problem and is discussed in the dissertation of Tarman.[43] In this instance it was necessary to introduce, in an ad hoc manner, an eddy viscosity to keep the number of equation down to a manageable size. Aubry et al[44] used a similar device in their remarkable model system for the boundary layer (also based on the empirical eigenfunctions).

A systematic development which leads to a generalized eddy viscosity is a consequence of the method of slaved variables.[45,46] In this method a portion of the system of governing dynamical equation is replaced by algebraic equations, thus *slaving* (in time) certain of the dependent variables to the others, which are governed by the prescribed evolutionary equations. A mathematical framework for rigorously treating these ideas is contained in the study of *inertial manifolds.*[47,48,49,50] By definition an inertial manifold is one towards which solutions tend exponentially. In the remainder of this section we discuss recent work connecting empirical eigenfunctions and inertial manifolds.[51] More specifically we deal with the idea of approximate inertial manifolds.[52,53]

If we write

$$(6.5) \qquad \mathbf{v}_N^c = (1 - P_N)\mathbf{v}$$

then

$$(6.6) \qquad \mathbf{v} = \mathbf{v}_N + \mathbf{v}_N^c.$$

The dynamical system (6.3) may be divided in a corresponding fashion,

$$(6.7) \qquad \frac{d}{dt}\mathbf{v}_N = P_N\mathbf{F}(\mathbf{v}_N + \mathbf{v}_N^c) = \mathbf{Q}_N(\mathbf{v}_N + \mathbf{v}_N^c),$$

$$(6.8) \qquad \frac{d}{dt}\mathbf{v}_N^c = (1 - P_N)\mathbf{F}_N(\mathbf{v}_N + \mathbf{v}_N^c) = \mathbf{R}_N(\mathbf{v}_N + \mathbf{v}_N^c).$$

Inspection of this system leads us to the following argument: If we seek a basis $\{\phi^{(n)}\}$ such that $\mathbf{R}_N \to 0$ as $t \uparrow \infty$ then we might hope that by taking the asymptotic initial data

$$(6.9) \qquad \mathbf{v}_N^c \,|_{t=0} = 0,$$

then $\mathbf{v}_N^c \equiv 0$, so that

$$(6.10) \qquad \frac{d\mathbf{v}_N}{dt} = Q_N(\mathbf{v}_N).$$

The last system is equivalent to (6.4), i.e. the Galerkin procedure.

In order to implement such a program we start with

$$(6.11) \qquad \| \mathbf{v} \|^2 = \| \mathbf{v}_N \|^2 + \| \mathbf{v}_N^c \|^2$$

and then search for a basis set $\{\phi^{(n)}\}$ which minimizes the *error*, $\| \mathbf{v}_N^c \|^2$ – more precisely we want to minimize the ensemble average of this quantity,

$$(6.12) \qquad \epsilon_N = <\| \mathbf{v}_N^c \|^2> .$$

The solution to the posed problem, *viz*, that the basis be optimal in the sense that (6.12) is minimized, leads precisely to the empirical eigenfunctions as the required basis set.

We might hope to improve on this procedure by retaining the influence of the neglected variables. Instead of outright neglect of $\mathbf{v}_N^c$ we can solve

$$(6.13) \qquad \mathbf{R}_N(\mathbf{v}_N + \mathbf{v}_N^c) = 0$$

for

$$(6.14) \qquad \mathbf{v}_N^c = \mathbf{S}(\mathbf{v}_N)$$

and substitute into (6.7) to obtain,

$$(6.15) \qquad \frac{d\mathbf{v}_N}{dt} = \mathbf{Q}_N(\mathbf{v}_N + \mathbf{S}(\mathbf{v}_N)).$$

This is the essence of the slaved variable method,[51] and results in an approximation to the inertial manifold.

Examination of the above discussion reveals that the strategy inplicit in arriving at a reduced system changed in passing from (6.10) to (6.15). In the first instance we sought a basis system, $\{\phi^{(n)}\}$, which minimized the average error (6.12). On the other hand to arrive at (6.15) we impose the condition, (6.13). The latter implies that we should attempt to minimize $\| \mathbf{R}_N \|$ or equivalently $\| \mathbf{v}_N^c \|$, instead of (6.12).

To develop this idea into a criterion, consider

$$(6.16) \qquad \| \dot{\mathbf{v}} \|^2 = \| \dot{\mathbf{v}}_N \|^2 + \| \dot{\mathbf{v}}_N^c \|^2 .$$

The desired criterion is that $<\| \dot{\mathbf{v}}_N^c \|^2>$ be minimized, or equivalently that

$$(6.17) \qquad E = <\| \dot{\mathbf{v}}_N \|^2>$$

be maximized. The objective is sufficiently close to the usual K-L formulation so that we can state the procedure to follow in order to solve the new optimization problem. First, form the *acceleration* covariance[51]

$$(6.18) \qquad \mathbf{L} = < \dot{\mathbf{v}} \otimes \dot{\mathbf{v}} >$$

where $\otimes$ denotes the outer product. The solution to the stated optimization problem is given by the eigenfunctions of $\mathbf{L}$, i.e. $\{\psi^{(n)}\}$ such that,

$$(6.19) \qquad \mathbf{L}\psi = \lambda\psi.$$

The operator can be shown to be hermitian, non-negative and in certain cases square integrable. It then follows from Mercers theorem[54] that $\{\psi^{(n)}\}$ form a complete orthonormal basis. We can then appeal to the Karhunen-Loeve framework[5] to show that

$$(6.20) \qquad \dot{\mathbf{v}} = \sum_n b_n \psi^n, \quad b_n = (\psi^{(n)}, \mathbf{v})$$

almost everywhere.

The basis set derived in this way, $\{\psi^{(n)}\}$, is optimal, by the above criterion, for use in the slaving method. To summarize, we split the system $(b_1, b_2, \ldots)$ into

(6.21) $$\mathbf{v}_N \leftrightarrow \mathbf{b} = (b_1, \ldots, b_N)$$

(6.22) $$\mathbf{v}_N^c \leftrightarrow \mathbf{b}^c = (b_{N+1}, \ldots)$$

according to some à priori criterion error bound,

(6.23) $$<\| \mathbf{b}^c \|^2> < \epsilon.$$

The underlying dynamical system may be written in the notation of (6.21,22) as

(6.24)
$$\begin{cases} \dfrac{d\mathbf{b}}{dt} = \mathbf{T}(\mathbf{b}, \mathbf{b}^c), \\[2mm] \dfrac{d\mathbf{b}^c}{dt} = \mathbf{T}^c(\mathbf{b}, \mathbf{b}^c). \end{cases}$$

This splitting depends on the error criterion, (6.23). The slaved system (approximate inertial manifold) is then obtained by neglecting the time derivative in the second part of (6.24),

$$\begin{aligned} \frac{d\mathbf{b}}{dt} &= \mathbf{T}(\mathbf{b}, \mathbf{b}^c) \\ 0 &= \mathbf{T}^c(\mathbf{b}, \mathbf{b}^c). \end{aligned}$$

The above illustrates a somewhat different approach to the use of the empirical eigenfunctions. It has been applied to the G-L equation[51], with an encouraging degree of success. Its application to other, more demanding cases, is an area of research.

Acknowledgement

The research here was supported by DARPA-URI under grant number N00014-86-K0754.

References

1. Lumley, J.L., *The structure of inhomogeneous turbulent flows*, In: Atmospheric Turbulence and Radio Wave Propagation (A.M. Yaglom and V.I. Tatarski, eds.) 166-178, Moscow: Nauka, (1967).

2. Lumley, J.L., *Stochastic Tools in Turbulence*, Academic Press, N.Y., (1970).

3. Lumley, J.L., *Coherent structures in turbulence*, In: Transition and Turbulence, (R.E. Meyer, ed.), 215-242, Academic Press, N.Y., (1981).

4. Preisendorfer, R.W. *Principal Component Analysis in Meteorology and Oceanography*, Elsevier (1988)

5. Ash, R.B. and M.F. Gardner, *Topics in Stochastic Processes*, Academic Press, NY, (1975).

6. Devijver, P.A. and J. Kittler *Pattern Recognition: A Statistical Approach* Prentice/Hall International (1982).

7. Breuer, K. and L. Sirovich *The use of the Karhunen-Loève procedure for the calculation of linear eigenfunctions*, Jour. Comp. Phys.(to appear 1991).

8. Kim, J., P. Moin and R. Moser, *Turbulence statistics in fully developed channel flow at low Reynolds number*, Jour. Flu. Mech., **177**, 133-166 (1987).

9. Drazin, P.G. and W.H. Reid, *Hydrodynamic Stability*, Cambridge University Press, (1981).

10. Chandrasekhar, *Hydrodynamic and Hydromagnetic Stability*, (Oxford University Press). (1961).

11. L. Sirovich, S. Balachandar and M. Maxey, *Simulations of turbulent thermal convection*, Phys. Fluids A, **1**, 1911-1914, (1989).

12. S. Balachandar, M. Maxey and L. Sirovich, *Direct numerical simulation of high Rayleigh number turbulent thermal convection*, Jour. Sci. Comp., **4**, No. 2, (1989).

13. Malkus, W.V.R., *Discrete transitions in turbulent convection*, Proc. Roy. Soc. of London, Ser. A, **225**, 185-212 (1954).

14. Priestly, C.H.B., *Turbulent Heat Transfer in the Lower Atmosphere*, Univ. Chicago Press (1959).

15. Castaing, B., G. Gunaratne, F. Heslot, L. Kadanoff, A. Libschaber, S. Thomae, X-Z. Wu, S. Zaleski and G. Zanetti, *Scaling of hard thermal turbulence in Rayleigh-Bénard Convection*, (submitted for publication).

16. Heslot, F., B. Castaing and A. Libschaber, *Transitions in helium gas,*, Phys. Rev. **A36**, 5870-5873 (1987).

17. Landau, L.D., *Turbulence*, Doklady AN SSR, **44**, 339-342, (1944).

18. Constantin P., C. Foiaş, O.P. Manley and R. Temam, *Determining modes and fractal dimension of turbulent flows*, J. Fluid Mech. **150**, 427-440 (1985).

19. Bergé, P., Y. Pomeau and C. Vidal, *Order in Chaos*, John Wiley and Sons (1984).

20. Schuster, G.S., *Deterministic Chaos: An Introduction*, Physik-Verlag, Weinheim, FRG, (1984).

21. Kaplan, J.L. and J.A. Yorke, *Chaotic behavior of in multi-dimensional difference equations* In Lecture Notes in Math, **730**, 204, Springer Verlag (1978).

22. Wolf, A., J. B. Swift, H.L. Swinney and J.A. Vastano, *Determining Lyaponuv exponents from a time series*, Physica **16D**, p. 285-317, (1985).

23. Keefe, L., *Comparison of calculated and predicted forms for Lyapunov spectral of Navier-Stokes Equations*, Bull. Am. Phys. Soc., **34**, 2296 (1989).

24. Deane, A. and L. Sirovich, *A computational study of Rayleigh-Bénard convection Part 1: Rayleigh number dependence* Jour. Flu. Mech. **222**, 231 (1990).

25. Sirovich, L. and A. Deane, *A computational study of Rayleigh-Bénard convection Part 2: dimension considerations*, Jour. Flu. Mech. **222**, 251 (1990).

26. Sirovich, L., M. Maxey and H. Tarman, *An Eigenfunction Analysis of Turbulent Thermal Convection*, Post-conference Proceedings (editor, B. Launder) Springer (1988).

27. Sirovich, L., M. Maxey and H. Tarman, *Analysis of turbulent thermal convection*, Proc. 6th Symposium on Turbulent Shear Flow (1987).

28. Tarman, H. and L. Sirovich, *An analysis of turbulent thermal convection*, (submitted) (1989).

29. Deane, A., and L. Sirovich, *Lyapunov Dimension of Rayleigh-Bénard Convection* (The Forum on Chaotic Flows Third Joint ASCE/ASME Mechanics Conference, July 1989) (to appear).

30. Kraichnan, R., *Turbulent thermal convection at arbitrary Prandlt number*, Phys. Flu. **5**, 1374-1389 (1962).

31. Payne, F.R., and J.L. Lumley, *Large eddy structure of the turbulent wake behind a circular cylinder*, Phys. Fluids **10**, S194-196, (1967).

32. Glauser, M.N., S.J. Lieb and W.K. George, *Coherent structure in an axisymmetric jet mixing layer*, Proc. 5th Symp. Turb. Shear Flow, Cornell University, Springer Verlag (1985).

33. Moser, R.D., P. Moin and A. Leonard, *A spectral numerical method for the Navier-Stokes equations with applications to Taylor-Couette flow*, J. Comput. Phys. **52**, No. 3, 524-544 (1983).

34. K.S. Ball, L. Sirovich and L.R. Keefe, *Dynamical eigenfunction decomposition of turbulent channel flow*, International Journal for Numerical Methods in Fluids (To appear, 1990).

35. Sirovich, L., *Turbulence and the dynamics of coherent structures, Pt. I: Coherent Structures*. Quar. Appl. Math., Vol XLV, No. 3, 561-571; *Turbulence and the dynamics of coherent structures, Pt. II: Symmetries and transformations*, Quar. Appl. Math., Vol. XLV, No. 3, 573-582; *Turbulence and the dynamics of coherent structures, Pt. III: Dynamics and scaling*, Quar. Appl. Math., Vol. XLV, No. 3, 583-590, (1987).

36. Sirovich, L. and H. Park, *Turbulent Thermal Convection in a Finite Domain: Theory*, Phys. Flu. A, **2**, 1649 (1990).

37. Park, H. and L. Sirovich, *Turbulent Thermal Convection in a finite domain: Numerical experiments and results,*, Phys. Flu. A, **2**, 1659 (1990).

38. Sirovich, L., M. Kirby and M. Winter, *An eigenfunction approach to large scale transitional structures in jet flow*, Phys. Fluids A **2** 127-136 (1990).

39. Sirovich, L., and J.D. Rodriguez, *Coherent structures and chaos: A model problem*, Physics Letters A, **120**, 211 (1987).

40. Sirovich, L., J.D. Rodriguez and B. Knight, *Two boundary value problem for Ginzburg Landau equation*, Physica D **43**, 63 (1990).

41. J.D. Rodriguez and L. Sirovich, *Low dimensional dynamics for the complex Ginzburg Landau Equation*, Physica D **43**, 77 (1990).

42. Sirovich, L., *Chaotic dynamics of coherent structures*, Physica D **37**, 126-145 (1989).

43. Tarman, H., *Analysis of turbulent thermal convection*, Thesis, Brown University (1989).

44. Aubry, N., P. Holmes, J.L. Lumley, and E. Stone, *The dynamics of coherent structures in the wall region of a turbulent boundary layer*, J. Flu. Mech. **192**, 115-173 (1988).

45. Haken, H., *Synergetics*, 3rd Edition, Springer-Verlag, (1983).

46. van Kampen, N.G., *Elimination of Fast Variables*, Phys. Rep. 124, 69-160, (1985).

47. Foias, C., G.R. Sella and R. Temam, *Inertial manifolds for nonlinear evolutionary equations*, J.D.E. **73**, 309-353, (1988).

48. Mallet-Paret, J. and G.R. Sell, *Inertial manifolds for reaction-diffusion equations in higher space dimensions*, Jour. A.M.S., Volume 1, Number 4, 805-866 (1988).

49. Constantin, P., C. Foias, B. Nicolenko and R. Temam, *Integral manifolds and inertial manifolds for dissipative partial differential equations*, Applied Math. Sci., Springer Verlag (1989).

50. Constantin, P., *Remarks on the Navier-Stokes Equations* (These proceeding).

51. Sirovich, L., B.W. Knight and J.D. Rodriguez, *Optimal low dimensional dynamical approximations*, Quar. Appl. Math **48**, 535 (1990).

52. Titi, E., *On approximate inertial manifolds to the Navier-Stokes equations*, Math. Sci. Inst. Rep. (Cornell), (1989).

53. Foias, D., O.P. Manley and R. Teman, *Sur l'interaction des petits et grands tourbillars dans les écoulements turbulents*, C.R. Acad. Sci. Paris, Serie I, **305**, 497-500, (1987).

54. Riesz, F. and B. Sz. Nagy, *Functional Analysis*, Ungar, N.Y., 1955.

6

Spatiotemporal Chaos in Interfacial Waves

J.P. Gollub and R. Ramshankar

Abstract

Spatiotemporal chaos occurs in many hydrodynamic systems. This state is distinguished from temporal chaos by at least partial loss of coherence between different spatial regions, and from fully developed turbulence by the absence of a cascade. Several experimental and theoretical examples of spatiotemporal chaos are briefly reviewed. Parametrically forced surface waves provide a favorable context for the study of spatiotemporal chaos since the number of active modes can be controlled by varying the driving frequency. In this chapter, we describe the onset and statistical properties of spatiotemporal chaos in surface waves. Of particular note is the presence of phenomena on scales much greater than the wavelength, so that the de-correlation that is one hallmark of spatiotemporal chaos is incomplete.

1. Introduction

The discovery of deterministic chaotic dynamics in hydrodynamic systems has caused various investigators to speculate about possible connections between chaotic dynamics and fully developed turbulence. For example, chaotic dynamics can be described by motion in phase space on an attractor of modest dimension, implying that only a few degrees of freedom are required to capture the dynamics. (We use the term "chaotic dynamics" or "chaotic flows" to refer to low-dimensional situations.) Many investigators have wondered whether any aspects of such a description can be carried over or extended to cover fully developed turbulence, which has many active degrees of freedom.

Turbulent flows differ from chaotic flows in several respects: they lack spatial coherence over distances comparable to the size of the system; they are believed not to be low-dimensional; and they have a wide range of length scales. There also appears to be an interesting intermediate case in which spatial coherence is destroyed or at least reduced, but there is still a dominant scale. The extent to which such situations may be described deterministically is unknown. Such states have been called "spatiotemporal chaos" or "weak turbulence," but a precise definition has not been given. One example that is often cited is the phenomenon of time-dependent Rayleigh-Bénard convection near threshold in a layer whose lateral dimension L is much larger than its depth d. A similar phenomenon occurs in a thin layer of nematic liquid crystal in the presence of a small electric potential difference across the layer. In both cases, linear structures (rolls) resulting from an initial instability become disordered through the generation of defects or dislocations, yielding a state that retains a dominant scale but does not have long range order. Temporal fluctuations in these states are broadband. Spatiotemporal chaos also occurs in hydrodynamic surface waves, and the primary purpose of this article is to elucidate the properties of this state.

Phenomena of this type have received less quantitative study than fully developed turbulence, but are worthy of attention because of their intermediate status between chaos and turbulence. In Section 2, we briefly review some experiments and model systems pertaining to spatiotemporal chaos. In Section 3, we introduce the phenomenon of parametrically forced surface waves, and show how temporal chaos arises from the interaction between several modes. In Section 4, we describe the onset of spatiotemporal chaos, and in Section 5, the statistical properties of this state.

2. Spatiotemporal Chaos: Background

2.1. MEANING OF SPATIOTEMPORAL CHAOS

In an effort to understand spatiotemporal chaos in extended systems and its relationship to fully developed turbulence, Hohenberg and Shraiman (1989) proposed that three characteristic length scales (and their ratios to the system size L) should be distinguished: a dissipation scale l_D, an excitation scale l_E, and a correlation length ξ. The latter is defined through a correlation function.

$$C(r) = \langle \Delta U(r,t) \Delta U(0,t) \rangle, \tag{1}$$

where $\Delta U(r,t)$ and $\Delta U(0,t)$ are the fluctuations of the dynamical variable $U(r,t)$ from the time average value at r and at the origin and $\langle \cdots \rangle$ denotes a long time average. (Translational invariance is assumed.) This quantity is expected to decay as

$$C(r) = \exp(-r/\xi), \tag{2}$$

where ξ is the correlation length.

When $\xi > L$, as occurs for small systems, then the fluctuations may be chaotic in time but are coherent in space. The regime of spatiotemporal chaos, on the other hand, is suggested to occur when $\xi \ll L$, so that fluctuations are incoherent, but the Reynolds number is sufficiently small that $l_E \sim l_D$ and there is no cascade. If the Reynolds number is higher, then $l_D \ll l_E$, and a wide range of length scales are present; this is the case of fully developed turbulence. It might be possible to regard strong turbulence as being in some sense a limiting case of spatiotemporal chaos.

What other properties are expected for a spatiotemporally chaotic state? The dynamical variables are nonperiodic in both space and time. In principle, a deterministic description should be possible, though in practice hard to establish. A statistical approach may be the only practical route to both experimental characterization and theoretical computation.

2.2. OTHER EXPERIMENTAL REALIZATIONS OF SPATIOTEMPORAL CHAOS

A variety of fluid dynamical experiments in the past few years have revealed states that are apparently spatiotemporally chaotic, although a complete characterization is often difficult to accomplish. In general, one looks for nearly two-dimensional situations, since experiments in three dimensions are more difficult. In addition, two-dimensionality tends to delay the transition to fully developed turbulence to higher values of the Reynolds number.

Ordinary Rayleigh-Bénard convection in layers of large horizontal extent provides one example of spatiotemporal chaos. The process appears to involve the repetitive nucleation and elimination of defects in the pattern of rolls (Ahlers, Cannell, and Steinberg, 1985; Pocheau, Croquette, and Le Gal, 1985; Heutmaker and Gollub, 1987). A great variety of types of defects have been noted, including disclinations, dislocations, and grain boundaries; examples are shown in Figure 1. An example of the time-dependence produced by their motion is shown in Figure 2. The dynamics of such processes is believed to involve a rather complex interplay between the defects on the one hand, and large scale flows which both move the defects and are influenced by them (Pocheau, 1989). For a general review of convective pattern dynamics, see Croquette (1989).

In hopes of studying a somewhat simpler situation involving only one space dimension, Ciliberto and Bigazzi (1988) and later Daviaud, Dubois, and Bergé (1989) considered the problem of convection in an annulus. The flow consisted of a pattern of oscillating azimuthal rolls. The system was much farther above the onset of convection than was the case for the two-dimensional convection of Figure 2, and the dynamics cannot be described by the motion of defects. Ciliberto and Bigazzi noted that the amplitude of oscillation was spatially inhomogeneous and temporally intermittent. The regions of high activity were localized and varied strongly in time.

Convection in binary mixtures provides another context in which spatiotemporally chaotic phenomena are observed (Ahlers, Cannell, and Heinrichs, 1987; Steinberg, Moses, and Fineberg, 1987; Kolodner *et al.*, 1988).

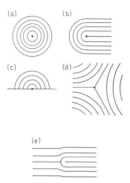

FIGURE 1: Examples of point defects that occur in Rayleigh-Bénard convection and contribute to the onset of spatiotemporal chaos: (a,b) a disclination at the center of a curved patch of rolls; (c) a disclination at the side wall bounding the fluid; (d) a disclination joining three patches of curved rolls; (e) an extra roll pair forming a dislocation (from Heutmaker, 1986).

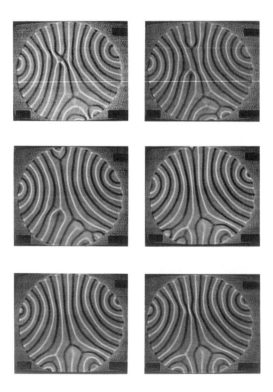

FIGURE 2: Time-dependent flow due to persistent defect motion far above the onset of convection in a circular cell. The pictures span an interval of approximately 30 minutes (from Heutmaker and Gollub, 1987).

FIGURE 3: An example of spatiotemporal chaos in electrohydrodynamic convection in a nematic material. The cellular pattern arises from a secondary instability of the rolls, and leads to chaotic time-dependence (from Albert, 1986).

The motivation for studying mixtures of fluids is that time-dependent phenomena including various types of propagating waves occur close to the onset of convection, where theoretical explanation based on model equations is practical.

Electrohydrodynamic instabilities in nematic materials provide a third example of spatiotemporal chaos (Kai and Hirakawa, 1978; Rehberg, Rasenat, and Steinberg, 1989). Here, the underlying motion, which resembles the Rayleigh-Bénard flow, results from the interaction of an applied electric field with the anisotropy of the dielectric constant and conductivity of the fluid. Again, time-dependent phenomena are associated with the generation and interaction of defects, or with secondary instabilities, depending on the experimental parameters. An example of a time-dependent pattern resulting from a secondary instability is shown in Figure 3.

In all of these systems, there is a well-defined length scale, and in contrast to the situation of fully developed turbulence, no wavenumber cascade mediated by vortex stretching. However, interesting scaling properties may well exist.

2.3. MODEL SYSTEMS

Many theoretical and numerical studies have dealt with model equations
that exhibit spatiotemporal chaos. These model systems are far simpler
than the full Navier-Stokes equations, but can be chosen to represent the
basic space-time symmetries of the problem properly They can often be
derived as low order approximations valid when gradients of the dynamical
variables are sufficiently small.

The Ginzburg-Landau equation has figured prominently in these discussions. It may be written in various forms, but the basic terms are generally
linear, cubic, and second order gradients of the dynamical field of interest:

$$dA/dt = \mu A + \alpha \nabla^2 A - \beta |A|^2 A, \tag{3}$$

where $A(r,t)$ is a complex two-dimensional field and the coefficients μ, α,
and β are complex constants. The amplitude A describes, for example, the
slow space and time modulations of a physical wave-field

$$T(r,t) = \text{Re}\{A(r,t)\exp[ik_0 x + \omega_0 t)]\}. \tag{4}$$

This equation, for suitable choices of parameters, contains topological defects whose dynamics can be quite complex (Coullet *et al.*, 1987; Coullet,
Gil, and Lega, 1989; Coullet, Gil, and Repaux, 1989; Bodenschatz *et al.*,
1989). A related approach aimed at understanding the interactions between
individual defects has been given by Goren *et al.* (1989).

Another well-studied type of model equation is the Kuramoto-Sivashinsky
equation

$$d\phi/dt + \eta\phi + d^2\phi/dx^2 + d^4\phi/dx^4 + \phi(d\phi/dx) = 0, \tag{5}$$

where ϕ is the field of interest and η is a control parameter. It has been
shown to be a suitable model for certain phenomena involving flame fronts,
liquid films, and chemical reaction-diffusion phenomena, and has been studied by several investigators (Manneville, 1981; Pomeau, Pumir, and Pelc,
1984; Pumir, 1985; Shraiman, 1986; Chat and Manneville, 1987). This equation displays complex spatiotemporal phenomena involving the disordered
motion of the *phase* of a cellular pattern.

Several investigators have considered a type of model system comprised
of a one- or two-dimensional lattice of coupled nonlinear maps, each of
which is capable of temporal chaos. Such systems were first studied by
Kaneko (1984) and Waller and Kapral (1984). These models have provided
interesting examples of spatiotemporal chaos, but are not readily applicable to the interpretation of hydrodynamic experiments. On the other
hand, discrete models having chaotic dynamics may be quite appropriate
in modeling neural networks (Sompolinsky, Crisanti, and Sommers, 1988).

3. Vertically Forced Surface Waves: Background

3.1. PARAMETRIC INSTABILITY

Michael Faraday (1831) observed that liquids placed on a vertically oscillating horizontal plate manifest waves at half the frequency of oscillation. This observation has given rise to a voluminous literature, rapidly expanding in the last decade. Much of it has been recently reviewed by Miles and Henderson (1989). Here we concentrate on those features relevant to understanding dynamical complexity.

Benjamin and Ursell (1954) first showed how Faraday's waves could be understood by linearizing the hydrodynamic equations (neglecting damping) and decomposing them into spatial Fourier modes. Individual modes were found to show a subharmonic instability when their wave frequencies are close to half the excitation frequency. The behavior of a single mode amplitude (in their approximation) is identical to that of a rigid pendulum subjected to vertical oscillation of the pivot.

A careful experimental study of the transient growth and eventual stabilization of these waves, and comparison with theoretical predictions incorporating dissipation, has recently been performed by Henderson and Miles (1989). This effort led only to partial agreement, perhaps because the proper treatment of boundary effects and dissipation is quite difficult and somewhat elusive.

It may be worth noting that the transient behavior of a single mode can be surprisingly complex. The surface displacement may be described by the following expression

$$Z_j(r,t) = [A_j(t)\cos(\omega_0 t/2) + B_j(t)\sin(\omega_0 t/2)]f_j(r), \qquad (6)$$

where $f_j(r)$ gives the spatial mode structure, and the coefficients A_j and B_j are mode amplitudes that in the weakly nonlinear regime vary on a timescale much longer than the inverse of the excitation frequency ω_0. At least in a square geometry, there are certain choices of driving amplitude and frequency for which there are three stable and two unstable solutions in the (A_j, B_j) phase plane. The stable solutions correspond to the flat surface and to two non-zero solutions with different phases with respect to the drive. The basins of attraction of these solutions are expected to be thin and intertwined (Gu, Sethna, and Narain, 1988), so that predictability is limited even for the transient behavior of these stable patterns. Similar behavior has been seen experimentally (Simonelli and Gollub, 1989). Thus, it is not necessary to look far to see the origins of dynamical complexity.

3.2. CHAOTIC DYNAMICS

Next, we consider situations leading to chaotic dynamics as a prelude to our discussion of spatiotemporal chaos in surface waves.

Chaos can occur for a single spatial mode as the excitation amplitude is increased. Phenomena of this type were documented by Gollub and Meyer (1983), who studied the axisymmetric modes in a cylindrical container. They observed a sequence of bifurcations leading first to a small displacement of the center of the pattern and a slow precession of the entire pattern, and then (at higher excitation amplitude) to azimuthally modulated waves with chaotic temporal dynamics. (However, Miles and Henderson (1989) have speculated that a second undetected mode may have been involved.)

Chaotic dynamics was later shown to arise even for small amplitude waves when two distinct spatial patterns or modes have nearly the same resonant frequency Ciliberto and Gollub, 1985). This situation can arise through an "accidental" near degeneracy. For example, in a circular geometry, the linear modes have the form

$$f_{lm}(r, \theta) = J_l(q_{lm}r)\sin(l\theta + \phi_{lm}), \tag{7}$$

where the J_l are Bessel functions of order l, and the wavenumber q_{lm} is determined by the vanishing of the derivative J_l' at the boundary of the cell. The modes may be labelled by the indices (l, m), where l is the number of angular maxima and $m - 1$ is the number of modal circles. Ciliberto and Gollub (1985) showed that the near degeneracy of the modes (4,3) and (7,2) gives rise to chaotic mode competition in the region near the intersection of the linear stability boundaries of these two modes. The properties of the resulting chaotic state were extensively studied in that work. Meron and Procaccia (1986) later formulated a low dimensional dynamical system to describe the dynamics. A fairly general approach to describing multimode dynamics in parametrically excited systems has been given by Meron (1987). The use of symmetry considerations in conjunction with bifurcation theory has also yielded insight into the bifurcation structure of this system (Crawford, Knobloch, and Riecke, 1989). Many of the phenomena are apparently not strongly dependent on the specific form of the hydrodynamic equations.

A second type of situation leading to chaos occurs when two modes have almost the same frequency due to symmetry. For example, in a rectangular geometry, the mode structure is of the form

$$f_{mn}(x, y) = \cos(m\pi x/L_x)\cos(n\pi y/L_y). \tag{8}$$

In a square geometry, the modes (m, n) and (n, m) are degenerate due to symmetry. When this symmetry is broken by making the cell dimensions L_x and L_y slightly different, a remarkable sequence of bifurcations leading from oscillations to chaos occurs (Simonelli and Gollub, 1989). The appropriate phase space is spanned by the four time-dependent mode amplitudes $A_{mn}, B_{mn}, A_{nm}, B_{nm}$. These were measured experimentally; an example showing a complex oscillation that is a precursor of chaos is shown in Figure 4. In this motion, the system cycles from a "pure mode" with only (3,2)

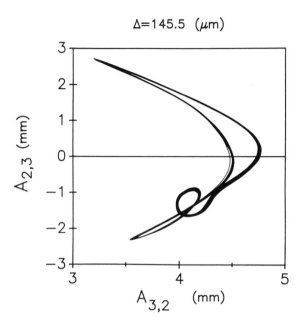

$\Delta = 145.5 \ (\mu m)$

FIGURE 4. Complex oscillation involving two interfacial wave modes. The coordinate axes are the amplitudes of the two modes (component in phase with the forcing).

present, to a "mixed mode" with all amplitudes non-zero, and back again. The pattern itself appears to rotate.

Chaotic surface waves have also been observed by Funakoshi and Inoue (1988) and Nobili *et al.* (1988), who studied the case of horizontal forcing. Comparison with a theoretical formulation proposed by Miles (1984) seems to be fairly successful. Chaotic dynamics is clearly a ubiquitous phenomenon for surface waves when several modes are simultaneously present on the surface.

4. Transition to Spatiotemporal Chaos in Surface Waves

The vertically forced surface wave system has the very attractive feature that, simply by increasing the drive frequency, the wavelength λ can be made small compared to the dimension L of the system. A regime of spatiotemporal chaos then occurs for driving amplitudes only slightly above that needed to produce waves. The appearance of spatiotemporal chaos in this regime may be viewed as a consequence of the fact that the resonant frequencies of the modes become quite closely spaced, so that many modes are nearly degenerate and can then have significant nonlinear coupling in the amplitude equations. In this sense, spatiotemporal chaos may perhaps be viewed as an extension of chaos due to modal degeneracy, to situations involving many modes or even a continuum. The resulting disordered waves have been noted by Ezerskii, Korotin, and Rabinovich (1985) and a quantitative study of the transition was undertaken by Tufillaro, Ramshankar, and Gollub (1989). Here we describe the salient features of the basic ordered state and the transition to the disordered state.

4.1. THE ORDERED STATE

The transition to spatiotemporally chaotic state was studied in a square plexiglas cell of dimensions 8 cm × 8 cm × 2 cm containing a 1 cm depth of n-butyl alcohol. This fluid was chosen because it wets the cell walls reproducibly and is not too volatile. The cell is mounted on an electromagnetic shaker driven by a frequency synthesizer and power amplifier. Harmonics in the driving amplitude are negligible. The excitation frequency of 320 Hz yields waves at 160 Hz, and a wavelength of 2 mm.

Shortly after the excitation is started, highly coherent and spatially ordered waves with $\xi > L$ occur for

$$\varepsilon = (A - A_c)/A_c \leq 0.1, \tag{9}$$

where A_c is the critical driving amplitude for the appearance of waves. This initial (but unsteady) state is shown in Figure 5a, which is a shadowgraph image formed by the projection of collimated light through the fluid surface and onto a diffusely scattering mylar sheet fastened to the top of the cell. The pattern is well described by a single mode with equal wavenumbers in the two directions. The nature of the selected states near onset in a fairly large cell is a subtle problem which has been studied by Douady and Fauve (1988). The meniscus at the lateral boundary can affect the selected pattern, eliminating or rendering less stable the modes with unequal wavenumbers in the two directions.

The simplest hypothesis for the form of the surface deformation in the ordered initial state is $f(x,y) = \cos(q_0 x) + \cos(q_0 y)$, i.e., a superposition of standing waves in the x and y directions (parallel to the cell walls) with wavenumber q_0. However, the optical imaging that produces the pattern shown in Figure 5(a) is nonlinear. A computation (Milner, private communication) of the expected light intensity on the mylar sheet surface shows that the observed image intensity $u(x,y)$ is given by

$$u(x,y) = b[\cos^2(q_0 x) + \cos^2(q_0 y)] + c[\cos(q_0 x)\cos(q_0 y)]$$
$$+ d[\cos^4(q_0 x) + \cos^4(q_0 y)] + e[\cos^2(q_0 x)\cos^2(q_0 y)]$$
$$+ f[\cos^2(q_0 x) + \cos^2(q_0 y)][\cos(q_0 x)\cos(q_0 y)] + \cdots, \tag{10}$$

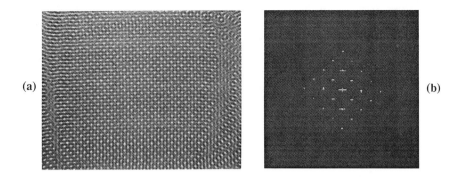

(a)   (b)

FIGURE 5. (a) Shadowgraph image of an ordered capillary wave pattern obtained initially after excitation in a square cell. The dimensionless driving amplitude ε is 0.07. The region shown corresponds approximately to $6.0 \times 4.8 \text{cm}^2$. (b) The spectral power $P(q_x, q_y)$ for the ordered state shown as a 2-dimensional image. The center of the image is the origin in the (q_x, q_y) space.

FIGURE 6. Shadowgraph image of the entire cell ($8.0 \times 8.0 \text{ cm}^2$) showing a peripheral region where the waves are aligned with the cell walls and a central region where the waves are oriented at an angle θ to the side walls at $\varepsilon = 0.07$. The pattern drifts slowly, leading to the creation and annihilation of wavefronts at the side walls of the cell, but θ remains approximately constant. The pattern becomes better aligned with the side walls as the excitation amplitude is increased. The apparent vanishing of the waves near the cell walls is an optical effect.

where b, c, d, e, and f are constants that depend on the system parameters (wavenumber, wave amplitude, refractive index, and distance to the screen). A computation of the two-dimensional spectra $P(q_x, q_y)$ for such a nonlinear imaging process results in peaks along the q_x and q_y axes at $|q| = 2q_0$ and peaks along the diagonal at $|q| = \sqrt{2}q_0$, and their harmonics. This is in at least qualitative agreement with the observed spectra in the ordered initial state shown in Figure 5(b). The data are consistent with the surface deformation field $f(x, y)$, but an additional smaller contribution proportional to $\cos(q_0 x)\cos(q_0 y)$ cannot be excluded.

This ordered initial pattern is not quite time-independent. Most strikingly, the waves become misaligned with respect to the side walls. The pattern (see Figure 6) eventually consists of a peripheral region several wavelengths in extent where waves are generally aligned with the boundaries, and an interior region oriented at an angle $\theta(\varepsilon)$ that depends on ε. The two-dimensional Fourier spectrum computed from the central region of the cell is also rotated, but is otherwise the same as in Figure 5(b). The rotated pattern is nearly, but not quite, time-independent. A slow drift of the pattern occurs, leading to creation and annihilation of wavefronts at the side walls of the cell. We have not yet established whether this symmetry-breaking drift is intrinsic or a result of imperfections in the geometry or in the imposed acceleration.

The pattern becomes *better* aligned with the cell boundary as ε is increased, as long as the threshold of spatiotemporal chaos is not exceeded. Figure 7 shows the variation of θ with ε. The alignment with the cell walls appears essentially complete (i.e., $\theta = 0$) just as the ordered pattern itself becomes unstable at $\varepsilon \approx 0.1$.

The initial pattern formation was found to be supercritical. This was confirmed by measuring the average spectral amplitude at the wavenumber $2q_0$ from images of the central 25% of the fluid surface as a function of the excitation amplitude. Figure 8 shows the variation of this quantity (obtained by integrating the two-dimensional wavenumber spectra over a circle of radius $2q_0$) for both increasing and decreasing values of ε. There is no hysteresis, indicating that the bifurcation is supercritical. However, the pattern does not fill the cell uniformly. Rather, it begins near the center and fills the entire cell only at $\varepsilon \approx 0.03$. Therefore, it is probably not meaningful to study the detailed shape of the function shown in Figure 8. We do not know whether an inhomogeneity in the excitation amplitude is present. If so, it would have to arise from the elasticity of the cell itself at the relatively high frequency of excitation.

Finally, we studied the transient decay of the patterns when the excitation was turned off. The decay of the wave amplitude is at least as fast as the response of the shaker table (approximately 6 periods of the excitation frequency). This places a lower bound on the dissipation experienced by the waves.

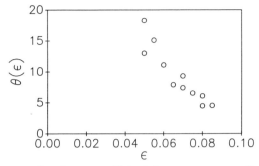

FIGURE 7. The angular deviation $\theta(\varepsilon)$ of the wave pattern (in degrees) in the central part of the cell from the side walls as a function of ε.

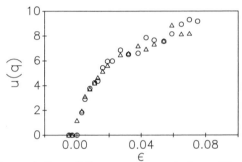

FIGURE 8. The variation of the average spectral amplitude (arbitrary units) at wavenumber $2q_0$ obtained from images of the fluid surface as a function of the excitation amplitude. The circles represent the variation of $u(2q_0)$ as ε is increased and triangles represent the variation of $u(2q_0)$ as ε is decreased. The absence of hysteresis indicates that the bifurcation is supercritical.

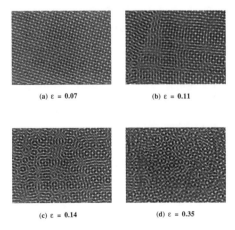

(a) $\varepsilon = 0.07$ (b) $\varepsilon = 0.11$

(c) $\varepsilon = 0.14$ (d) $\varepsilon = 0.35$

FIGURE 9. A sequence of surface wave patterns for four different values of ε spanning the onset of spatiotemporal chaos. The region shown is approximately 3.6×2.9 cm^2.

4.2. TRANSITION TO THE SPATIOTEMPORALLY CHAOTIC STATE

The first evidence of time dependence is observed just below $\varepsilon = 0.1$, in the form of small translational oscillations of the wave peaks. This is followed by a well-defined transition to incoherent waves with $\xi \cong \lambda \ll L$ for $\varepsilon \geq 0.1$. The transition is illustrated by the sequence of patterns shown in Figure 9. Only the center 20% of the cell area is shown in these photographs. The transition is mediated by long wavelength modulations which result from a secondary instability of the pattern (Ezerskii *et al.*, 1986).

To measure the correlation length ξ, two-dimensional autocorrelation functions of the images were computed. The envelopes of these functions decay as a function of distance r, as shown for several values of ε in Figure 10(a). The decay is fitted satisfactorily by an exponential function, from which the correlation length ξ can be obtained. The remarkable feature of $\xi(\varepsilon)$, as shown in Figure 10(b), is the well-defined transition near $\varepsilon = 0.1$, beyond which $\xi \cong \lambda \ll L$. (We note parenthetically that the decay of correlation is not isotropic, so that the use of a single length is an oversimplification.)

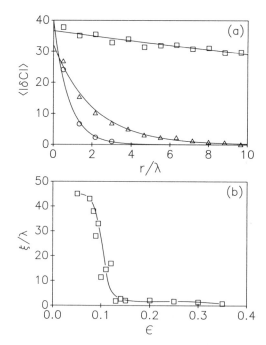

FIGURE 10. (a) Decay of the envelope of the autocorrelation function for three values of ε as a function of $r/\lambda(\varepsilon = 0.07$, squares; $\varepsilon = 0.15$, triangles; $\varepsilon = 0.35$, circles). (b) Correlation length ξ as a function of ε from exponential fits to the curves in (a).

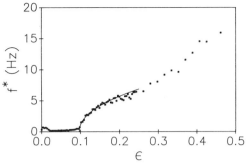

FIGURE 11. Characteristic fluctuation frequency f^* of the pattern, as a function of ε. Note the sharp increase in f^* at the onset of spatial disorder.

This sudden decrease in the correlation length marks the onset of spatiotemporal chaos. The temporal aspect of the transition may be seen by examining the *characteristic frequency* f^* of the time-dependence. It may be determined either from the first moment of the power spectrum, or from the decay time to half-maximum of the temporal autocorrelation function. The latter is plotted in Figure 11. The sudden rise in f^* at $\varepsilon = 0.1$ coincides with the transition noted in the correlation length.

5. Statistical Properties of Spatiotemporal Chaos in Surface Waves

In this Section, we describe a number of measurements aimed at elucidating the statistical properties of spatiotemporal chaos in surface waves. Naturally, we hope that some of these will turn out to be generic properties.

5.1. FLUCTUATIONS IN FOURIER SPACE

First, we consider the question of whether the fluctuations of the Fourier modes at low q may exhibit Gaussian fluctuations. Behavior of this type was suggested by Hohenberg and Shraiman (1989) to be an expected consequence of the central limit theorem when ξ is small, since many coherence volumes contribute to each Fourier mode.

To answer this question, it is necessary to determine the Fourier amplitudes at time intervals short compared to the typical times for fluctuations. In practice, an interval of about 0.05 s is required. To accomplish this using two-dimensional images is computationally demanding. We instead utilized a fast line scan camera consisting of an array of 512 photodiodes. With an appropriate lens, a region scanning approximately 20 wavelengths can be studied. The optical intensity is averaged over a few cycles of the fast wave oscillation, and all 512 elements are read out nearly simultaneously (actually, at 400 kHz). The integration and readout process is repeated at a

(a)   (b)

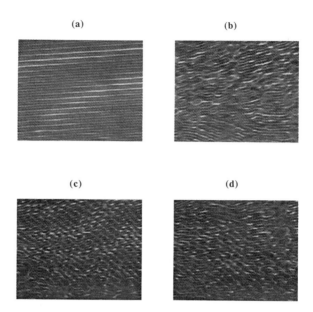

(c)   (d)

FIGURE 12. Examples of the variation of $U(x,t)$ shown as a two-dimensional image. The intensity at a given point is proportional to $U(x,t)$. The interval shown corresponds to approximately 3.8 cm (ordinate) and 12.8 s (abcissa). (a) $\varepsilon = 0.07$, ordered state; (b,c,d) $\varepsilon = 0.15$, disordered state at different times.

rate of 20 Hz, for 512 scans in 25.6 s. Half of the resulting array $U(x_i, t_j)$ is displayed in Figure 12 for the ordered and chaotic states. Bright regions correspond to antinodes of the fast waves, or maxima of their envelope.

A few slowly evolving defects may be seen in the ordered state. In contrast, the regime of spatiotemporal chaos shows considerable complexity. Interestingly, there is evidence in some cases (Figure 12(c)) for structures that are large compared to the wavelength (the bright swaths). Propagating waves are also apparent, as may be seen from the weak parallel lines normal to the diagonal in Figure 12(c). Their intensity and direction of travel fluctuate in time.

In order to study the fluctuations in Fourier space, we compute the Fourier amplitudes $U(q_x, t)$ for each q_x up to Nyquist spatial frequency, and the fluctuations $\Delta U(q_x, t) = U(q_x, t) - \langle U(q_x) \rangle$. This is a complex quantity whose real and imaginary parts are shown for a typical case in Figure 13. We have chosen $q_x/q_0 = 0.4$, where $q_0 = 2\pi/\lambda$, but similar results are obtained for all $q_x/q_0 \leq 2$. (Recall that the imaging process causes the apparent wavenumber to be $2q_0$.) The fluctuations are found to have a *Gaussian probability distribution*, and both real and imaginary parts have the same variance. These distributions are shown in Figure 14.

The fluctuations are Gaussian only for $\varepsilon > 0.1$. For smaller ε, the fluctuations are much slower, but even when sampled over a very long time do not

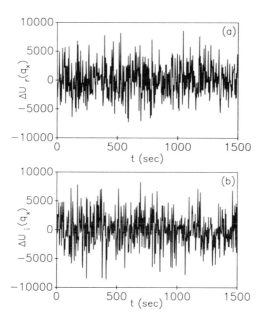

FIGURE 13. Real and imaginary parts of the fluctuations of $U(q_x)$ (in arbitrary units) for $q_x = 0.4q_0$ and $\varepsilon=0.15$.

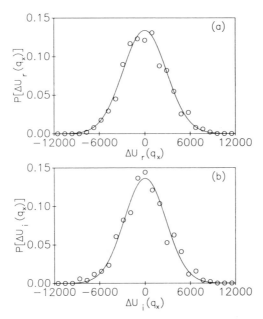

FIGURE 14. The probability distribution of the fluctuations $\Delta U(q_x)$ from figure 13. (a) Real part; (b) the imaginary part. Gaussian fits (solid lines) are satisfactory and the variances of the two sets are the same within 10% .

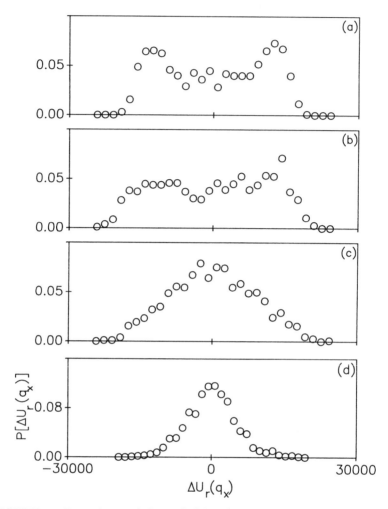

FIGURE 15. Dependence of the probability distributions on driving amplitude ε: (a) $\varepsilon=0.05$; (b) $\varepsilon=0.07$; (c) $\varepsilon=0.10$; (d) $\varepsilon=0.14$. In all cases $q_x = q_0$. The distribution becomes Gaussian at the transition to spatiotemporal chaos.

exhibit a Gaussian distribution. This is shown in Figure 15, where distributions for various ε are presented. The Gaussian distribution is associated only with the spatiotemporally chaotic state.

It is also worth noting that probability distributions for the *local image intensity* fluctuations $\Delta U(x,t)$ are *non-Gaussian* for all ε. This may be in part a consequence of imaging nonlinearity, but it would also probably be characteristic of the local wave height fluctuations as well.

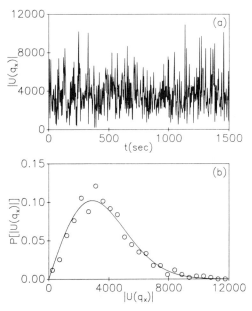

FIGURE 16. (a) The fluctuations of the absolute amplitude $|U(q_x)|$, whose real and imaginary parts are shown in Figure 12. (b) The probability distribution of $|U(q_x)|$. The solid curve shows the Rayleigh density, and the good fit implies that the real and imaginary parts are independent random variables.

One might wonder whether the fluctuations in the real and imaginary parts of $U(q_x,t)$ are caused by phase fluctuations alone. This is easily demonstrated it not to be the case. The magnitude $|U(q_x,t)|$ fluctuates substantially and follows a Rayleigh distribution:

$$P[|U(q_x)|] \sim |U(q_x)| \exp(-|U(q_x)|^2/2\sigma^2), \qquad (11)$$

where σ^2 is found to be independent of wavenumber for $q_x/q_0 < 2$ and is the same as the variance of the real and imaginary parts separately. A fit to the Rayleigh distribution is shown in Figure 16. This is the expected form for the distribution of a quantity whose real and imaginary parts are *independent* Gaussian random variables with equal variance.

The Gaussian character of the fluctuations and the short correlation length leads to the tempting hypothesis that the dynamical behavior can be completely confined to the scale of the wavelength. However, such a conclusion would be false; there are significant long-range features. These were already noted as bright swaths in $U(x,t)$. We can also see from images $U(x,y)$ (see Figure 17(a)) that there are domains much larger than λ that appear to have correlated orientation.

(a)

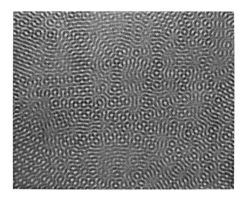

(b)

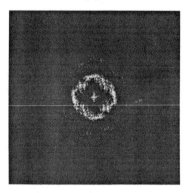

FIGURE 17. (a) A shadowgraph image of the capillary wave pattern obtained in a square cell for driving amplitude $\varepsilon=0.15$, in the spatiotemporally chaotic regime. The pattern shows spatial disorder on the average although locally correlated regions are clearly visible. The region shown corresponds to approximately 6.0 x 4.8 cm^2. (b) The spectral power $P(q_x, q_y)$ for this state shown as a 2-dimensional image. The center of the image represents the origin in the (q_x, q_y) space.

5.2. TWO DIMENSIONAL SPECTRA

As we mentioned in Section 4.1, two dimensional spectra $P(q_x, q_y)$ can be obtained from images of the flow. For $\varepsilon < 0.1$ the spectrum is a lattice of discrete spots as shown in Figure 5(b). For $\varepsilon > 0.1$, the bulk of the power is concentrated on a four-lobed ring which connects the peaks along the axes to those on the diagonals (see Figure 17(b)). The distribution of power $P(q_x, q_y)$ along the axial and the diagonal directions represents the major features of this function. Scans along these two directions are shown in Figure 18 for the ordered and disordered states, averaged over several

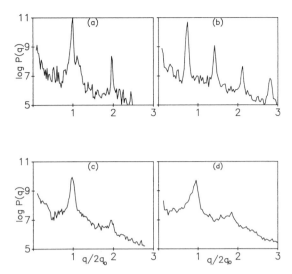

FIGURE 18. The variation of the spectral power $P(q_x, q_y)$ (in arbitrary units) along the axial and diagonal directions. (a) $\varepsilon=0.07$, axial direction; (b) $\varepsilon=0.07$, diagonal direction; (c) $\varepsilon=0.15$, axial direction; (d) $\varepsilon=0.15$, diagonal direction.

sets of data. The ordered state shows sharp, discrete peaks at $q = 2q_0$ along the axial direction and at $q = \sqrt{2}q_0$ along the diagonal direction. The disordered state is characterized by peaks that are not as high as in the case of the ordered state and are significantly broader.

Since the width of the peaks of the spatial spectra should be approximately proportional to the inverse of the correlation length of the pattern, the spectrum in principle provides an alternate way of estimating the spatial correlation length. We obtained the fractional spectral widths $2\sigma/q_0$ by computing the second moment σ^2 of $P(q_x, q_y)$ about each peak, integrating out to the point at which the data falls below the peak by two orders of magnitude. For the ordered state, we find that $2\sigma/q_0$ is approximately 0.05. The spectral width is nearly the same for scans along the axial and diagonal directions in q space. In the disordered state, the widths are larger, and differ by a factor of two in the axial and diagonal directions (0.13 and 0.28 respectively). The smaller axial width indicates larger residual translational correlation along directions parallel to the cell walls. The anisotropy is noticeable even at $\varepsilon=0.3$, where the average width is 0.4. The actual numerical values cited in this paragraph are somewhat dependent on the cutoffs used.

5.3. ORIENTATIONAL CORRELATIONS

Locally, the patterns have a tendency toward fourfold symmetry. This prompted us to compute a fourfold orientational correlation function $G_4(r)$

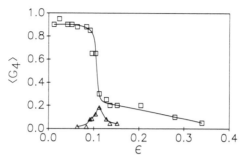

FIGURE 19. Extent of fourfold orientational correlation at large distances, $G_4(\infty)$, as a function of ε (squares). The triangles show the rms fluctuation in the orientational correlation as a function of ε. There is a substantial decline in orientational disorder near $\varepsilon=0.10$ as in figure 9. However, some residual orientational order remains.

(Tufillaro, Ramshankar, and Gollub, 1989) that measures the extent to which the local orientation at one point persists to another at a distance r away. We find that for large distances $r \sim 10\lambda$, $G_4 \cong 0.9$ for $\varepsilon < 0.1$, where $G_4 = 1$ for perfect orientational order. The extent of correlation declines sharply near $\varepsilon=0.1$. However, there is still a residual long-range orientational correlation of roughly 0.1–0.2 for $\varepsilon > 0.1$. These features are demonstrated in Figure 19.

The observation of persistent orientational correlation at large distances further confirms that the translational correlation function (averaged over angles) is not sufficient to indicate the extent to which different regions fluctuate coherently.

5.4. TEMPORAL SPECTRA

The shape of the temporal spectrum in the disordered state provides another useful characterization of the chaotic fluctuations. Figure 20(a) shows the spectrum of the local fluctuations of the image intensity at a typical point. Frequencies are normalized to the basic wave frequency $\omega_0/2$. A power law tail is evident for frequencies such that $(2\omega/\omega_0) > 10^{-2}$, where

$$P(\omega) \sim (\omega)^{-3.7\pm0.2}. \tag{12}$$

This exponent does not seem to depend significantly on ε.

Another way to characterize the temporal fluctuations is to utilize the $x - t$ data of Figure 12 to obtain a frequency spectrum at a selected value of q_x. An example obtained at $q_x/q_0 = 0.65$ is shown in Figure 20(b). The result is essentially the same as for the local signal, and is independent of

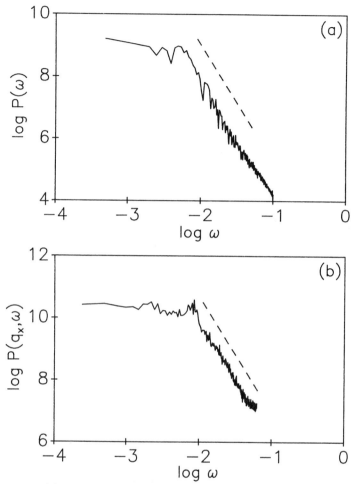

FIGURE 20. (a) The spectrum of local fluctuations obtained by measuring the light intensity at a single point. (b) The wavenumber-selected spectrum $P(q_x, \omega)$ (in arbitrary units) for $q_x/q_0 = 0.65$ obtained from the $x - t$ scans. In both cases, the horizontal axis denotes frequency normalized by the basic wave frequency (160 Hz). The power law fall off is indicated by the dashed lines.

wavenumber for $q_x/q_0 < 2$. In summary, both the local and q-dependent spectra reveal significant fluctuations at frequencies two orders of magnitude below the basic wave frequency. The spectrum of these fluctuations is flat at low frequencies and falls off as a power law at frequencies comparable to or higher than f^*.

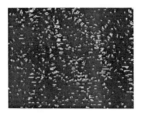

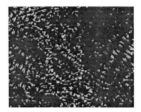

FIGURE 21. Difference image formed by subtraction of two images of the waves in the chaotic state, separated by 0.15 sec. Two examples are shown. The spatial inhomogeneity of the active regions is evident.

5.5. SPATIOTEMPORAL INTERMITTENCY

As noted in other studies of spatiotemporal chaos, the time dependence of interfacial waves is localized and intermittent. This property may be seen most easily from the *difference* between two images of the waves separated by a short time interval (0.1–0.2 s). Figure 21 shows examples of such a difference image. It is apparent that the time dependence of the pattern at this short time scale is highly inhomogeneous. This inhomogeneity is related to the domains of similarly oriented waves, and provides an additional example of correlations extending beyond the scale of the wavelength. The active regions are typically the boundaries between the domains.

5.6. ROLE OF THE BOUNDARIES

The lateral boundaries may have a significant effect on the onset and statistical properties of spatiotemporally chaotic waves in a finite size system. The basic ordered wave pattern is apparently independent of the global geometry, and has fourfold symmetry even in a circular cell (Ezerskii, Korotin, and Rabinovich, 1985; Tufillaro, Ramshankar, and Gollub, 1989). We also find that the onset of spatiotemporal chaos, as measured for example by the characteristic frequency $f^*(\varepsilon)$ of chaotic fluctuations, is quite similar

in both cases. (Earlier preliminary work incorrectly suggested a more grad-
ual transition in the circular case, probably due to surface contamination.)
The dependence of the statistical properties in the disordered regime on
the geometry requires further investigation.

5.7. ADDITIONAL OBSERVATIONS

Some measurements were made at excitation frequencies a factor of two
higher and lower than 320 Hz. Although the *transition process* is somewhat
dependent on the excitation frequency, the properties of the disordered
state seem to be robust. The nature of the working fluid could also be
significant. In fact, Ezersky and Rabinovich (1990) have found that the
properties of the spatiotemporally chaotic state do depend somewhat on the
working fluid. We found, for example, that the transition to the disordered
state is not as sharp for water as it is for butanol, which has a lower
surface tension. We have not undertaken a thorough investigation of these
questions.

6. Discussion and Conclusion

After an introduction and brief review of several examples of spatiotemporal chaos, we have focussed on the problem of parametrically forced interfacial waves. Chaotic time dependence due to modal interactions is evident
for certain driving frequencies even when only two modes are accessible.
For higher excitation frequencies, the wavelength is much shorter than the
lateral dimension of the system, and spatiotemporal chaos occurs for *any*
choice of excitation frequency.

Our emphasis has been on thorough characterization of the disordered
wave state. We noted that the translational correlation length is of the order
of the wavelength, and that, as a result, the Fourier amplitudes have Gaussian fluctuations. However, some orientational order persists in the spatiotemporally chaotic regime and the decay of correlation is not isotropic.
The 2-dimensional spatial spectrum in the choatic state is characterized by
a broad anisotropic ring. We also note the power law decay in both the local
and wavenumber-selected power spectra $P(\omega)$ at frequencies comparable to
or higher than the characteristic fluctuation frequency $f *$. Difference images reveal inhomogeneous and intermittent fluctuations on scales larger
than the wavelength.

There have been several efforts to provide a quantitative explanation
of the phenomena described here. One of the earliest proposals has been
given by Ezerskii *et al.* (1986) to explain their observations. They argue that
the essential phenomena can be described in only one space dimension and
time, and propose a generalization of the Ginzburg-Landau equation which
they suggest is generic for parametric forcing. The differential equation for

the wave amplitude $u(\eta, \tau)$ expressed in terms of non-dimensional space and time variables has the form

$$\partial u / \partial \tau - i \partial^2 u / \partial \eta^2 + u = i \eta u^* + i u |u|^2 - i \sigma u. \tag{13}$$

This equation yields the slow space and time modulations of each traveling wave component of the standing wave. Numerical simulations indicate that the equation has several different chaotic attractors that are accessible from different initial conditions. However, our experiments do not indicate any dependence on initial conditions in the chaotic regime. Furthermore, the orientational correlations visible in Figure 17 and measured in Figure 19 would seem to require a two-dimensional description. Still, the relevance of model equations to this problem, as in other cases involving the disordering of cellular or roll-like structures, seems fairly clear.

Another theoretical effort has been undertaken by Milner (1989) who derives amplitude equations to describe the evolution of the wave pattern. Unlike Ezerskii et al., Milner's calculations include nonlinear damping terms, and consider the problem of arbitrary patterns in two dimensions. It turns out that the nonlinear damping terms are essential to account for the symmetry and the stabilization of the amplitude of the regular pattern. For excitation amplitudes just above the onset of waves, a reduction of the number of marginal modes (from traveling capillary waves to standing waves) is possible, leading to a simpler set of amplitude equations. This simpler set of amplitude equations has a Lyapunov functional, from which one obtains the wavenumber and the pattern symmetry. Milner's calculations show that the square symmetry of the pattern as observed in the experiments is the most stable uniform state. However, the original amplitude equations have a secondary instability to transverse amplitude modulations (TAM). That is, the uniform square pattern collides with the TAM instability boundary in parameter space, leading to amplitude modulation of the standing waves.

Theoretical estimates of the pattern amplitude and the wavenumber of the secondary instability are in good agreement with the experimental observations (see Table I). However, the predicted onset of the secondary instability occurs at $\varepsilon = 0.01$, considerably below the observed transition at $\varepsilon = 0.10$. Milner suggests that a factor of three increase in the computed nonlinear damping coefficients, which are hard to estimate, would account for this shift. A further test of the theory may be made by measuring the change of the onset of the secondary instability with frequency or surface tension.

We end with several remarks on spatiotemporal chaos and its relationship to temporal chaos and fully developed turbulence. The loss of spatial coherence is the most dramatic way in which the phenomena described here differ from temporal chaos. In some systems, a relatively low-dimensional description is still useful even in the presence of spatiotemporal complexity (Aubry et al., 1988; Sirovich, 1991). It is not clear whether such an

Table I. A comparison of the experimental observations (present work) and theoretical predictions of Milner (1989).

	Measured value	Theoretical prediction
Threshold acceleration	7.2 g	8 g
amplitude for wave onset (n-butanol at 30 C, ω_0=320 Hz.)		
Onset ε for secondary instability	0.10	0.01
Pattern amplitude at secondary instability onset	$\sim 100\mu$m	95 μm
Wavenumber of secondary instability	$\sim 8q_0$	$6.6q_0$

approach might be possible for wave dynamics when the system is large compared to the basic wavelength.

In both spatiotemporal chaos and fully developed turbulence a statistical approach is useful as a practical matter. Careful measurement and computation of various correlation functions provide at least some of the required tools for characterization. In both cases large scale structures are often involved and intermittent behavior is apparent. In fully developed turbulence a scaling range exists, bounded at small q by the coherent structures and at large q by viscous dissipation. Though we found evidence for correlations on scales large compared to the wavelength in disordered surface waves, spatial scaling behavior was not observed.

Acknowledgements

We appreciate helpful discussions with P. Hohenberg and S. Milner. This work was supported by the University Research Initiative program under Contract No. DARPA/ONR N00014-85-K-0759.

References

Ahlers, G., Cannell, D.S., and Steinberg, V., 1985, "Time dependence of flow patterns near the convective threshold in a circular cylinder,"Phys. Rev. Lett. 54, 1373-1376.

Ahlers, G., Cannell, D.S., and Heinrichs, R., 1987, "Convection in a binary mixture," Nuc. Phys. B (Proc. Suppl.) 2, 77-86.

Albert, B.S., 1986, "Pattern formation in a nematic liquid crystal undergoing an electrohydrodynamic instability in the presence of a spatially periodic forcing," Haverford College, B.S. thesis (unpublished).

Aubry, N., Holmes, P., Lumley, J.L., and Stone, E., 1988, "The dynamics of coherent structures in the wall region of a turbulent boundary layer," J. Fluid Mech. 192, 115-173.

Benjamin, T.B. and Ursell, F., 1954, "The stability of the plane free surface of a liquid in vertical periodic motion," Proc. R. Soc. Lond. A225, 505-515.

Bodenschatz, E., Kaiser, M., Kramer, L., Pesch, W., Weber, A., and Zimmermann, W., 1989, "Pattern and defects in liquid crystals," in *New Trends in Nonlinear Dynamics and Pattern Forming Phenomena: The Geometry of Nonequillibrium* (eds. P. Coullet and P. Huerre), Plenum Press.

Chaté, H. and Manneville, P., 1987, "Transition to turbulence via spatiotemporal intermittency," Phys. Rev. Lett. 58, 112-115.

Ciliberto, S. and Gollub, J.P., 1985, "Chaotic mode competition in parametrically forced surface waves," J. Fluid Mech. 158, 381-398.

Ciliberto, S. and Bigazzi, P., 1988, "Spatiotemporal intermittency in Rayleigh–Bénard convection," Phys. Rev. Lett. 60, 286-289.

Coullet, P., Elphick, C., Gil, L., and Lega, J., 1987, "Topological defects of wave patterns," Phys. Rev. Lett. 59, 884-886.

Coullet, P., Gil, L., and Repaux, D., 1989, "Defects and subcritical bifurcations," Phys. Rev. Lett. 62, 2957-2960.

Coullet, P., Gil, L., and Lega, L., 1989, "A form of turbulence associated with defects," Physica D 37, 91-103.

Crawford, J.D., Knobloch, E., and Riecke, H., 1989, "Mode interactions and symmetry," in *Proceedings of the International Conference on Singular Behavior and Nonlinear Dynamics*, Samos, Greece, World Scientific.

Croquette, V., 1989, "Convective pattern dynamics at low Prandtl number," Contemp. Phys. 30, 113-133 (Part I); Contemp. Phys. 30, 153-171 (Part II).

Daviaud, F., Dubois, M., and Bergé, P., 1989, "Spatio-temporal intermittency in quasi one-dimensional Rayleigh-Bnard convection," Europhys. Lett., 9, 441-446.

Douady, S. and Fauve, S., 1988, "Pattern selection in Faraday instability," Europhys. Lett. 6, 221-226.

Ezerskii, A.B., Korotin, P.I., and Rabinovich, M.I., 1985, "Random self-modulation of two-dimensional structures on a liquid surface during parametric excitation," Pis'ma Zh. Eksp. Teor. Fiz. 41, 129-131; JETP Lett. 41, 157-160.

Ezerskii, A.B., Rabinovich, M.I., Reutov, V.P., and Starobinets, I.M., 1986, "Spatiotemporal chaos in the parametric excitation of a capillary ripple," Zh. Eksp. Teor. Fiz. 91, 2070-2083; Sov. Phys. JETP 64, 1228-1236.

Ezersky, A.B. and Rabinovich, M.I., 1990, "Nonlinear wave competition and anisotropic spectra of spatio-temporal chaos of Faraday ripples," Europhys. Lett. 1990, 243-249.

Faraday, M., 1831, "On a peculiar class of acoustical figures, and on certain forms assumed by groups of particles upon vibrating elastic surfaces," Phil. Trans. R. Soc. Lond. 121, 299-340.

Funakoshi, M. and Inoue, S., 1987, "Chaotic behavior of resonantly forced surface water waves," Phys. Lett. A. 121, 229-232.

Gollub, J.P. and Meyer, C.W., 1983, "Symmetry-breaking instabilities on a fluid surface," Physica 6D, 337-346.

Goren, G., Procaccia, I., Rasenat, S., and Steinberg, V., 1989, "Interactions and dynamics of topological defects", Phys. Rev. Lett. 63, 1237-1240.

Gu, X.M., Sethna, P.R., and Narain, A., 1988, "On three-dimensional non-linear subharmonic resonant surface waves in a fluid. Part I: Theory," Trans ASME E: J. Appl. Mech. 55, 213-219.

Henderson, D.M. and Miles, J.W., 1990, "Single-mode Faraday waves in small containers," J. Fluid Mech. 213, 95-109.

Heutmaker, M.S., 1986, "A quantitative study of pattern evolution and time dependence on Rayleigh-Bénard convection," University of Pennsylvania, Ph.D. thesis (unpublished).

Heutmaker, M.S. and Gollub, J.P., 1987, "Wave-vector field of convective flow patterns," Phys. Rev. A 35, 242-259.

Hohenberg, P.C. and Shraiman, B.I., 1989, "Chaotic Behavior of an Extended System," Physica D 37, 109-115.

Kai, S. and Hirakawa, K., 1978, "Successive transitions in electrohydrodynamic instabilities of nematics," Prog. Theor. Phys. Suppl. 64, 212- 243.

Kaneko, K., 1984, "Period-doubling of kink-antikink patterns, quasiperiodicity in anti-ferro-like structures and spatial intermittency in coupled logistic lattice," Prog. Theor. Phys. 72, 480-486.

Kolodner, P., Bensimon, D., and Surko, C.M., 1988 "Traveling wave convection in an annulus," Phys. Rev. Lett. 60, 1723-1726.

Manneville, P., 1981, "Statistical properties of chaotic solutions of a one-dimensional model for phase turbulence," Phys. Lett. 84A, 129-132.

Meron, E. and Procaccia, I., 1986, "Low-dimensional chaos in surface waves: theoretical analysis of an experiment," Phys. Rev. A 34, 3221-3237.

Meron, E., 1987, "Parametric excitation of multimode dissipative systems," Phys. Rev. A 35, 4892-4895.

Miles, J.W., 1984, "Resonantly forced surface waves in a circular cylinder," J. Fluid Mech. 149, 15-31.

Miles, J. and Henderson, D., 1990, "Parametrically forced surface waves," Ann. Rev. Fluid Mech. 22, 143-165.

Milner, S., 1989, "Square patterns and secondary instabilities in driven capillary waves," preprint.

Nobili, M., Ciliberto, S., Cocchiaro, B., Faetti, S., and Fronzoni, L., 1988, "Time-dependent surface waves in a horizontally oscillating container," Europhys. Lett. 7, 587-592.

Pocheau, A., 1989, "Phase dynamics attractors in an extended cylindrical convective layer," J. Physique 50, 25-69.

Pocheau, A., Croquette, V., and Le Gal, P., 1985, "Turbulence in a cylindrical container of Argon near threshold for convection," Phys. Rev. Lett. 55, 1094-1097.

Pomeau, Y., Pumir, A., and Pelcé, 1984, "Intrinsic stochasticity with many degrees of freedom," J. Stat. Phys. 37, 39-49.

Pumir, A., 1985, "Statistical properties of an equation describing fluid interfaces," J. Physique 46, 511-522.

Rehberg, I., Rasenat, S., and Steinberg, V., 1989, "Traveling waves and defect-initiated turbulence in electroconvecting nematics," Phys. Rev. Lett. 62, 756-759.

Shraiman, B.I., 1986, "Order, disorder and phase turbulence," Phys. Rev. Lett. 57, 325-328.

Simonelli, F. and Gollub, J.P., 1989, "Surface wave mode interactions: effects of symmetry and degeneracy," J. Fluid Mech. 199, 471-494.

Sirovich, L., 1991, "Empirical Eigenfunctions and Low Dimensional Systems," in New Perspectives in Turbulence, L. Sirovich, Ed., Springer-Verlag. (This volume.)

Sompolinsky, H., Crisanti, A., and Sommers, H.J., 1988, "Chaos in random neural networks," Phys. Rev. Lett. 61, 259-262.

Steinberg, V., Moses, E., and Fineberg, J., 1987, "Spatio-temporal complexity at the onset of convection in a binary fluid," Nuc. Phys. B (Proc. Suppl.) 2, 109-124.

Tufillaro, N.B., Ramshankar, R., and Gollub, J.P., 1989, "Order-disorder transition in capillary ripples," Phys. Rev. Lett. 62, 422-425.

Waller, I. and Kapral, R., 1984, " Spatial and temporal structures in systems of coupled nonlinear oscillators," Phys. Rev. A 30, 2047-2055.

7

The Complementary Roles of Experiments and Simulation in Coherent Structure Studies

Mogens V. Melander, Hyder S. Husain and Fazle Hussain

Introduction

The past two decades' vigorous studies of coherent structures (CS) have failed to produce a consensus on what CS are, let alone a CS-based turbulence theory or even an objective, mathematical definition of CS. What started out as a promise for a mechanistic explanation for fluid turbulence—as researchers found or reinvented CS in their 'search for order in disorder' and presumed to have discovered a deterministic, tractable route to turbulence phenomena—has unfolded itself as a Pandora's box. Successive studies of CS continue to raise more questions than they answer. Thus, even though we understand more about turbulence via CS concepts, we have become painfully even more aware of how complex turbulence is. CS are not the panacea they were initially presumed to be, nor are they as simple as we all had hoped. Despite some progress through CS research, the secrets of turbulence remain ever impenetrable.

If every flow geometry has its own brand of CS, or worse yet, if every flow has a large (if not infinite) variety of CS, then is the effort required to unravel the anatomical details of every CS worthwhile? CS studies will then involve monumental efforts, not only because the statistics of so many structures will have to be measured, but also because intricate detection criteria will have to be devised to discriminate among different categories of structures. Even if every detail of all CS were known, it is unclear how this information would be utilized in a theory, predictive model, or application. A CS-based turbulence theory has not yet been proposed, nor is there any in sight. CS continue to remain almost exclusively in the arena of experimental studies. Unfortunately, research in this field is still in its infancy, at the information gathering stage; comprehensive analysis and synthesis are still to follow.

Such morphological cataloging, although a hallmark of science (particularly in chemistry and biology), would be virtually endless. No doubt, CS is a rich subject, inviting potentially fruitful explorations. But we have chosen for our own work at the University of Houston a much more limited agenda. What are the generic features, if any, of CS in different turbulent flows? Can we learn about CS interactions in turbulent flows and hence about turbulence physics via studies of a few paradigms? Can/should CS control produce desired turbulence management? The last item, which is primarily technologically-motivated, although extensively studied by us, will not be addressed here. This note is neither a review of the field of CS, nor a summary of all of our works in CS. Comprehensive reviews of CS research (e.g., Kline *et al.*1967; Crow & Champagne 1971; Brown & Roshko 1974; Coles 1981; Lumley 1981; Ho & Huerre 1984; Wallace 1985, Fiedler 1988; Bridges et al 1989; Kline & Robinson 1989) as well as summaries of our own works (Hussain 1981, 1986) have appeared before.

Flow visualization pioneered the discovery of CS and has remained the mainstay for many researchers, as if showing a mere suggestion of a large-scale organized event in turbulent flow is still a challenge and constitutes a study of CS dynamics! What little quantitative CS data there are came from measurements at a point or in a plane. We need to know the time-evolving 3D topology and dynamics of CS and their interactions. Unfortunately, state-of-the-art measurement technology is too limited in this respect and is not likely to improve soon.

The advent of supercomputers has presented an alternative method, namely direct numerical simulation (DNS). Although limited to low Re (in the absence of modeling) and simple geometries—future developments of supercomputers will surely relax this limitation—DNS can provide complete 3D instantaneous flow details. (Incidentally, the same computer limitations—memory and speed—which are the pacing factors for DNS are also so for modern measurement techniques such as 3D holographic particle velocimetry). We are thus forced to forge a partnership between the two: DNS and experiments, one complementing the other. The advantages and complementary roles of experiments and numerical simulation have been re-iterated (e.g., Hussain 1986). While experiments are unavoidable at high Re and in complex flows, DNS has decided advantages in parametric studies of effects of initial and boundary conditions, controlled excitation, freestream turbulence, etc., which are not independently controllable in experiments. DNS (including LES) is the only way by which one can study 2D flows (including 2D vortex dynamics and turbulence). Experiments and flow visualization can be used to identify the most interesting CS in a flow and their key interactions that need to be pursued further through idealized numerical simulation. DNS can, in addition, be used to discover new mechanisms and quantify them, as well as suggest new experiments for verification and further exploration. The combined use of experiments and DNS seems to be the logical path to follow, and has therefore been the thrust of our own research.

Definition, role and eduction of CS

Turbulent flows, characterized by random 3D vorticity fluctuations, are often regarded as an evolving tangle of vortex lines (Tennekes & Lumley 1972). This idea has, in fact, been used directly by Schwarz (1983), who has employed tangles of vortex filaments to simulate turbulence in superfluids. Because we view CS as the underlying organized motions in turbulence, it is natural to define CS in terms of vorticity. Thus we defined a CS as a large-scale motion with instantaneous phase-correlated (i.e., coherent) vorticity over its spatial extent; that is, underlying the random 3D vorticity fluctuations, that we call turbulence, there is a vorticity field which is instantaneously coherent over the extent of the structure (Hussain 1981).

Based on this definition, a general scheme has been developed to educe CS even in fully-developed turbulent flows (Hussain 1986). This eduction scheme entails averaging an ensemble of properly aligned flow realizations (instantaneous vorticity fields showing large-scale coherence). A coherent structure is a statistical measure of organized structures of one kind (i.e., shape, size, strength, orientation, etc.) and ensemble averaging allows one to separate the incoherent turbulence form the coherent component. Joint probability statistics, involving shape, size, strength, orientation, etc. as variables, reveal the dominant modes (the preferred modes) of vortical structures in a particular flow, which warrant detailed investigation. This iterative method of CS eduction does not assume any structure mode, as in pattern recognition and template matching techniques (Mumford 1983). Central to the CS studies is the concept of the preferred mode, without which CS is not a useful idea (Hussain 1981).

The concept of CS helps one to explain many important features and peculiarities of turbulent shear flows, and allows a better understanding of the flow physics which are not at all discernible either from voluminous instantaneous data or from time-average data. Furthermore, the CS approach promises not only to provide insight into the mechanisms of turbulence, but also to allow these mechanisms to be controlled. The role of CS in the transports of heat, mass, and momentum, which in turn affect mixing, chemical reaction, drag, aerodynamic noise, etc. can be controlled by modifying the generation, evolution, and interaction of CS. Such control is not likely to be possible in a turbulent flow having no CS.

CS and the role of vortex dynamics

Our definition of CS in terms of organized vorticity is especially useful when viewed in the light of vortex dynamics; the topology of CS during their evolution and interaction is tractable through vortex dynamics. When we decompose the vorticity field into coherent and incoherent (random) parts by

$$\omega = \omega_c + \omega_r, \tag{1}$$

the Biot-Savart law,

$$\mathbf{u}(\mathbf{r}) = \frac{1}{4\pi} \int \frac{\omega \times (\mathbf{r} - \mathbf{r}')}{|\mathbf{r} - \mathbf{r}'|} d\mathbf{r}' + \nabla \phi, \tag{2}$$

produces two velocity fields corresponding to ω_c and ω_r, i.e.,

$$\mathbf{u} = \mathbf{u}_c + \mathbf{u}_r. \tag{3}$$

These two velocity fields have different long-range asymptotic properties. This is most easily seen when we consider $|\mathbf{r} - \mathbf{r}'|$ to be large and expand (2) in moments of the vorticity distribution about some point r_0 inside a CS. The induced velocity field (2) thereby takes the form of an infinite series in inverse powers of $|\mathbf{r} - \mathbf{r}_0|$, thus constituting a systematic description of the influence of a CS and its internal small-scale structure on the far field. The $\mathbf{u}_c$-component induced by the coherent vorticity tapers off at large distances as $1/|\mathbf{r} - \mathbf{r}_0|$ or $1/|\mathbf{r} - \mathbf{r}_0|^2$ (depending on whether the flow is 2D or 3D). On the other hand, random vorticity components will average to zero when integrated over a volume of the size of a CS. Therefore, we expect $\mathbf{u}_r$ to tend to zero as a high power of $1/|\mathbf{r} - \mathbf{r}_0|$, or perhaps even faster. Hence, the coherent vorticity has long-range effects and accounts for the far-field, while the random vorticity has only short-range, local influence. However, the local influence of the random vorticity should not be underestimated, as it can cause local alterations of CS and thereby indirectly influence the far field through the vorticity moments of the CS. This is evident from the phase averaged vorticity transport equation,

$$\frac{D}{Dt} \langle \omega \rangle = \langle \omega \rangle \cdot \nabla \langle \mathbf{u} \rangle + \frac{1}{\text{Re}} \nabla^2 \langle \omega \rangle$$

$$+ \langle \omega_r \cdot \nabla \mathbf{u}_r \rangle - \langle \mathbf{u}_r \cdot \nabla \omega_r \rangle, \tag{4}$$

where the third and fourth terms of the RHS show the effects of stretching and advection of incoherent vorticity by the incoherent velocity on the organization of the coherent vorticity field (Hussain 1986). Evaluation of these two nonlinear terms constitutes the prime limitation of our ability to predict the evolution of CS.

We may put these ideas into a formal computational framework by modifying the definition of the instantaneous coherent vorticity field by putting,

$$\omega_c(\mathbf{r}) = \omega_{c,\sigma}(\mathbf{r}) = \int \omega(\mathbf{r}_0) K_\sigma(\mathbf{r} - \mathbf{r}_0) d\mathbf{r}_0. \tag{5}$$

Here K_σ is a smoothing kernel, which smooths fine-scale structures based on the length-scale σ. The smoothing kernel should be a C^∞ test-function for the Dirac delta function, as that ensures $\omega_{c,\sigma} \to \omega$ as $\sigma \to 0$. This definition is not ideal because it is not clear exactly how σ should be selected. If we choose σ too large, a coherent feature such as a rib in a mixing layer will be represented as being too fat. If, on the other hand, σ is chosen too small, then $\omega_{c,\sigma}$ will include part of the random vorticity. Nevertheless, this definition is very helpful as it is mathematically precise and also correctly reproduces the far field induced by CS. Furthermore, it reduces to a simple filtering in wavenumber space and is therefore computationally handy.

Note that, in our eduction scheme, we do not assume any smoothing function; the ensemble average takes care of the smoothing operation. However, in general, in direct numerical simulations we do not have the luxury of generating a large number of sample functions. Experimental data may guide one to construct proper smoothing functions to be used in numerical simulations.

The true significance of (5) and its physical moment expansion lie in the fact that the long-range interaction of two or more CS can be expressed in terms of a few low-order moments. This idea has been used with success in 2D vortex dynamics (Melander *et al.* 1986). Besides elucidating the physics of well separated coherent structures, the moment expansion also has a significant computational potential. Since the CS interactions are described by only a few degrees of freedom, which evolve on a much slower timescale than do the incoherent vorticity, the moment approach opens up an attractive avenue for parallel computation. Local CS interaction can be allocated to individual processors, with the communication between processors taking place through the exchange of low order moments, thus yielding a very low communication to computation ratio. On the other hand, the moment expansion is seldom of any use when we consider the interaction of CS in very close proximity or the self-deformation of a single CS (a notable exception being the pairing of 2D-vortices (Melander *et al.* 1988, 1987)). The reason for this is the simple fact that the expansion parameter $1/|\mathbf{r} - \mathbf{r}_0|$ is no longer small. Likewise, we cannot formulate equations for close interactions entirely in terms of $\omega_{c,\sigma}$, as any such attempt inevitably results in the classical closure problem. Thus, for close CS interactions we must retain the full Navier-Stokes equations.

With the long-range interaction described by a simple mathematical framework, we concentrate on the close interactions. These interactions can be divided into the following categories: (1) self-deformation of a single isolated CS, including the effects of turbulence it may generate; (2) the interaction of a single isolated CS with a background potential flow; (3) the interaction of a single isolated CS with a turbulent background; (4) the isolated interaction of two CS in very close proximity; and (5) the isolated interaction of two CS in the presence of a turbulent background.

These close interactions and self-deformations are by themselves intriguing facets of fluid mechanics. Such problems as vortex breakdown, vortex pairing, and vortex reconnection are examples of fundamental interactions which fall into the above categories. These problems are so complex that, even when studied in isolation from other flow effects, our present understanding of their dynamics does not usually allow simple predictive models. Perhaps an even greater difficulty is that we do not currently have a generic catalog of the key interactions. Thus, to further our understanding of CS interactions and their roles in turbulent flows, we must first identify the basic CS interactions primarily via experiments, and then make a detailed study of the dynamics of each interaction through idealized direct numerical simulation.

Only a thorough insight into the dynamics of key vortex interactions can further the present level of understanding of turbulence and the role of CS. The theory of turbulence in 2D flows illustrates this point clearly. In 2D flows, fundamental vortex interactions, such as axisymmetrization, merger or pairing, and dipole formation provide the mechanisms for the emergence of large-scale coherent structures (McWilliams 1984). These structures in turn invalidate earlier statistical theories based on random phases (Babiano et al. 1987). The dynamics of each of these fundamental 2D vortex interactions are now well understood (Melander et al. 1987, 1988). Axisymmetrization, the process by which a single isolated 2D vortex becomes circular, serves as a simple example of the importance of interaction between large and small scales. Although an initially non-circular vortex may not have any small scales, it immediately spins off small-scale spiral filaments. In turn, these filaments, which are advected by the flow induced by the vortex core, influence the vortex core in such a way as to drive the core towards circular symmetry on a convective timescale (Melander et al. 1987).

Morphology of CS: the role of experiments

We will now discuss the topology and a number of key interactions involving CS which have been revealed through experiments. For each of these, a large body of experimental data is available, which calls for future complementary simulation, visualization, and mechanism extraction.

GENERAL TOPOLOGY OF CS

Centers and saddles

Our detailed studies of the spanwise cross-section of CS in jets, wakes, and mixing layers revealed that the topology of various coherent structure properties is very similar irrespective of flow geometry and structure morphology. Examples of various structure geometries are: axisymmetric CS in the near field of a circular jet (Hussain & Zaman 1980), helical CS in the far-field of a circular jet (Tso & Hussain 1989), and quasi-two-dimensional

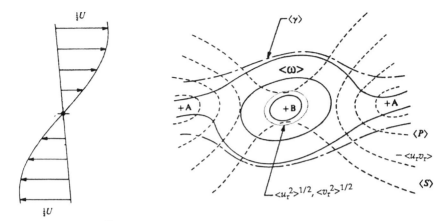

FIGURE 1. Topological features of a coherent structure showing approximate contours of various coherent structure properties: coherent vorticity $\langle\omega\rangle$; coherent intermittency $\langle\gamma\rangle$; incoherent turbulence intensities $\langle u_r^2\rangle^{1/2}$, $\langle v_r^2\rangle^{1/2}$; incoherent Reynolds stress $-\langle u_r v_r\rangle$; coherent strain rate $\langle S\rangle$; coherent production $\langle P\rangle$.

CS in plane wakes and mixing layers (Hayakawa & Hussain 1989; Metcalfe *et al.* 1987). The distributions of various coherent structure properties in the spanwise plane is schematically shown in figure 1. Coherent structures are characterized by two critical points, the saddle (A) and center (B). These are associated with the minimum and maximum spanwise coherent vorticity respectively. The saddle region is also characterized by longitudinal vorticity, peak levels of incoherent Reynolds stress, coherent strain rate and hence coherent turbulence production. Using the theory of critical points, basic topological elements of coherent structures (namely nodes, saddles, and foci) have been studied (Perry & Fairlie 1974; Perry & Chong 1984); however, the dynamical significance of these critical points has been revealed only after quantitative measurements of various CS properties (Hussain 1981).

As an example, educed structure properties, namely coherent vorticity $\langle\omega_Z\rangle/S_M$, incoherent Reynolds stress $\langle u_r v_r\rangle/U_0^2$, and coherent turbulence production $\langle P\rangle/U_0^2 S_M$, in the self-preserving region of a single-stream plane mixing layer are shown in figure 2. In these figures, coordinates are nondimensionalized by θ and U_c, i.e., $Y = y/\theta$ and $T = tU_c/\theta$, where U_0 is the free-stream velocity, U_c is the structure convection velocity, θ is the local momentum thickness, and S_M is the maximum of the mean velocity gradient. Measurements were made with a linear array of eight X-wire probes. The spanwise vorticity ω_z was calculated from eight pairs of u and v signals using the Taylor hypothesis. The educed CS is based on an ensemble average of 700 realizations. These data show that the peaks of $\langle u_r v_r\rangle$ and $\langle P\rangle$ indeed occur near the saddle region of the structure.

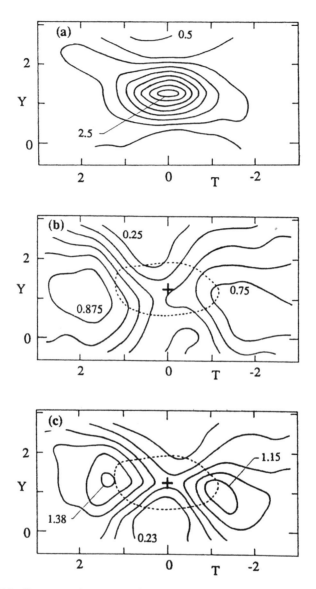

FIGURE 2. Contours of various coherent structure properties in a single-stream plane mixing layer at $x/\theta_e = 2000$. (a) Coherent vorticity $\langle \omega_z \rangle / S_M$; contour levels are: 2.5, 2.25, 2, 1.75, 1.5, 1.25,1.0, 0.75, 0.5. (b) Incoherent Reynolds stress $\langle u_r v_r \rangle / U_0^2$; contour levels are: 0.875, 0.75, 0.625, 0.5, 0.375, 0.25 ($\times 10^{-2}$). (c) Coherent turbulence production $\langle P \rangle / U_0^2 S_M$; contour levels are: 1.38, 1.15, 0.92, 0.69, 0.46, 0.23 ($\times 10^{-2}$).

Ribs and their role in turbulence production

Since vortex stretching is the most likely mechanism of turbulence production, it was argued that the saddle region, which is characterized by very low spanwise vorticity and high turbulence production, must consist of a series of longitudinal vortices aligned with the diverging separatrix. It was also suggested that such stretching of predominantly longitudinal vortices was the dominant mechanism of production in most turbulent flows (Hussain 1984). We have named these longitudinal vortical substructures connecting larger CS as *ribs*. The ribs have been visualized in a number of laboratories including ours (e.g., Bernal & Roshko 1986; Jimenez, Cogollos & Bernal 1985; Lasheras *et al.* 1986).

During the evolution of spanwise large-scale structures (rolls), ribs connecting two adjacent rolls are wrapped around the rolls and subjected to continual stretching. Ambient fluid entrained by the induction of ribs acquires vorticity and is transported away from the saddle region towards the roll. When this entrained fluid reaches the rolls, the interaction of the two orthogonal vorticities produces three-dimensional vorticity fluctuations, causing enhanced mixing. The direct role of ribs in mixing and sustaining chemical reaction has been demonstrated by Metcalfe & Hussain (1989). Due to the interaction of ribs with the spanwise structure, turbulence produced in the saddle region is advected towards the structure center, resulting in high values of incoherent turbulence intensities there (Hussain 1981).

Obviously, to capture the topology of ribs and their interactions with spanwise rolls is extremely difficult. The measurement of all three components of vorticity simultaneously at a number of points in space is still beyond current measurement technology. However, with our limited experimental resources, we have been able to capture the signature of ribs in wake flows by measuring the transverse component of vorticity, ω_y (Hayakawa & Hussain 1989). Because the ribs are inclined in the (x, y)-plane, intersecting the (y, z)-plane between successive rolls, their signature should appear in the (x, z) plane as ω_y concentrations. A conceptual sketch of two rolls and a rib connecting them is shown in figure 3. For simplicity, this figure includes only one side of the wake, and the rolls are drawn as two-dimensional structures.

A question arises as to how to discriminate ω_y concentrations due to ribs from those due to distorted rolls, and how to recognize the two different coherent structures without having to simultaneously measure spanwise and normal vorticities. To identify the signature of ribs, one needs to carefully examine simultaneous records of $\omega_y(z, t)$ and $u(z, t)$ (figure 4). Elongated u-contours in the z-direction and simultaneous absence of significant ω_y are indications of advection of (locally) two-dimensional rolls under the hot-wire rake. Paired ω_y-concentrations with opposite circulations (area surrounded by dotted lines) associated with lower-speed u-contours (dot-

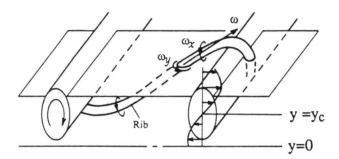

FIGURE 3. Conceptual sketch of the rib-roll structure in a plane wake. For simplicity, only one side of the wake is drawn.

filled contours) between the ω_y pairs are believed to be due to distorted rolls. This is because the induced velocity between the legs of the contorted roll (figure 5) is in the upstream direction (opposite to the main flow). Concentrations of ω_y (hatched contours) between elongated u-contours indicate ribs.

Further insight into the interaction of ribs and rolls, particularly the details of the vortex line topology and the winding of ribs in a roll, requires the complete flow field data, which can presently be provided only by direct numerical simulations. Such simulations have been performed by Metcalfe et al. (1987), Grinstein et al. (1989) and others (e.g., Rogers & Moser 1989, Comte et al. 1989). As expected, the occurrence of ribs has been conclusively demonstrated, and further studies are under way in collaboration with Metcalfe and Grinstein.

Coherent Structure Interactions

Following our initial success in educing CS in various flows, we have begun to classify CS interactions by the morphology of structures involved and to examine in detail the various CS properties during a particular type of interaction, and the associated effects on entrainment, mixing, combustion, enstrophy cascade, and the generation of helicity and aerodynamic noise. Some CS interactions that we find interesting are pairing and tearing of like-signed vortices, merger of opposite-signed vortices involving parallel vortical structures, entanglement involving like-signed nonparallel vortices, interaction of vortices whose vorticities are orthogonal to each other (e.g., rib-roll interaction), and reconnection of antiparallel vortices.

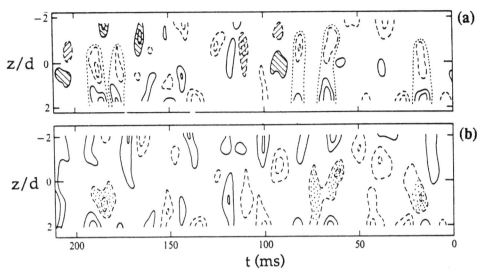

FIGURE 4. Measured, instantaneous ω_z- and u-maps at $x/d = 20$; contours levels of (a) $\omega_z/S_M = +3, +1.5$; (b) $(u - U_c)/U_0 = \pm 0.125, \pm 0.075$. Here d is the cylinder diameter, S_M is the the local maximum time-mean shear rate, U_c is the convection speed and U_0 is the free stream velocity.

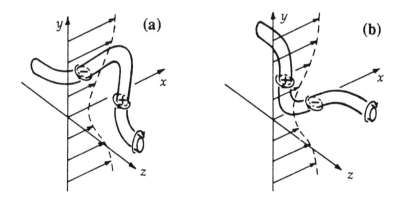

FIGURE 5. Conceptual picture showing the induced velocity associated with deformed rolls having (a) outward and (b) inward kink.

Pairing and tearing

After the initial rollup of a vortex sheet into discrete vortices in a plane mixing layer or an axisymmetric jet, the most common interaction a shear layer undergoes is pairing of two neighboring vortices. This interaction has been treated as a nonlinear instability problem (Kelly 1967; Monkewitz 1988) involving the resonant growth of the subharmonic component. The interaction between a wave of frequency f and a subharmonic wave of frequency $f/2$ produces an $f/2$ component, which is capable of reinforcing the subharmonic depending upon the phase angle between the two waves. While pairing occurs over a wide range of phase, tearing—the shredding of a vortex by its neighboring vortices—occurs over an extremely narrow range of phase (Husain & Hussain 1989). Vortex pairing plays the key role in large-scale mixing and growth of shear layers, and can be an important source of aerodynamic noise generation. Space here disallows a detailed discussion of pairing dynamics. CS properties during pairing have been studied extensively by Hussain & Zaman (1980) in a circular jet and by Husain (1984) in an elliptic jet. Effects of pairing on aerodynamic noise generation, first proposed by Laufer (1974), have been studied by Zaman (1985) and Bridges & Hussain (1987).

Vortex pairing is a consequence of mutual induction between like-signed adjacent vortices. Merger of opposite-signed vortices is anomalous, but can happen between two vortices of different strengths. When equally strong, opposite-signed vortices form a dipole, which is a prototypical flow module of fundamental interest. In the presence of viscosity, such a dipole plays the key role in reconnection (discussed below) and can undergo decay by mutual annihilation due to cross-diffusion (Stanaway et al. 1988). Merger of opposite-signed vortices has been observed (through flow visualization) in the transitional region of a plane jet with a fully developed laminar velocity profile. Plane jets with an underdeveloped exit velocity profile form symmetric vortices, and pairing of like-signed vortices takes place independently in the two shear layers (figure 6a). There is no direct interaction of the two shear layers until after transition and breakdown of paired structures. However, for a plane jet with a fully-developed laminar plane Poiseuille profile, the vortices roll up antisymmetrically. Adjacent like-signed vortices pair independently on either side of the jet. The resulting paired vortex P, with double the circulation of a single vortex S, pulls part or all of vortex S (of opposite sign) from the other shear layer across the jet centerline and wraps around itself (figure 6b). The antisymmetric mode of vortex formation and merger of opposite-signed vortices (an example of counter-gradient transport of momentum) are found to produce increased jet spreading and intense mixing.

FIGURE 6. Coherent structure interaction in the near field of a plane jet; (a) top-hat initial velocity profile; (b) fully-developed initial velocity profile.

INTERACTIONS OF 3D STRUCTURES: ELLIPTIC JETS

In general, interactions involving 2D (i.e., axisymmetric or roller type) structures are rare in turbulent flows. Structures are typically 3D and undergo complex motions because of azimuthally nonuniform self- and mutual-inductions. Elliptic vortex rings are a natural choice for fundamental studies of 3D CS. With this in mind, we have explored the CS dynamics in the near field of elliptic jets of various aspect ratios (ratio of the major to minor axes) and initial conditions (initially laminar and fully developed turbulent boundary layer at the jet exit plane) (Husain 1984).

Switching of axes

Elliptic vortex rings have complicated motions due to curvature-dependent self-advection, which cause them to undergo 3D deformation and change their aspect ratio as they evolve. The deformation of an isolated vortex filament, simulated numerically (Bridges 1988), is shown in two views in figure 7(a). Figures 7(b, c) are flow visualization pictures taken in an elliptic jet of aspect ratio 2:1; figure 7(b) shows the deformation as viewed simultaneously from the major-axis plane (top sequence) and minor-axis plane (bottom sequence), and figure 7(c) shows the front view of an advecting structure in the jet. Initially, when an elliptic vortex ring rolls up, it is in a plane parallel to the nozzle exit plane. Due to higher curvature, the major-axis sides move ahead of the minor-axis sides. The forward inclination of the major-axis sides forms folds in the initial minor-axis sides. As a result, the induced velocity on the minor-axis sides, now directed outward, increases, and the minor-axis sides move outward. Consequently, the vortex again takes an elliptic shape; but now the major-axis is in the plane of the

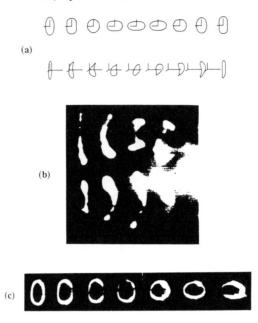

(a)

(b)

(c)

FIGURE 7. (a) Deformation of an elliptic vortex filament computed by numerical simulation. (b) Flow-visualization picture viewed simultaneously normal to the major-axis plane (top) and the minor-axis plane (bottom). (c) Flow visualization picture in a view aligned with the jet axis.

initial minor axis, that is, the axes have switched. In contrast with the deformation of an isolated inviscid elliptic vortex ring (Dhanak & Berdardinis 1981; Bridges 1988), the switching of axes does not continue indefinitely in a jet. Viscous as well as turbulent diffusion causes the core radius to increase, producing diminishing self-induced velocity and deformation, and delay of the axis-switching process.

The deformation of structures by itself can cause enhanced mixing in elliptic jets: the major-axis sides induce ambient fluid to move toward the jet axis, while jet fluid is ejected by the outward motion on the minor-axis sides. By using excitation to increase the strength and regularity of the CS, turbulence and mixing can be further enhanced. Figure 8 shows the mass flux in a jet as a function of downstream distance for unexcited and excited circular and elliptic (2:1) jets. For self-excited jets whistler nozzles were used (Hasan & Hussain 1982). The knowledge of CS suggested that entrainment in elliptic jets will be higher than in circular jets. Excitation can further enhance this effect. Self-excitation by whistler nozzles produces further enhancements in an economic way. We thus predicted that an elliptic whistler jet will produce the maximum entrainment rate, as was indeed subsequently proven by the data (figure 8).

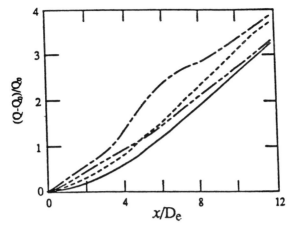

FIGURE 8. Mass entrainment in different jets as function of downstream distance. —, unexcited circular jet; — - - —, unexcited 2:1 elliptic jet; - - - - , self-excited (whistler) circular jet; — - —, self-excited 2:1 elliptic jet.

Pairing through entanglement

Expecting that the pairing mechanism of deformed elliptic structures to be quite different from that of planar or axisymmetric structures, we have studied the details of pairing in an elliptic jet of 2:1 aspect ratio. Controlled periodic excitation was used to stabilize the interaction process in space, thus allowing phase-locked measurements to educe structure details during various phases of pairing. The jet was excited at the stable pairing mode frequency of the jet column (i.e., $St_{D_e} = fD_e/U_e = 0.85$, where f is the excitation frequency, D_e is the equivalent diameter and U_e is the exit velocity). Coherent azimuthal vorticity contours $\langle \omega_z \rangle / f$ in both major-axis and minor-axis planes for five sequential phases during pairing are shown in figure 9(a). Because of nonplanar and nonuniform self-induction and the consequent effect on mutual induction, merging of elliptic vortices does not occur uniformly around the entire perimeter. Merger occurs only on the initial major-axis sides, while on the initial minor-axis sides, the trailing vortex rushes through the leading vortex without pairing and then breaks down violently.

Since pairing occurs over a small segment on the major-axis sides, pairing in an elliptic jet is morphologically different form that in a circular jet. In a circular jet, pairing takes place through a leapfrog motion of two vortices, while pairing over small portions of elliptic vortices starts as an entanglement rather than a leapfrogging process. Entangled vortices soon merge into a single vortex through diffusion. This interaction process is schematically shown in a perspective view in figure 9(b).

Interactions of 3D structures in an elliptic jet produce considerably greater entrainment and mixing than in circular or plane jets. We have studied pair-

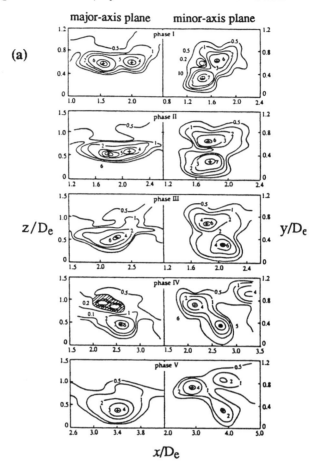

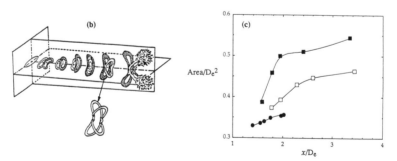

FIGURE 9. (a) Vorticity contours in the major- and minor-axis planes in an elliptic jet during pairing. (b) Schematics of the pairing process in an elliptic jet. (c) Area enclosed by the vorticity level $\langle\omega_z\rangle/f = 1$; □ , major-axis plane; ■ , minor-axis plane; ●, circular jet.

ing in a circular jet under initial conditions identical to those of the elliptic jet, which allowed us to compare the effects of three dimensionality on various structure properties. As an example, let us consider vorticity diffusion during pairing as a way of understanding entrainment. For this, structure boundaries defined by a low-level vorticity contour (e.g., $\langle \omega_z \rangle / f = 1$) are compared during the interaction of structures in circular and elliptic jets. The area enclosed by the vorticity level $\langle \omega_z \rangle / f = 1$ (figure 9c) shows that the structure cross-section increases at a much higher rate in the elliptic jet than in the circular jet. This indicates that more (nonvortical) ambient fluid is engulfed by the elliptic jet than by the circular jet; that is, elliptic jets produce more rapid mixing than circular jets.

Cut-and-connect interaction: jet bifurcation

Vortex reconnection is a frequent event in turbulent flows and presumably plays crucial roles in mixing and generation of helicity, enstrophy, and aerodynamic noise (Hussain 1986). Also called 'cut-and-connect' or 'cross-linking', this phenomenon has recently received considerable attention because it is an example of non-preserving topology. It has also been argued that vortex reconnection is a prime candidate for a finite-time-singularity of the Navier-Stokes equations (Siggia & Pumir 1985; Kerr & Hussain 1989).

Vortex reconnection occurs randomly in space and time in a turbulent flow. Thus, details of the reconnection mechanism would be hard to capture in a turbulent flow field in the laboratory or in a DNS of a turbulent flow field. Special situations need to be devised in both experiments and numerical simulations in order to capture the reconnection details. We were able to stabilize the reconnection process of vortical structures in elliptic jets (for aspect ratios $> 4 : 1$) under excitation at the preferred mode frequency, which enabled us to study the effects on time-average turbulence measures (Hussain & Husain 1989). This is an example where flow visualization and concepts of vortex dynamics have allowed us to understand the peculiarities of time-average measures resulting from the deformation of elliptic vortical structures and the subsequent cut-and-connect interaction.

For low-aspect-ratio jets (below about 3.5:1), switching of axes occurs a few times. For jets of higher aspect ratios, after the first switching, the major-axis sides continue to move in towards the jet centerline instead of reversing direction to switch axes again. This causes the original major-axis sides to come into contact with each other. The result is two nearly circular vortices produced by a cut-and-connect interaction (discussed next). Such a jet then spreads more in the initial minor-axis plane due to the presence and growth of two side-by-side structures. A sequence of this bifurcation process in an elliptic vortex ring is shown in figure 10(a). Figure 10(b) shows flow visualization of the deformation and bifurcation of an elliptic vortical structure in a 4:1 jet; (i)–(iii) denote three phases of bifurcation

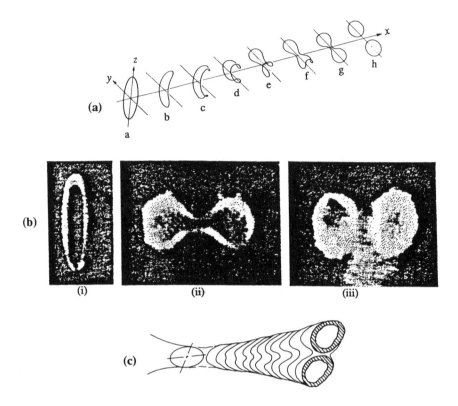

FIGURE 10. (a) Schematics of the cut-and-connect process in an elliptic vortex ring. (b) Flow visualization picture of the bifurcation of an elliptic vortical structure in the 4:1 jet. (i), (ii), and (iii) denote three phases of bifurcation recorded at successive x-locations. (c) Schematic of the bifurcation of an elliptic jet.

recorded at successive x-locations. The resulting bifurcation of the elliptic jet is shown schematically in figure 10(c).

Reconnection: an example of the complementary approach

Vortex reconnection is one of the fundamental 3D vortex interactions, not only between CS, as exemplified by its occurrence in the bifurcating elliptic jet, but also a frequently occurring phenomenon in fully-developed turbulent flows. The latter point is clearly illustrated by the superfluid turbulence simulation of Schwarz (1983). In fact, Schwarz' calculations, which are based on vortex filaments, clearly show that if the reconnection process is artificially suppressed, then the measurable global quantities are in error. Furthermore, Siggia (1984) showed that two nonparallel vortex filaments in close proximity align locally in an antiparallel configuration. Thus, there is good reason to nominate the reconnection of two antiparallel vortex tubes as a canonical fundamental CS interaction.

Vortex reconnection has been simulated numerically by, among others, Siggia & Pumir (1985), Winckelmans & Leonard (1988), Pumir & Kerr (1987), Ashurst & Meiron (1987), Kida & Takaoka (1987), Melander & Hussain(1988), Zabusky & Melander (1989), Kida et al. (1989), and Kerr & Hussain (1989). A number of numerical techniques and initial conditions have been employed, and not all simulations have been performed with the same objective. Our objective was to reveal the underlying physical mechanism of vortex reconnection, and we have made significant progress in that respect (Melander & Hussain 1988). It is, however, important to emphasize that the so-called *bridging* mechanism can operate in different ways depending on the mean flow in which the reconnection is embedded. The simulation by Kida et al. (1989) shows this point clearly by considering the two successive reconnections that result when two vortex rings collide. The mean flow changes direction relative to the local mutual induction of the antiparallel vortices, and consequently, the end products of the first and second reconnection differ considerably. One should keep this in mind when comparing the results of simulations like Melander & Hussain (1988) and Kerr & Hussain (1989) with the experimental results of Schatzle (1987).

Figure 11 shows three views of the vortex pair for a symmetric initial condition consisting of two antiparallel vortices with inclined sinusoidal perturbations. Wire-frame plots of the vorticity norm $|\omega|$ during the interaction are shown in figure 12 at $t = 0, 1, 3.5, 4.25, 5, 6$, and 9. In figures 12(a-f), the level surface represents 30% of the initial peak vorticity $\omega_{max}(0)$, while in figure12(g), the wire-plot is for 10% of $\omega_{max}(0)$ at $t = 9$. Here t denotes time t^* nondimensionalized by $\omega_{max}(0)$ such that $t = t^*|\omega_{max}(0)|/20$. The transport of scalar is shown in figure 13 for $t = 0, 2.75, 3.5, 4.25$, and 5. The scalar, at a Schmidt number of unity, marks only one vortex. The circulation in the symmetric plane is shown in figure 14. For the sake of clear identification, let us distinguish the two planes of symmetry: we will call the xy-plane the *symmetric plane* and yz-plane the *dividing plane*. It is a simple consequence of Gauss's theorem that the sum of the circulations in half of the symmetric plane and half of the dividing plane is conserved in time. Hence, the degree of reconnection is measured by the decrease in the circulation in the symmetric plane.

It is evident that the reconnection process has three phases. The first phase is largely an inviscid process wherein the vortices are advected upwards by mutual induction, while the perturbation grows due to local self-induction and differential mutual induction. As the cores approach each other in the symmetric plane, they flatten and the second phase, which is characterized by a rapid circulation transfer from the symmetric plane

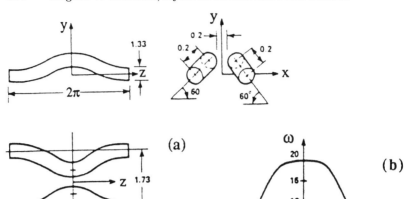

FIGURE 11. (a) Schematic of the initial vortex tubes and coordinates. (b) Vorticity distribution in the vortex cores.

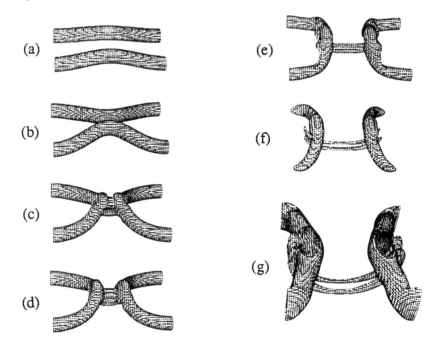

FIGURE 12. $|\omega|$ surface at 30% of the initial peak. (a) $t = 0$; (b) $t = 2.75$; (c) $t = 3.5$; (d) $t = 4.25$; (e) $t = 5$; (f) $t = 6$. (g) $|\omega|$ surface at 10% of the initial peak at $t = 9$.

to the dividing (yz) plane, begins. The transfer of circulation is caused by vorticity annihilation in the contact zone. This phase is called *bridging*, as the reconnected vortex lines accumulate into characteristically shaped *bridges* located in front of and orthogonal to the annihilating remnants of the original counter-rotating vortices. The bridges play a significant role in the dynamics of reconnection, as they are responsible for halting the circulation transfer and the cross-linking of vortex lines. As the bridges accumulate circulation, the remnants of the original counter-rotating vortices in the contact zone get swept backwards relative to the main structure by the velocity field induced by the bridges. This results in a curvature reversal in the contact zone. Due to this curvature reversal, the original vortices reverse self-induction in the contact zone such that they no longer are pressed against each other; on the contrary, they separate from each other at a slow rate. The mechanism for rapid vorticity annihilation in the contact zone is therefore no longer present; consequently, the circulation transfer stops short of completion. The transfer arrest marks the beginning of the third phase of the reconnection process, characterized by bridges with a pair of vorticity threads in between. Although the threads diffuse they are also stretched axially by the flow induced by the bridges. Much as for a Burger's vortex, a balance forms between spreading of the vortex cores by diffusion and contraction through axial stretching. The threads therefore have a very long lifetime and remain an important coherent feature of the vortex structure. In fact they were still present in basically the same shape and strength at $t = 18$ (not shown here) when the calculation was stopped. Kida *et al.* (1989) also found persistent threads in their simulation of the collision of two vortex rings. The long-lasting threads in their case are produced by the second reconnection, not the first. The threads formed in their first reconnection collapse on the vortices and get eliminated by cross-diffusion.

The reconnection process, particularly in the elliptic jet, vividly illustrates the limitation of flow visualization (Hussain 1981, 1986). Not only does the experimentalist face large Schmidt numbers, which suggest that the marker boundaries are different from the vorticity boundaries due to different diffusion rates, but there is also the inherent difficulty that vorticity is amplified through vortex stretching precisely where the marker is diluted (see also Bridges *et al.* 1989). The last problem is perhaps the most serious one as it is present for all values of the Schmidt number, even at *unity* Schmidt number. In order to highlight this problem, the numerical simulation was performed with a passive marker. The Schmidt number was unity, and the marker was initially distributed such that its concentration exactly equalled the vorticity amplitude of one vortex. In spite of these ideal conditions, the marker quickly left the contact zone as seen in figure 13 and got concentrated in regions of little interest. We emphasize that even though the marker contour amplitude is very low, threads are completely invisible in the marker panels of figure 13(e) during the third phase. Furthermore, the marker does not faithfully reproduce the head-tail

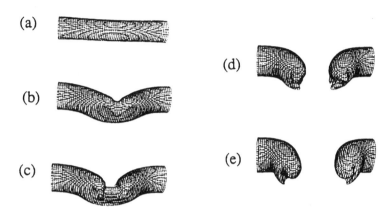

FIGURE 13. Scalar at 5% of the initial peak. (a) $t = 0$; (b) $t = 2.275$; (c) $t = 3.5$; (d) $t = 4.25$; (e) $t = 5$.

structure of the vorticity (i.e., vortex dipole) in the contact zone during the second phase. This is because the tail is stretched significantly more than the head. In light of the simulation results, it is not surprising that flow visualization of the bifurcating elliptic jet as well as collision of two vortex rings failed to reveal the presence of the threads.

When the jet experiments were first performed, it was generally believed that vortex reconnection would generally result in a quick transfer of all circulation from the symmetric plane into the dividing plane. This belief was the basis for the first attempt to model the process mathematically (Takaki & Hussain 1985). When this assumption is imposed on the bifurcating elliptic jet, one would expect a velocity profile in the dividing plane with a velocity peak located inside the newly formed vortex rings, as illustrated in figures 15(a, b). In reality, measurements (Hussain & Husain 1989) show that the largest velocity occurs at the intersection of the symmetry planes (figure 15d). The threads revealed by the numerical simulation account for the correct velocity peak, as illustrated in figure 15(c).

Topological properties

The direct simulation provides considerable amounts of spatial data inaccessible experimentally with current measurement technology. These provide valuable information on the topology and dynamics of interacting

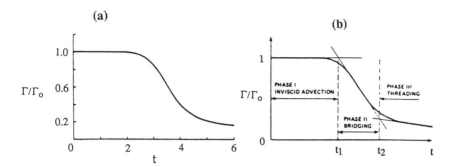

FIGURE 14. (a) Circulation as a function of time in one half of the symmetric plane; (b) phases of timescales of reconnection.

vortices. While post-processing and visual examination on a color graphics workstation prove extremely illuminating, this visual information is typically overwhelming and must be properly synthesized and interpreted. The approach undertaken here evolved from that employed to study coherent structures in turbulent shear flows (Hussain 1981, 1986). Of the variety of properties that can be useful in studying the topology and dynamics, we limit our attention to scalar intensity, enstrophy production, dissipation and helicity density. We have selected a few planar cross-sections which capture striking dynamical and topological aspects of the reconnection process. Figures 16-18 show planar cross sections of vorticity norm $|\omega|$, scalar s, dissipation $\varepsilon = 2\nu s_{ij}s_{ij}$, helicity $h = u_i\omega_i$, enstrophy production $P_\omega = \omega_i s_{ij}\omega_j$ and relative helicity $h_r = u_i\omega_i/(|u||\omega|)$ in a plane parallel to the dividing plane, but at a distance 0.196 from it (this plane essentially goes through the threads). Three instants were selected: $t = 3.5$ at the middle of the circulation transfer, $t = 4.75$ at the time of peak vorticity in the bridges and $t = 6$ representing the slowly evolving post-reconnection vortical structure.

Throughout the first phase (the inviscid advection), the flow has little or no helicity, and the vortex lines inside each vortex are almost parallel. During and especially towards the end of the bridging phase, high helicity develops locally on each side of the dividing plane (figures 19a,b). Far away from the dividing plane the helicity remains low and the vortexlines are untwisted. However, in the threading phase, the peak helicity decays and the helicity distribution gradually spreads outward from the dividing plane. The helicity peaks remain in the threads after the bridging phase and are the most dominant feature of the helicity distribution (figure 19b). The reason is that while the bridges build up circulation, the weaker threads

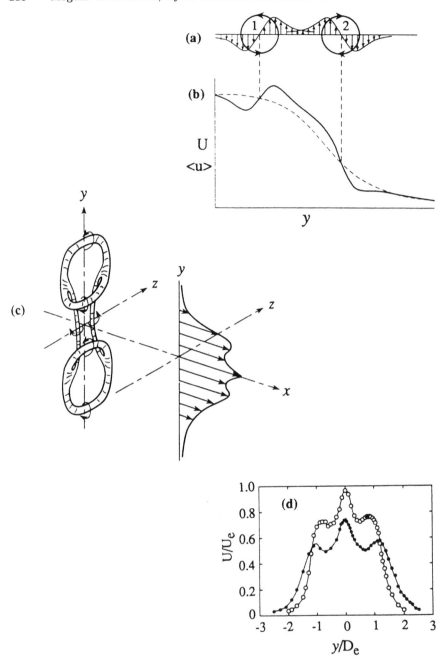

FIGURE 15. Schematics, showing the modification of the longitudinal velocity profile due to jet bifurcation. (a,b) Modification due to bridges only; (c) modification due to bridges and threads; (d) experimental $U(y)$ profiles at $x/D_e = 3.5$ (open symbol) and $x/D_e = 5$ (solid symbol) in the minor plane of the 4:1 jet.

are being stretched and aligned orthogonal to the bridges (figure 12f). The high helicity in the threads is a consequence of this near orthogonality of threads and bridges—the strong swirling velocity of the bridges nearly aligning with the threads.

The helicity distribution after the reconnection (e.g., $t = 4.75$ and $t = 6$) is particularly interesting, as it consists of intertwined regions of positive and negative helicity. Apart from the threads, where helicity stands out clearly (figures 19a,b), the structure is an enigma until one examines bundles of vortexlines (figures 20 a-c). Figures 20(b,c) show that vortexlines emerging from the threads do not wrap all the way around the bridges as one would expect from $|\omega|$ surfaces, e.g., expanded views of figures 12(d,e). Instead, they form asymmetric hairpin-like structures (see A and B, figure 20b). The vorticity is high in one hairpin leg and low in the other. Vortical structures of this kind are unfortunately not faithfully reflected in the surface plots of $|\omega|$. The intense leg (A) is the extension of the thread (C) and its high vorticity makes its helicity stand out in figure 19. The diffuse leg is difficult to identify in the helicity distribution as the $|\omega|$ there is low and the angle between vorticity and velocity is larger. The intense leg induces a flow away from the diving plane on the outside of the bridges, causing vortex stretching in the outer part of the bridge cores (see point A in figures 17e,18e). The relative helicity also reflects this axial flow away from the dividing plane (see point b in figures 17f,18f).

Inside the bridge the vortex lines have a slow helical twist near the dividing plane (barely discernible in figure 20c). Contrary to the intense hairpin leg, the twist produces an axial flow toward the diving plane at the center of the bridge. Since this axial flow is toward the diving plane from both sides, it results in a negative enstrophy production (dotted lines in figures 17e,18e; see also figure 19e). The twisting of vortexlines is the result of a skewed (nonconcentric) vorticity distribution in the bridges cores. The peak vorticity in the bridge is not at the location of the vorticity centroid, but it is far away from the contact zone, where vorticity is lower than in bridges—a consequence of earlier stretching by the dipole during bridging. Figures 20(b,c) show that a vortexline which neatly follows the centerline of the vortex far away from the dividing ceases to do so in the bridge. This effect clearly results in vortexline twisting and thus induced axial flow (figure 20d). The direction of the axial flow is determined by the orientation of the twist, which in turn is determined by the fastest swirling velocity along the vortex. The bridges, having the highest peak vorticity, also have the highest swirling velocity. Hence the axial flow is toward the dividing plane from both sides, resulting in vortex compression (i.e., negative P_ω in figures 16–18) which decreases the peak vorticity and thereby also the twisting rate. This constitutes a new inviscid mechanism, clearly distinct from diffusion, for smoothing the vorticity intensity along the reconnected vortex lines.

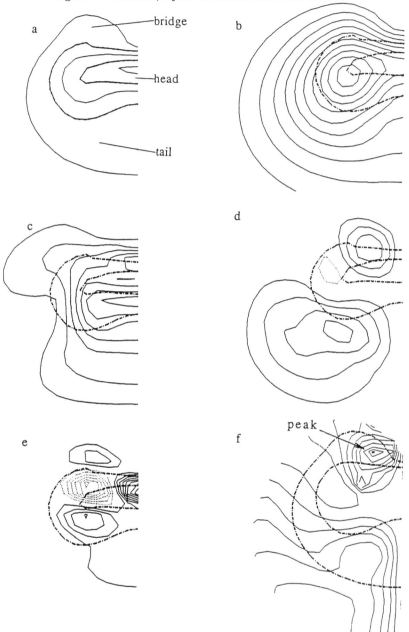

FIGURE 16. Plane cross-section at $t = 3.5$ parallel to the dividing plane but removed $\Delta x = 0.196$. Contour levels in panels (a,b) are scaled by the peak values in the plane. (a) $|\omega|$; (b) s; (c) ε; (d) h; (e) P_ω; (f) h_r contours giving the angle between vorticity and velocity vectors in increments of $\Delta\theta = 10°$. Solid lines represent positive values, dotted lines are negative values.

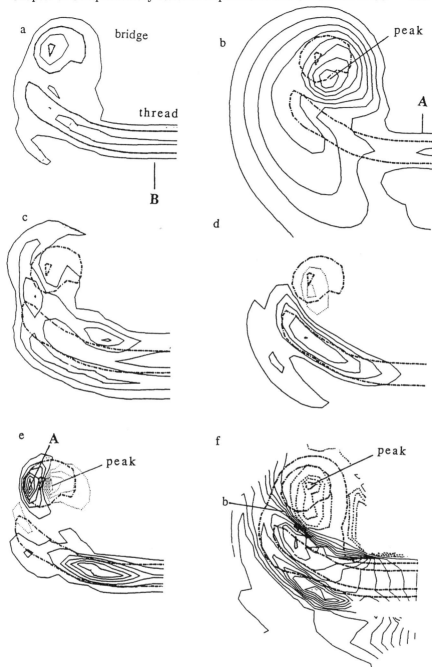

FIGURE 17. Plane cross-sections at $t = 4.75$ (see caption from figure 16).

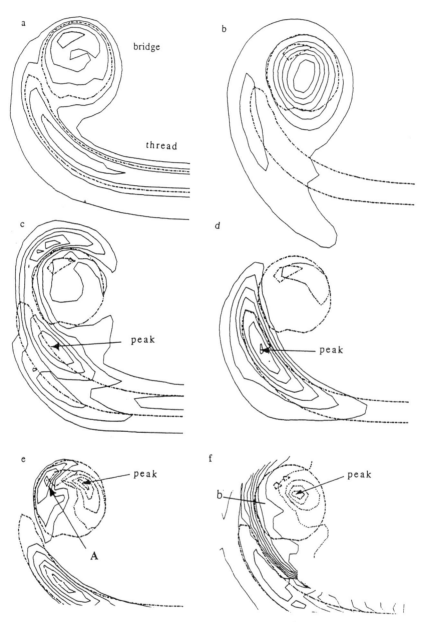

FIGURE 18. Plane cross-sections at $t = 6$ (see caption from figure 16).

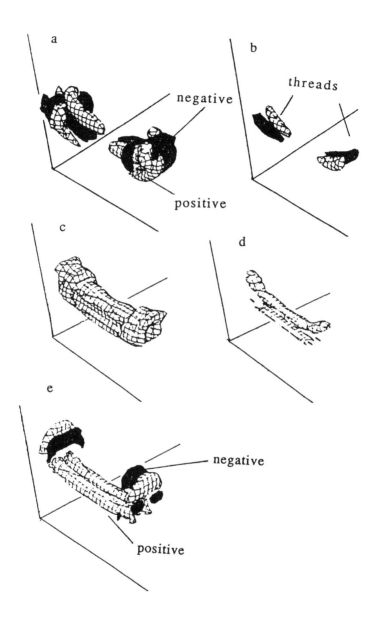

FIGURE 19. (a) h at $t = 4.5$, 20% of peak value; (b) h at 50% level; (c) ε at 20% level; (d) ε at 50% level; (e) P_ω at 20% of positive and negative peak values.

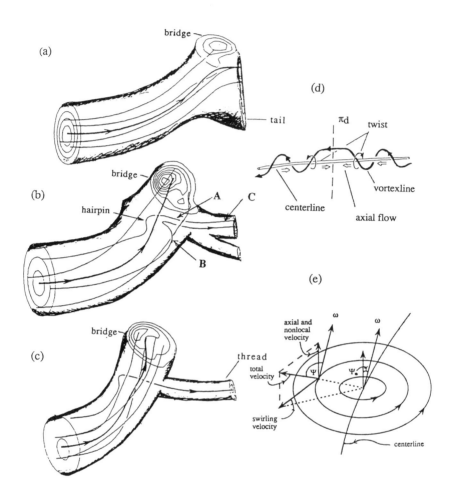

FIGURE 20. Vortex lines in a quarter of the computational domain. Also shown are the plane cross-sections of the vorticity norms in symmetric and dividing planes and in the box side. A sketch of the vortex has been overlaid in order to orient the reader. (a) $t = 3.5$; (b) $t = 4.75$; (c) $t = 6.0$; (d) axial flow due to vortexline twist; (e) explanation for higher h_r on vortex centerline.

The relative helicity is sensitive to noise level amplitudes and must be treated with care. Interestingly enough, numerical noise in almost all irrotational regions exhibits large relative helicity. Therefore, it is essential to condition the relative helicity by both vorticity and velocity amplitude thresholds. One thus has a physically meaningful quantity as a measure for the angle between velocity and vorticity vectors. In studying the present simulation we found the *conditional relative helicity* to be useful for identifying sharp turns in vortex lines, such as those occurring at the tip of hairpin-like structures. Near such structures the relative helicity exhibits a characteristic sharp variation from positive to negative values. The conditional relative helicity thereby highlights asymmetric hairpin structures which elude other diagnostics. Furthermore, the relative helicity clearly marks the centerline of a vortex. This is a simple consequence of the fact that the swirling velocity vanishes at the centerline of a vortex whereas other components of the velocity such as axial and nonlocal components in general do not; figure 20(e) illustrates that the angle Ψ_0 on the vortex axis is lower than the angle Ψ away from the axis. In figures 16(f), 17(f) and 18(f) one therefore observes a sharp peak in relative helicity near the vortex center.

Until the threading is well under way, ε is high in the center of the contact zone and the peak is located in the diving plane. Since, as a simple consequence of the symmetry, neither of the symmetry planes have helicity h, the peaks of h and ε remain spatially exclusive during the bridging phase. However, during the threading phase the peak of ε bifurcates into two peaks, which move away from each other in the dividing plane. Thereby, the peaks of ε locate themselves between two peaks of h, one on each side of the dividing plane (figures 19a-d). The spatial exclusiveness remains, but is not as clear as at earlier times; for example in the plane cross-sections (figures 18c,d) the peaks almost coincide. We have found no unique correspondence between dissipation and helicity.

Concluding remarks

Our understanding of turbulence physics is greatly dependent on our understanding of CS, their interactions, and interactions between CS and incoherent turbulence. To understand the dynamics of CS and their significance in turbulent flows, we need to know the topological details of various coherent structure properties. While a variety of approaches, mostly flow visualization, have been employed to study CS, a discussion of their dynamical role must involve quantitative data.

Our general-purpose eduction scheme has proven successful in educing structures in fully developed turbulent flows. Rakes of hot-wires can provide important information regarding structure topology and mode of interactions in laboratory flows. However, to obtain details of structure interaction with a finer spatial resolution and to obtain quantities which are not possible to measure in the laboratory, we need to resort to numerical simulation. In fact, we must combine the powers of both methods to reveal the flow physics that either method alone is inadequate to provide. Through the use of specific examples, we have illustrated how experimental observations and numerical simulations are best used in complementary roles. We have demonstrated why the deeper understanding of the role and dynamics of CS hinges on a better understanding of fundamental interactions of vortices in close proximity. Furthermore, we have included a comprehensive list of interactions, which have been identified experimentally, but need to be simulated numerically under idealized conditions as the internal topology and dynamics can not be adequately uncovered by current experimental techniques.

Acknowledgments

This research has been supported by Office of Naval Research Grant N00014-87-K-0126 and Department of Energy Grant DE-FG05-88ER13839.

References

Ashurst, W. T. & Meiron, D. I. 1987 *Phys. Rev. Lett.* 58, 1632.

Babiano, A., Basdevant, G., Legras, B. & Sadourny, R. 1987 *J. Fluid Mech.* 183, 379.

Bernal, L. P. & Roshko, A. 1986 *J. Fluid Mech.* 170, 499.

Bridges, J. 1988 (personal communication).

Bridges, J. & Hussain, A. K. M. F. 1987 *J. Sound & Vibration* 117, 289.

Bridges, J., Husain, H. S. & Hussain, F. 1990 in *Turbulence at the Crossroads: Whither Turbulence?* (ed. J.L. Lumley), p. 132, Springer-Verlag.

Brown, G. L. & Roshko, A. 1974, *J. Fluid Mech.* 64, 775.

Coles, D. 1981 *Proc. Indian Acad. Sci.* 4, 111.

Comte, M., Lesieur, M. & Fouillet, Y. 1989 in *Topological Fluid Mech.* (eds. H.K. Moffatt & A. Tsinober), p.649. Cambridge U. Press.

Crow, S. C. & Champagne, F. H. 1971 *J. Fluid Mech.* 48, 547.

Dhanak, M. R. & Barnardinis, B. 1981 *J. Fluid Mech.* 109, 189.

Fiedler, H. E. 1988 *Prog. Aerospace Sci.* 25, 231.

Grinstein, F. F., Hussain, F. & Oran, E. S. 1989 AIAA paper No.89-0977.

Hasan, M. A. Z. & Hussain, A. K. M. F. 1982 *J. Fluid Mech.* 115, 59.

Hayakawa, M. & Hussain, F. 1989 *J. Fluid Mech.* 206, 375

Ho, C. M. & Huerre, P. 1984 *Ann. Rev. Fluid. Mech.* 16, 365.

Husain, H. S. 1984 Ph.D. Thesis, U. of Houston

Husain, H. S. & Hussain, F. 1989 in *Advances in Turbulence 2* (eds. H.H. Fernholz & H.E. Fiedler) p. 96, Springer-Verlag.

Hussain, A. K. M. F. 1981 *Proc. Indian Acad. Sci. (Eng. Sci.)* 4, 129.

Hussain, A. K. M. F. 1984 in *Turbulence and Chaotic Phenomena in Fluids* (ed. T. Tatsumi), p. 453.

Hussain, A. K. M. F. 1986 *J. Fluid Mech.* 173, 303.

Hussain, A. K. M. F. & Zaman, K. B. M. Q. & 1980 *J. Fluid Mech.* 101, 493.

Hussain, F. & Husain, H. S. 1989 *J. Fluid Mech.* 208, 257.

Jimenez, J., Cogollos, M. & Bernal, L. P. 1985 *J. Fluid Mech.* 152, 125.

Kelly, R. E. 1967 *J. Fluid Mech.* 27, 667.

Kerr, R.M. & Hussain, F. 1989 *Physica D* 37, 474.

Kida, S., Takaoka, M. 1987 *Phys. Fluids* 20, 2911.

Kida, S., Takaoka, M. & Hussain, F. 1989 *Phys. Fluids* A1, 630.

Kline, S. J., Reynolds, W. D., Schraub, F. A. & Runstadler, P. W. 1967 *J. Fluid. Mech.* 30, 741.

Kline, S. J. & Robinson, S. K. 1989 Second IUTAM Symp. on Structure of Turbulence & Drag Reduction, Federal Institute of Tech., Zurich, Switzerland.

Lasheras, J. C., Cho, J. S. & Maxworthy, T. 1986 *J. Fluid Mech.* 172, 231.

Laufer, J. 1974 *Omaggio Carto Ferrari*, p. 451

Lumley, J. L. 1981, in *Transition and Turbulence* (ed. R.E. Meyer), p. 215, Academic.

McWilliams, J. C. 1984 *J. Fluid Mech.* 146, 21.

Melander, M. V., Zabusky, N. J. & Styczek, A. S. 1986 *J. Fluid Mech.* 167, 95.

Melander, M. V., McWilliams, J. C., Zabusky, N. J. 1987 *J. Fluid Mech.* 178, 137.

Melander, M. V., McWilliams, J. C., Zabusky, N. J. 1987 *Phys. Fluids* 30, 2610.

Melander, M. V, Zabusky, N. J. & McWilliams, J. C. 1988 *J. Fluid Mech.* 195, 303.

Melander, M. V. & Hussain, F. 1988 CTR Report No. CTR-588, p. 257

Melander, M.V. & Hussain, F. 1989 in *Topological Fluid Mech.* (eds. H.K. Moffatt & A. Tsinober), p. 485. Cambridge U. Press.

Metcalfe, R. W., Hussain, A. K. M. F., Menon, S. & Hayakawa, M. 1987 in *Turbulent shear flows vol. 5* (eds. F. Durst, *et al.*), 110. Springer.

Metcalfe, R.W., Hussain, F 1989 in *Topological Fluid Mech.* (eds. H.K. Moffatt & A. Tsinober), p. 659. Cambridge U. Press.

Monkewitz, P. A. 1988 *J. Fluid Mech.* 18, 223.

Mumford, J. C. 1982 *J. Fluid Mech.* 118, 241

Perry, A. E. & Fairlie, B. D. 1974 *Adv. Geophys.* B18, 299.

Perry, A. E. & Cong, M. S. 1984 *Ann. Rev. Fluid Mech.* 19, 125.

Pumir, A. & Kerr, R. M. 1987 *Phys. Fluids* 58, 1636.

Rogers, M. M. & Moser, R. D. 1989 (private communication)

Schatzle, P. 1987 Ph.D. Thesis, California Institute of Tech.

Schwarz, K. W. 1983 *Phys. Rev. Lett.* 50, 364.

Siggia, E. 1984 *Phys. Fluids* 28, 794.

Siggia, E. & Pumir, A. 1985 *Phys. Rev. Lett.* 55, 1749.

Stanaway, S., Shariff, K. & Hussain, F. 1988 CTR Report No. CTR-S88, 287.

Takaki, R. & Hussain, F. 1985 Turb. Shear Flow V, Cornell U. 3.19

Tennekes, H. & Lumley, J. L. 1972 *A First Course in Turbulence*, MIT press.

Tso, J. & Hussain, F. 1989 *J. Fluid Mech.* 203, 425.

Wallace, J. M. 1985 *Lect. Notes Phys.* 235, 253.

Winckelmans, G. & Leonard, A. 1989 in *Mathematical Aspects of Vortex Dynamics*, p. 25, Siam.

Zabusky, N. J. & Melander, M. V. 1989 *Physica D* 37, 555.

Zaman, K. B. M. Q. 1985 *J. Fluid Mech.* 152, 83.

Chapter 8
REMARKS ON THE NAVIER–STOKES EQUATIONS

Peter Constantin

Abstract

Necessary and sufficient conditions for the absence of singularities in solutions of the three dimensional Navier-Stokes equations are recalled. New global weak solutions are constructed. They enjoy the properties that the spatial integral of the vorticity magnitude is a priori bounded in time and that the space and time integral of the $\frac{4}{3+\varepsilon}$ power of the magnitude of the gradient of the vorticity is a priori bounded. Vortex sheet, vortex line and even more general vortex structures with arbitrarily large vortex strengths are initial data which give rise to global weak solutions of this type of the Navier-Stokes equations. The two dimensional Hausdorff measure of level sets of the vorticity magnitude is studied and a priori bounds on an average such measure, $< \mu >$ are obtained. When expressed in terms of the Reynolds number and the Kolmogorov dissipation length η, these bounds are

$$< \mu > \leq \frac{L^3}{\eta} \left(1 + Re^{-\frac{1}{2}} \right)^{\frac{1}{2}}$$

The area of level sets of scalars and in particular isotherms in Rayleigh-Benard convection is studied. A quantity, $< \mu >_{r,t} (x_0)$, describing an average value of the area of a portion of a level set contained in a small ball of radius r about the point x_0 is bounded from above by

$$< \mu >_{r,t} (x_0) \leq C\kappa^{-\frac{1}{2}} r^{\frac{5}{2}} < v(x_0) >^{\frac{1}{2}}$$

where κ is the diffusivity constant and $< v >$ is the average maximal velocity. The inequality is valid for r larger than the local length scale $\lambda = \kappa < v >^{-1}$. It is found that this local scale is a microscale, if the statistics of the velocity field are homogeneous. This suggests that 2.5 is a lower bound for the fractal dimension of an (ensemble) average interface in homogeneous turbulent flow, a fact which agrees with the experimental lower bound of 2.35.

A similar quantity, $< \mu >_{\delta,t}$, representing an average value of the area of a portion of a level set contained in the region of space $D_\delta = \{x \in D; dist(x, \partial D) > \delta\}$, where D is the domain of aspect ratio of order one and diameter L where the convection takes place, satisfies the inequality

$$< \mu >_{\delta,t} \leq C\left(L^3\delta^{-1} + L^{\frac{5}{2}}\delta^{-\frac{1}{2}}Ra^{\frac{1}{2}}\right)$$

where Ra is the Rayleigh number.

1. INTRODUCTION

The aim of this lecture is threefold. First, to recall some (mostly well-known) results concerning the possibility of development of singularities in solutions of the three dimensional Navier-Stokes equations. Secondly, to present a new large class of global weak solutions which possess a somewhat richer structure than the previously known ones. Thirdly, to describe a general analytical method for the study of the area of dynamic interfaces and apply it to vorticity level sets in decaying turbulence and to isotherms in Rayleigh-Benard convection. I will endeavor to keep the style of this lecture as accessible and informal as possible. The complete mathematical statements and proofs of most of the previously unpublished results presented here can be found in [C].

The second section of this work is devoted to the the problem of singularities in the three dimensional incompressible Navier-Stokes equations. The existing mathematical body of knowledge leaves open the possibility of the spontaneous generation of singularities in some solutions of the three dimensional incompressible Navier-Stokes equations. By singularities I mean infinite velocities. In view of the parabolic aspects of the Navier-Stokes equation, if the velocities are finite, then all the other hydrodynamic variables (vorticity, pressure, vorticity gradients, etc) are finite, too. This and similar statements will be formulated in a precise quantitative way in the second section. By spontaneous generation I mean the appearance in finite time of a singularity, given smooth initial data, boundary conditions and body forces. A theorem ([C-F-T]) states that if there exists an initial datum which generates a solution whose enstrophy becomes infinite in infinite time then there exist initial data which generate solutions whose enstrophy become infinite in finite time.

The possibilty of singularity formation is present even in very idealized circumstances, in particular, in the absence of boundaries. Therefore, the study of this possibility in the absence of boundaries, (i.e. the cases of the fluid occupying the whole of space and the fluid variables either decaying at infinity or spatially periodic), has the merit of isolating the main difficulty of the problem. Singularities cannot appear in two dimensional or nearly two-dimensional incompressible flow nor in low Reynolds number three dimensional situations. Also, the solution of the three dimensional incompressible Navier-Stokes equation is guaranteed to exist, be smooth and uniquely determined by the initial data for short times, for all Reynolds numbers and all admissible data. Unlike, for instance, the case of the Kelvin-Helmholtz instability, for the Navier-Stokes equation the size alone of a (smooth) disturbance determines a minimal existence time of the solution.

J. Leray ([L]) published in 1934 his famous result establishing the existence of global weak solutions for arbitrary initial data with finite energy and arbitrary Reynolds number. By global it is meant "for arbitrarily long time". The solutions are weak in the sense that they are given by generalized functions of a certain type which satisfy the Navier-Stokes equations as distributions. It is not known if the initial datum determines the Leray solution uniquely. Thus, different constructive procedures might conceivably lead to different solutions. The basic property of these weak solutions is that they satisfy an energy inequality

(1.1)
$$\frac{1}{2} \int |u(x,t)|^2 dx + \nu \int_0^t \int |\nabla u(x,s)|^2 dx ds$$

$$\leq \frac{1}{2} \int |u_0(x)|^2 dx + \int_0^t \int F(x,s) \cdot u(x,s) dx ds$$

In (1.1) u is the three-component, divergence-free velocity vector; ν the kinematic viscosity coefficient; u_0 the initial velocities; and F, the body forces. Periodic boundary conditions are assumed, for simplicity. The spatial integrals are taken over a standard cube of periodicity $[-\frac{L}{2}, \frac{L}{2}]^3$. It follows from (1.1) that the space and time integral of the square of the magnitude of gradient of the velocities is a well-defined quantity, bounded a priori in terms of the initial data, F and, of course ν. The vorticity, $\omega = curl u$ satisfies the equation

(1.2)
$$\left(\frac{\partial}{\partial t} + u \cdot \nabla - \nu \Delta\right)\omega = \omega \cdot \nabla u + f$$

where $f = curl F$. In view of the energy inequality, the right hand side of (1.2) is a priori integrable in space and time provided F and f are sufficiently regular. This suggests that

$$sup_{0 \leq t} \int |\omega(x,t)| dx$$

is a priori bounded if the initial data are not too singular. This fact and related matters are also the object of Section 2. We consider initial data such that the velocity u_0 is square integrable and divergence-free and such that the vorticity ω_0 is a finite Radon measure on the torus. Expressed in terms of the initial vorticity alone, these conditions are first, that ω_0 is a finite divergence-free Radon measure on the torus and secondly, that ω_0

belongs also to the space V'. V' is the dual of the space V of square integrable, divergence-free functions on the torus whose first order derivatives are periodic and square integrable. This is a maximal degree of generality consistent with the requirement of finite kinetic energy and structured vorticity. Under these assumptions, together with appropriate regularity requirements on F, f we prove that the Navier-Stokes equations possess global Leray weak solutions for which

$$ess.sup_{0 \leq t} \int |\omega(x,t)| dx$$

and

$$\int_0^t \int |\nabla \omega(x,s)|^{\frac{4}{3+\varepsilon}} dx ds$$

are a priori bounded in terms of the data. The proof of this fact uses a change of dependent variable for the vorticity of certain approximate solutions. The approximation procedure we chose is the one employed in [C-K-N] for the construction of suitable weak solutions (that is, solutions which satisfy a local energy estimate). The reason for this choice is the fact that these approximate solutions obey a vorticity equation sufficiently similar to (1.2) for the present method to succeed. Incidentally, we obtain thus suitable weak solutions and therefore the 1-dimensional Hausdorff measure of their space and time singular set is zero ([C-K-N]). The class of initial vorticities for which this result is applicable contains the physically interesting examples of vortex sheets and vortex lines. There are results in the literature concerning the case of small Reynolds number. For instance, in [G-M] it is shown that if ω_0 belongs to the Morrey class $\frac{3}{2}$ and is small then there exist unique, globally smooth solutions which decay to zero (the result is for $F = 0$). Similar results are stated in [C-S]. In the present case, large Reynolds numbers are permitted, allowing for complex dynamical behavior. Moreover, dimensional analysis suggests that measures which are concentrated on certain sets of Hausdorff dimension $\geq \frac{1}{2}$ and are smooth with respect to Hausdorff measure, give rise to vortex structures for which the present results are applicable. The result is applicable for initial velocities which are square integrable, divergence-free and of bounded variation. However, it is not known whether $\int |\nabla u(x,t)| dx$ is bounded a priori in time for general data. What can be proved is that there exists an absolute constant c such that

$$\int_0^t exp\left(c \int |\nabla u(x,s)| dx \right) ds$$

is bounded a priori in terms of the data. The proof of this bound uses the kinematic fact that the map $\omega \to \nabla u$ is of weak-type (1,1) and the dynamical fact [F-G-T] that

$$\int_0^t ds \|\nabla u(\cdot, s)\|_{L^\infty}^{\frac{1}{2}}$$

is a priori bounded.

The results of [F-G-T] (see also [D] for the case of bounded domains) are the only ones that I am aware of, in which the behavior of high order derivatives for weak solutions is investigated. For instance, it is proven in [F-G-T] that

$$\int_0^t ds \left(\int |\Delta u(x,s)|^2 dx \right)^{\frac{1}{3}}$$

is a priori bounded in terms of the initial data, for arbitrary L^2 data. The corresponding result in this work is the boundedness of

$$\int_0^t \int |\Delta u(x,s)|^{\frac{4}{3+\varepsilon}} dx ds$$

Neither result implies the other and neither seems sufficient for global regularity. It may be useful to mention here that a sufficient condition for global regularity is the boundedness of

$$\int_0^t ds \left(\int |\Delta u(x,s)|^2 dx \right)^{\frac{2}{3}}$$

(see [C-F]. A simple modification of the argument on p. 79 yields the result above.)

In Section 3 we consider the problem of estimating a priori, in terms of the Reynolds number, the average area of a level set of the vorticity magnitude. Thus, we consider the level sets

$$A_\gamma = \{x; |\omega(x,t)| = \gamma\}$$

If ω is smooth, a well-known result ([M]) states that for almost all values of γ the sets A_γ are smooth 2-dimensional manifolds without boundary on the torus. (Thus, fractal sets do not occur typically as individual level sets of smooth fields.) We denote their 2-dimensional Hausdorff measure (area in this case) by

$$\mu(\gamma, t) = \int_{A_\gamma} dH^{(2)}(x)$$

and the time average of this quantity by

$$\mu_T(\gamma) = T^{-1} \int_0^T \mu(\gamma, t) dt$$

The main result of Section 3 is an a priori bound on

$$\int_0^\infty \mu_T^2(\gamma) d\gamma$$

in terms of the data. In the particular case of $F = 0$ (corresponding to decaying turbulence) this a priori bound can be written in terms of the Kolmogorov dissipation length η and the Reynolds number Re. It states that the average

$$< \mu >_T = \left(\gamma_{max}^{-1} \int_0^{\gamma_{max}} \mu_T^2(\gamma) d\gamma \right)^{\frac{1}{2}}$$

satisfies the inequality

$$< \mu >_T \leq \frac{L^3}{\eta} \left(1 + Re^{-\frac{1}{2}} \right)^{\frac{1}{2}}$$

The relation between T, Re and the average energy dissipation rate ε is the one appropriate for decaying turbulence ([B])

$$Re = T^2 \varepsilon \nu^{-1}$$

that is, T represents a time scale large enough for the turbulence to display universal statistical features at small scales but shorter than the time scale $L^2 \nu^{-1}$ of viscous decay.

Section 4 is devoted to the study of level sets of scalars and, in particular, to isotherms in turbulent Rayleigh-Benard convection. The scalar T (no relation to the time T above) satisfies the equation

(1.3)
$$\left(\frac{\partial}{\partial t} + u \cdot \nabla - \kappa \Delta \right) T = 0$$

in an open bounded set D with smooth boundary ∂D included in $\mathbb{R}^3$. The time dependent function u is assumed to vanish on ∂D and be divergence-free. The equation (1.3) is supplemented with initial data and boundary conditions which imply a maximum principle

(1.4)
$$0 \leq T(x,t) \leq M$$

valid for all x, t. Well-posedness of (1.3) and the bound (1.4) are the only consequences of the boundary conditions on T used in this work. Sometimes these facts can be justified on physical grounds in a more convincing manner than the boundary conditions on T can be.

We study the areas of portions of level sets of T which are situated at a positive distance from ∂D.

For any fixed x_0 in D we obtain the bound

$$< \mu >_{r,t} (x_0) \leq C\kappa^{-\frac{1}{2}} r^{\frac{5}{2}} \left(|u|_{2r,t} + \sigma \right)^{\frac{1}{2}} + Cr^2$$

for the quantity

$$< \mu >_{r,t} (x_0) = \left(\tau^{-1} \int_0^B \mu_{r,t}^2(\beta)d\beta \right)^{\frac{1}{2}}$$

where

$$\mu_{r,t}(\beta) = t^{-1} \int_0^t ds \int_{\{x;|x-x_0|\leq r, T(x,s)=\beta\}} dH^{(2)}(x),$$

and B, τ are respectively the maximal value and the maximal oscillation of T in the box of radius $2r$ around x_0 :

$$B = sup_{0\leq s\leq t, |x-x_0|\leq 2r} T(x,s),$$

and

$$\tau = \sup_{x,y\in B(x_0,2r), 0\leq s\leq t} |T(x,s) - T(y,s)|.$$

The coefficient σ is a small velocity (see 4.25A) and $|u|_{2r,t}(x_0)$ is the average of the velocity magnitude on the ball $B(x_0, 2r)$ and in the time interval $[0, t]$. It can be majorized by the Hardy-Littlewood maximal function

$$< |u|(x_0) >= \sup_{0<2r\leq\delta} |u|_{2r,t}(x_0).$$

One defines the local length scale

$$\lambda(x_0) = \kappa < |u|(x_0) >^{-1} .$$

For $r > \lambda$ the coefficient of $r^{2.5}$ in the inequality exceeds the coefficient of r^2. If we assume that $< |u| >$ can be evaluated using statistical selfsimilar families of solutions of the Navier-Stokes equations ([FMT]) then we obtain the familiar result

$$< |u| >= (\varepsilon\nu Re)^{\frac{1}{4}}$$

and using this value for $< |u| >$ in the definition of λ we obtain

$$\lambda = Pr^{-1}Re^{-\frac{1}{4}}\eta.$$

Clearly, in this case λ is a microscale and the average area is bounded by a multiple of $r^{2.5}$. Another possible way of evaluation of $< |u| >$ is suggested in [Chi]. In that work the velocity u_c in the central region of the convection experiment is deduced by a heuristic argument and equals

$$u_c = \kappa L^{-3}l^2 Ra$$

where l is the width of the thermal boundary layer and Ra is the Rayleigh number. One of the main issues in convection is the way l scales with Rayleigh number, i.e., what is the value of the constant β in the scaling law

$$l = LRa^{-\beta}.$$

The classical value is $\beta = \frac{1}{3}$ and the value suggested in [Chi] (see also the paper by L. Kadanoff in this volume) is $\beta = \frac{2}{7}$. If we replace $< |u| >$ by u_c in our definition of the local scale λ we obtain

$$\lambda = L\, Ra^{2\beta - 1}.$$

The classical value of β is precisely the one for which $l = \lambda$; the local scale λ is a microscale (smaller the the width of the thermal boundary layer) if and only if $\beta < \frac{1}{3}$, a relation verified by the value $\frac{2}{7}$.

The power 2.5 of r which appears in the upper bound agrees with the experimental evidence ([S]) which seems to indicate that turbulent interfaces have a Hausdorff dimension larger or equal to 2.35.

2. WEAK SOLUTIONS, STRONG SOLUTIONS AND SMOOTH SOLUTIONS

We consider the Navier-Stokes equation

$$(2.1) \qquad \frac{\partial u}{\partial t} + u \cdot \nabla u - \nu \Delta u + \nabla p = F$$

in three dimensions, with periodic boundary conditions

$$(2.2) \qquad u(x + Le_j, t) = u(x, t)$$

where $e_j = (\delta_{jk}), j, k = 1, 2, 3$ is the standard basis in $\mathbb{R}^3$,

$$(2.3) \qquad divu = 0$$

and initial conditions

$$(2.4) \qquad u(x, 0) = u_0(x)$$

The vorticity, $\omega = curlu$ satisfies the equation

$$(2.5) \qquad \frac{\partial \omega}{\partial t} + u \cdot \nabla \omega - \nu \Delta \omega = \omega \cdot \nabla u + f$$

where $f = curlF$.

We restrict ourselves to the space-periodic case only to fix ideas; most of the results and all the concepts remain unchanged in a general setting. We assume also that the spatial integrals of the velocities are zero. For the issues addressed here the presence of a smooth non-zero mean flow introduces absolutely no difficulty and only minor changes. This comment is valid for the other sections of this work also. A weak solution of the Navier-Stokes equations is a function u satisfying the following requirements ([C-F] Ch.8): it is divergence-free, spatially square-integrable for almost all times, the space integral of its scalar product with any smooth, divergence-free function is continuous as a function of time, its gradient is square-integrable in space and time, its time derivative is a generalized function of a certain type $(L^1(0, T; V'))$ and it satisfies the Navier-Stokes equations when multiplied scalarly with an arbitrary smooth time independent divergence-free function and integrated spatially. A Leray weak solution is a weak solution satisfying the energy inequality

$$\frac{1}{2} \int |u(x,t)|^2 dx + \nu \int_{t_0}^t \int |\nabla u(x,s)|^2 dx ds \leq$$

$$\leq \frac{1}{2} \int |u(x,t_0)|^2 dx + \int_{t_0}^t \int F(x,s) \cdot u(x,s) dx ds$$

for t_0 in a set of times which contains 0 and whose complement has Hausdorff dimension not larger than $\frac{1}{2}$ and all $t \geq t_0$.

The pressure corresponding to a weak solution is a generalized function which is obtained by the usual requirement that the divergence be maintained zero. Loosely speaking, a Leray weak solution is a solution for which the total kinetic energy and the energy dissipation rate are well defined quantities, but not much more is known. The result of Leray ([L]) states that such a solution exists for all square-integrable initial data, practically all F, all ν and all T. Very few time evolution nonlinear equations of mathematical physics are known to possess global weak solutions for all large data. If the initial data and F are smooth then it is well-known that the solution exists, is uniquely determined by the data and is smooth for at least a short time. (Even if the initial data are not smooth the solution becomes instantly smooth because of the parabolic smoothing effect; but this is a secondary matter). The time of guaranteed existence, uniqueness and smoothness depends on the size of the initial data and their derivatives measured by spatial integrals. The larger are the initial data, the shorter is the guaranteed existence time. If the initial data are small enough, (low Reynolds number case) then this guaranteed existence time is infinity. If the initial data are large then there exists a Leray weak solution corresponding to them which is defined for large times and coincides with the smooth solution as long as the latter is guaranteed to exist. All these are well known results (see for instance [C-F]). There exist weak Leray solutions of the Navier-Stokes equations which are not smooth ([Sc]) but they are generated by singular Fs. The problem of formation of singularities from smooth data is open.

A strong solution is a Leray weak solution which has the spatial integral of the square of the vorticity magnitude bounded uniformly in time and the gradients of the vorticity square-integrable in space and time. Strong solutions are an important auxiliary concept. If a solution is strong on a time interval $[0, T]$ then it is actually as smooth as F permits ([C-F] Ch.

10) on that time interval. In particular, if F is smooth then this statement constitues a first criterion for the absence of singularities. Actually, an apparently less stringent condition expressed in terms of the vorticity implies that the solution is strong and consquently smooth:

Condition I. *If*

$$\int_0^T dt \left(\int |\omega(x,t)|^2 dx \right)^2 < \infty$$

then the solution is smooth and uniquely determined by its initial data for $0 < t \leq T$.

This result goes back to J. Leray. It is important to note that the condition requires the quantity to be finite but its size is not restricted in any way; in particular it is allowed to become indefinitely large when the Reynolds number is increased indefinitely. The gap between this condition and the corresponding known information on the vorticity of Leray solutions is evident:

Known. *Any Leray weak solutions satisfies*

$$\int_0^T \int |\omega(x,t)|^2 dx dt < \infty$$

Another well-known necessary and sufficient condition for the absence of singularities is expressed in terms of the velocities:

Condition II. *If*

$$\int_0^T dt \left(\int |u(x,t)|^p dx \right)^{\frac{2}{p-3}} < \infty$$

for some p, $3 < p \leq \infty$ then the solution is smooth and uniquely determined by the initial data for $0 < t \leq T$.

This condition follows from the work of G. Prodi, J-L. Lions and J. Serrin (see for instance [Li]). The meaning of the condition in the case $p = \infty$ is

$$\int_0^T sup_x |u(x,t)|^2 dt < \infty$$

By contrast, the known corresponding result is

Known. *Any Leray weak solution satisfies*

$$\int_0^T dt \left(\int |u(x,t)|^p dx \right)^{\frac{4}{3(p-2)}} < \infty$$

for any p, $2 \leq p \leq 6$.

The result above is obtained immediately by interpolation from the fact that Leray solutions have uniformly bounded in time L^2 norms and space and time square-integrable gradients. Note, of course, that there is a gap between the known result and Condition II. The known inequality can be written down explicitly in terms of the data.

A condition expressed in terms of the pressure is the following

Condition III. *If*

$$\int_0^T sup_x (u \cdot \nabla p)_-(x,t)dt < \infty$$

then the solution is smooth and uniquely determined by the initial data for $0 < t \leq T$.

We recall that $f_-(x)$ equals $-f(x)$ if $f(x) \leq 0$ and 0 otherwise.

Finally, a condition expressed in terms of the gradients of the vorticity is the following

Condition IV. *If*

$$\int_0^T dt \left(\int |\nabla \omega(x,t)|^2 dx \right)^{\frac{2}{3}} < \infty$$

then the solution is smooth and uniquely determined for $0 < t \leq T$.

The known result is, in this case

Known. *Any Leray solution satisfies*

$$\int_0^T dt \left(\int |\nabla \omega(x,t)|^2 dx \right)^{\frac{1}{3}} < \infty$$

This result can be found in [F-G-T].

Now we are going to describe the construction of a class of Leray solutions which enjoy the properties

(2.6) $$ess.sup_{t \in (0,T]} \int |\omega(x,t)| dx < \infty$$

(compare to Condition I) and

(2.7) $$\int_0^T \int |\nabla \omega(x,t)|^{\frac{4}{3+\varepsilon}} dx \, dt < \infty$$

(compare to Condition IV).

Let us consider an arbitrary smooth function

$$q : \mathbb{R}^3 \to \mathbb{R}$$

multiply the i-th equation (2.5) by $\tau \frac{\partial q}{\partial y_i}$ computed at $y = \tau w(x,t)$ and add. The constant τ is an arbitrary positive constant which has the dimension of time so that y is dimensionless. It follows that the scalar dimensionless function

$$w(x,t) = q(\tau w(x,t))$$

satisfies the equation

(2.8) $$\left(\frac{\partial}{\partial t} + u \cdot \nabla - \nu \Delta\right)w + \nu \tau^2 \frac{\partial^2 q}{\partial y_i \partial y_l} \frac{\partial w_i}{\partial x_j} \frac{\partial w_l}{\partial x_j} = R$$

with

$$R = \tau \omega_j \frac{\partial u_i}{\partial x_j} \frac{\partial q}{\partial y_i} + \tau f_i \frac{\partial q}{\partial y_i}$$

Now, clearly, if the function q has bounded gradient

$$|\nabla q| \le C$$

uniformly for all y in $\mathbb{R}^3$ then

$$|R| \le \tau C(|\omega||\nabla u| + |f|)$$

and thus R is a priori integrable in space and time. One can choose q such that $|\nabla q|$ is bounded and also q is strictly convex. In view of the mean value theorem one sees that this is possible only if the Hessian of q decays at least as does $|y|^{-1}$ for y large. If we seek q of the form

$$q = h(1 + |y|^2)$$

we deduce

$$\frac{\partial^2 q}{\partial y_i \partial y_j} \xi_i \xi_j = 2[|\xi_\perp|^2 h'(\alpha) + |\xi_\parallel|^2 (h'(\alpha) + (2\alpha - 2)h''(\alpha))]$$

where $\alpha = 1 + |y|^2$ and $\xi_\parallel$, $\xi_\perp$, are the decomposition of the arbitrary vector ξ in vectors parallel to and perpendicular to y.

We choose q given by

$$q(y) = (1 + |y|^2)^{\frac{1}{2}} - \frac{1}{2(1 - \varepsilon)}(1 + |y|^2)^{\frac{1 - \varepsilon}{2}}$$

for any positive $\varepsilon \leq \frac{1}{2}$. It follows by direct computation that

$$\frac{\partial^2 q}{\partial y_i \partial y_j} \xi_i \xi_j > \frac{\varepsilon}{2}(1 + |y|^2)^{-\frac{1+\varepsilon}{2}} |\xi|^2$$

Also,
$$|\nabla q| \leq 1$$

and, clearly,

$$\frac{1 - 2\varepsilon}{2(1 - \varepsilon)}(1 + |y|^2)^{\frac{1}{2}} \leq q \leq (1 + |y|^2)^{\frac{1}{2}}$$

With this choice of q it follows from that any classical solution of the Navier-Stokes equation satisfies

$$\frac{1 - 2\varepsilon}{2(1 - \varepsilon)} \int (1 + \tau^2 |\omega(x, t)|^2)^{\frac{1}{2}} dx+$$

$$\frac{\varepsilon}{2}\nu\tau^2 \int_0^t \int (1 + \tau^2 |\omega(x, s)|^2)^{-\frac{1+\varepsilon}{2}} |\nabla\omega(x, s)|^2 dx ds \leq \gamma$$

Where γ is given by

$$\gamma = \tau \int_0^t \int |\nabla u(x, s)||\omega(x, s)| dx ds + \tau \int_0^t \int |f(x, s)| dx ds$$

$$+ \int (1 + \tau^2 |\omega_0(x)|^2)^{\frac{1}{2}} dx$$

As it is well known, γ is a priori bounded because of the energy inequality

$$\int_0^t \int |\nabla u(x, s)|^2 dx ds \leq \nu^{-1} \int |u_0(x)|^2 dx + L^2 \nu^{-2} \int_0^t \int |F(x, s)|^2 dx ds$$

In view of the energy inequality it follows that

$$\int_0^t \int |\nabla\omega(x, s)|^{\frac{4}{3+\varepsilon}} dx ds$$

is a priori bounded, for all t positive. Indeed, we write

$$|\nabla\omega(x, s)|^{\frac{4}{3+\varepsilon}} =$$

$$\left(|\nabla\omega(x, s)|^{\frac{4}{3+\varepsilon}}(1 + \tau^2 |\omega(x, s)|^2)^{-\frac{1+\varepsilon}{3+\varepsilon}}\right)\left((1 + \tau^2 |\omega(x, s)|^2)^{\frac{1+\varepsilon}{3+\varepsilon}}\right)$$

and use a Hölder inequality in the space and time integral raising the first paranthesis to $\frac{3+\varepsilon}{2}$ and the second one to the conjugate power, $\frac{3+\varepsilon}{1+\varepsilon}$. We obtain

$$\int_0^t \int |\nabla\omega(x,s)|^{\frac{4}{3+\varepsilon}}dxds$$

$$\leq (2\gamma\varepsilon^{-1}\nu^{-1}\tau^{-2})^{\frac{2}{3+\varepsilon}}\left(\int_0^t\int(1+\tau^2|\omega(x,s)|^2)dxds\right)^{\frac{1+\varepsilon}{3+\varepsilon}}$$

In view of the Calderon-Zygmund inequality

$$\int|\nabla v(x)|^p dx \leq C_p\int|curlv(x)|^p dx$$

valid for divergence free functions and $1 < p < \infty$ it follows that

$$\int_0^t\int|\nabla\nabla u(x,s)|^{\frac{4}{3+\varepsilon}}dxds$$

is bounded a priori in terms of the data. On the other hand, the a priori boundedness of

$$\int|\nabla u(x,t)|dx$$

does not follow from the preceding considerations. If one tries to imitate the proof of boundedness of the spatial L^1 norm of the vorticity for the full gradient one encounters the difficulty that the Hessian of the pressure is not estimated a priori in L^1 (space and time) in terms of the data. Of course these remarks concern the general case of large data and large T. The previous arguments provide a priori bounds for the quantities in the left hand sides of (2.6), (2.7) for as long as the solution is smooth. If we would know, from the boundedness of these quantities, how to deduce smoothness and uniqueness of solutions then we would be able to conclude that no singularities can occur in the solutions. On the other hand, the previous arguments cannot be applied as they are to Leray solutions because they are not smooth enough. In order to obtain (2.6) and (2.7) for weak solutions we approximate the Navier-Stokes equations by a sequence of equations

$$\left(\frac{\partial}{\partial t} - \nu\Delta + S_\delta(v)\cdot\nabla\right)v + \nabla p = S_\delta(F)$$

$$divv = 0$$

$$v(x,0) = u_{0,\delta}$$

where δ is a small number and $S_\delta(v)$, $S_\delta(F)$ are obtained from v, F by a convolution

$$S_\delta(v)(x,t) = \delta^{-4}\int_{-\infty}^{\frac{t}{T}}\int_{\mathbb{R}^3}\psi(\frac{y}{\delta},\frac{s}{\delta})v(x-Ly,t-Ts)dyds$$

The smooth function ψ is assumed to be non-negative with compact support in $|y| < 1, 1 < s < 2$ and have space and time integral equal to one. Note that if v is space periodic with period L so is $S_\delta(v)$. Moreover, the values of $S_\delta(v)$ at time t depend only of the values of v at positive times in the interval $(t - 2\delta T, t - \delta T)$. Also, clearly, if v is divergence-free, so is $S_\delta(v)$. The initial data are also mollified

$$u_{0,\delta}(x) = \delta^{-3} \int_{\mathbb{R}^3} \phi(\frac{y}{\delta}) u_0(x - Ly) dy$$

where the smooth function ϕ is non-negative and has space integral equal to one. This procedure of constructing weak solutions is well-known. It was used in [C-K-N] to construct suitable weak solutions (that is solutions which obey a local energy estimate) of the Navier-Stokes equations. Thus, incidentally, the solutions that we construct are suitable weak Leray solutions and consequently, the space-time 1 dimensional Hausdorff measure of their singular set is zero [C-K-N]. The solution v is unique, global and smooth if δ is fixed and small enough. This can be proved using a standard Galerkin approximation of the equation and the fact that S_δ is smoothing. Moreover, from energy estimates of the solution v it follows that

$$sup_{0 \leq t \leq T} \int |v(x,t)|^2 dx$$

and

$$\int_0^T \int |\nabla v(x,s)|^2 dx ds$$

are uniformly bounded, independently of δ for small δ. The time derivatives of the solutions v are uniformly bounded in $L^{\frac{4}{3}}(0,T;V')$. Therefore by a well-known result (see, for instance, [C-F]) the v-s belong to a compact set in L^2 in space and time. These are standard results, and would have been true for other approximation methods, such as, for instance, the usual Galerkin method. The only assumption on u_0 needed for this is that it be square-integrable. We chose the retarded mollification approximation because the curls of the v can be bounded uniformly in L^1. Indeed the equation for $\omega_\delta = curl v$ is

$$\left(\frac{\partial}{\partial t} - \nu \Delta + S_\delta(v) \cdot \nabla\right)\omega_\delta = S_\delta(f) - \varepsilon_{\cdot jk} S_\delta\left(\frac{\partial v_l}{\partial x_j}\right)\frac{\partial v_k}{\partial x_l}$$

where ε_{ijk} is the usual antisymmetric tensor. The right hand side is uniformly bounded, independently of δ in L^1 in space and time. Moreover, assuming that $curl u_0$ is a Radon measure, the initial curls $\omega_\delta(x,0)$ are uniformly bounded, independently of δ in L^1 in space. Because the solutions v are smooth we are allowed to perform the change to w where $w = q(\tau \omega_\delta)$ and q is as before. We deduce that

$$sup_{t < T} \int |\omega_\delta(x,t)| dx$$

and

$$\int_0^T \int |\nabla \omega_\delta(x,s)|^{\frac{4}{3+\varepsilon}}\, dx ds$$

are uniformly bounded, independently of δ. Consequently, these uniform bounds will be inherited by the weak limit. The theorem one can prove is

Theorem. *Let T, L, $\nu > 0$ be arbitrary. Assume that the initial data u_0 is a square-integrable, L-periodic, divergence-free function whose curl, ω_0 is a finite Radon measure on the torus. Assume that the body forces F and their curl f are L-periodic and that*

$$\int_0^T \int |F(x,t)|^2 dx dt < \infty$$

and

$$\int_0^T \int |f(x,t)| dx dt < \infty$$

Then there exists a Leray weak solution of the Navier-Stokes equations which satisfies

$$ess.sup_{t \le T} \int |\omega(x,t)| dx < \infty$$

$$\int_0^T \int |\nabla \omega(x,t)|^{\frac{4}{3+\varepsilon}}\, dx dt < \infty$$

for $0 < \varepsilon \le \frac{1}{2}$ and

$$\int_0^T \int |\nabla \nabla u(x,t)|^{\frac{4}{3+\varepsilon}}\, dx dt < \infty.$$

We end this section with a few comments. Because uniqueness of Leray weak solutions is not known, the question arises whether weak solutions obtained by different approximation procedures from the same initial data enjoy the properties of the solution obtained by the present procedure.

It would be interesting if a concrete characterization of the class of initial vorticities allowed by the Theorem would be available. While vortons are too singular even for this problem, dimensional analysis suggests that vorticities which are concentrated on certain sets of Hausdorff dimension $\ge \frac{1}{2}$ and are smooth with respect to $\frac{1}{2}$-dimensional Hausdorff measure would be in this class.

If the initial velocities belong to BV, are square-integrable and divergence-free then our construction is applicable but it does not guarantee that the BV norm of the solution is bounded a priori in time. However, the L^∞ bound in time for the BV norm of the solution fails logarithmically. By this we mean that one can prove that

$$\int_0^T exp\left(c \int |\nabla u(x,s)|dx\right)ds$$

is a priori bounded for a certain $c > 0$. The proof of this statement uses the fact ([F-G-T]) that

$$\int_0^T \|\nabla u(\cdot,s)\|_{L^\infty}^{\frac{1}{2}} ds$$

is bounded a priori and the fact that, because the operator $\omega \to \nabla u$ is of weak- type $(1,1)$ one has a logarithmic extrapolation inequality

$$\int |\nabla u(x)|dx \le \tau L^3 + c_1\left(log_+(\frac{\|\nabla u\|_{L^\infty}}{\tau})\right)\int |\omega(x)|dx$$

In the above inequality $\tau > 0$ is arbitrary and c_1 is independent of τ and is the the weak-type $(1,1)$ norm of the map $\omega \to \nabla u$. The constant c above is related to c_1 by $c = (2c_1)^{-1}$. Strictly speaking these operations need to be done for each approximate solution v, but the argument remains the same.

3. THE AREA OF LEVEL SETS OF THE VORTICITY

In this section we will consider the problem of estimating a priori in terms of the Reynolds number the typical area of a level set of the vorticity magnitude. We consider an arbitrary solution of the Navier-Stokes equations (2.1)-(2.4). For any positive number γ we consider the level sets

$$(3.1) \qquad A_\gamma = \{x; |\omega(x,t)| = \gamma\}$$

If ω is a smooth function then it is well-known ([M]) that for almost all γ the sets A_γ are smooth 2-dimensional manifolds without boundary in T_L^3. We denote their 2-dimensional Hausdorff measure (area in this case) by $\mu(\gamma,t)$:

$$(3.2) \qquad \mu(\gamma,t) = \int_{A_\gamma} dH^{(2)}(x)$$

We are interested in estimating the average $< \mu >$ in terms of the Reynolds number. Our method works also for local averages, that is averages of the area of portions of level sets confined to some fixed region of space.

We consider the function

$$(3.3) \qquad w(x,t) = (\tau^{-2} + |\omega(x,t)|^2)^{\frac{1}{2}}$$

Then w satisfies (2.8) with R bounded independently of τ :

$$(3.4) \qquad |R| \le |\omega||\nabla u| + |f|$$

Moreover, direct computation of shows that

$$(3.5) \qquad (\frac{\partial}{\partial t} + u \cdot \nabla u - \nu\Delta)w \le |\omega||\nabla u| + |f|$$

The sets A_γ are level sets of w:

$$(3.6) \qquad A_\gamma = \{x; w(x,t) = (\tau^{-2} + \gamma^2)^{\frac{1}{2}}\}$$

We take a positive, increasing function of one variable $G(\beta)$ and multiply the inequality (3.5) by $G'(w(x,t))$ where G' is the derivative of G. We obtain

$$(3.7) \qquad (\frac{\partial}{\partial t} + u \cdot \nabla - \nu\Delta)g + \nu G''|\nabla w|^2 \le |R|G'$$

In (3.7) we denoted

$$(3.8) \qquad g(x,t) = G(w(x,t))$$

and G', G'' were computed at $w(x,t)$. Integrating in (3.7) we obtain

$$(3.9) \quad \nu \int_0^T \int G''|\nabla w|^2 dx dt \le \int_0^T \int G'|R| dx dt + \int (g(x,0) - g(x,T)) dx$$

We will specify G in the following way. We take an arbitrary smooth function of one variable $\phi(\beta)$ compactly supported in the interval (τ^{-1}, ∞) and define

$$(3.10) \qquad G(\beta) = \int_{\tau^{-1}}^{\beta} (\beta - \lambda)\phi^2(\lambda) d\lambda$$

We observe that

(3.11)
$$0 \le G(\beta) \le \beta \|\phi\|^2$$

where

(3.12)
$$\|\phi\|^2 = \int_{-\infty}^{\infty} \phi^2(\lambda) d\lambda$$

and

(3.13)
$$G'(\beta) = \int_{\tau^{-1}}^{\beta} \phi^2(\lambda) d\lambda$$

Consequently,

(3.14)
$$0 \le G'(\beta) \le \|\phi\|^2$$

Moreover,

(3.15)
$$G''(\beta) = \phi^2(\beta)$$

Using (3.10)-(3.15) in (3.9) we obtain

(3.16)
$$\nu \int_0^T \int \phi^2(w(x,t)) |\nabla w(x,t)|^2 dx\, dt \le \|\phi\|^2 \left(\int_0^T \int |R| dx\, dt + \int w(x,0) dx \right)$$

and, as a consequence

(3.17)
$$T^{-1} \int_0^T \int \phi(w(x,t)) |\nabla w(x,t)| dx\, dt \le \|\phi\| \mathcal{E}(T)$$

where

(3.18)
$$\mathcal{E}(T) = \nu^{-\frac{1}{2}} L^{\frac{3}{2}} \left(T^{-1} \int_0^T \int |R| dx\, dt + T^{-1} \int w(x,0) dx \right)^{\frac{1}{2}}$$

Now we will use an important tool of geometric measure theory, established by Kronrod ([K]) in a special case and by Federer ([F]) in the general case. A proof can be found in [M] also.

Theorem 3.1. *Let ψ be a Borel measurable nonnegative function on D and let h be a Lipschitz continuous function in D, where D is an open subset of $\mathbb{R}^n$. Then*

$$(3.19) \qquad \int_D \psi(x)|\nabla h(x)|dx = \int_0^\infty d\beta \int_{A_\beta} \psi(x)dH^{(n-1)}(x)$$

where $H^{(n-1)}$ is the $(n-1)$-dimensional Hausdorff measure and

$$(3.20) \qquad A_\beta = \{x \in D; |h(x)| = \beta\}$$

Theorem 3.1 is local and therefore it is valid for periodic functions, also. We will use it in (3.17), with $h = w$ and $\psi(x) = \phi(w(x,t))$. We obtain

$$(3.21) \qquad T^{-1}\int_0^T dt \int_{\tau^{-1}}^\infty \phi(\beta)d\beta \int_{A_\beta} dH^{(2)}(x) \leq \|\phi\|\mathcal{E}(T)$$

Now, in view of (3.1), (3.6) and (3.20) we deduce

$$(3.22) \qquad A_\beta = A_{(\beta^2 - \tau^{-2})^{\frac{1}{2}}}$$

and therefore the inequality (3.21) can be re-written as

$$(3.23) \qquad T^{-1}\int_0^T \int_{\tau^{-1}}^\infty \phi(\beta)\mu((\beta^2 - \tau^{-2})^{\frac{1}{2}}, t)d\beta dt \leq \|\phi\|\mathcal{E}(T)$$

Let us define μ_T by

$$(3.24) \qquad \mu_T(\gamma) = T^{-1}\int_0^T \mu(\gamma, t)dt$$

Because ϕ is arbitrary we obtain from (3.23) that

$$(3.25) \qquad \int_{\tau^{-1}}^\infty \mu_T^2((\beta^2 - \tau^{-2})^{\frac{1}{2}})d\beta \leq (\mathcal{E}(T))^2$$

We perform the change of variable

$$(3.26) \qquad (\beta^2 - \tau^{-2})^{\frac{1}{2}} = \gamma$$

in the left hand side of (3.25). We obtain

$$(3.27) \qquad \int_0^\infty \mu_T^2(\gamma) \frac{\gamma}{(\gamma^2 + \tau^{-2})^{\frac{1}{2}}} d\gamma \le (\mathcal{E}(T))^2$$

In view of the definition (3.18) of $\mathcal{E}(T)$ and (3.3) of w we see that

$$(3.28) \qquad lim_{\tau \to \infty} \mathcal{E}(T) = \mathcal{R}(T)$$

with

$$(3.29) \qquad \mathcal{R}(T) = \nu^{-\frac{1}{2}} L^{\frac{3}{2}} \left(T^{-1} \int_0^T \int |R| dx dt + T^{-1} \int |w_0(x)| dx \right)^{\frac{1}{2}}$$

Passing to the limit of $\tau \to \infty$ we obtain the inequality

$$(3.30) \qquad \int_0^\infty \mu_T^2(\gamma) d\gamma \le (\mathcal{R}(T))^2$$

In a completely analogous fashion, passing to $limsup_{T \to \infty}$ in (3.23) we obtain

$$(3.31) \qquad \int_0^\infty \mu_\infty^2(\gamma) d\gamma \le \mathcal{R}_\infty^2$$

where

$$(3.32) \qquad \mu_\infty(\gamma) = limsup_{T \to \infty} T^{-1} \int_0^T \mu(\gamma, t) dt$$

and

$$(3.33) \qquad \mathcal{R}_\infty^2 = \nu^{-1} L^3 limsup_{T \to \infty} T^{-1} \int_0^T \int |R| dx dt$$

In view of the inequality (3.4) we proved the following theorem

Theorem 3.2. *Let ω be the curl of a classical $(\mathcal{C}^3)$ L-periodic solution of the Navier-Stokes equation (2.1) with smooth periodic driving forces F. Assume that the solution is defined for $0 \leq t \leq T_1$ where $T_1 \leq \infty$. Then the area*

$$(3.34) \qquad \mu(\gamma, t) = \int_{\{x; |\omega(x,t)| = \gamma\}} dH^{(2)}(x)$$

satisfies, for any $0 < T < T_1$

$$(3.35) \quad \int_0^\infty \mu_T^2(\gamma) d\gamma \leq \nu^{-1} L^3 T^{-1} \left(\int_0^T \int (|\omega(x,t)|^2 + |curl F(x,t)|) dx\, dt \right.$$

$$\left. + \int |\omega_0(x)| dx \right)$$

where μ_T is the time average of μ defined in (3.24).

Moreover, if $T_1 = \infty$ then

$$(3.36) \qquad \int_0^\infty \mu_\infty^2(\gamma) d\gamma \leq$$

$$\nu^{-1} L^3 limsup_{T \to \infty} T^{-1} \int_0^T \int (|\omega(x,t)|^2 + |curl F(x,t)|) dx\, dt$$

where $\mu_\infty = limsup_{T \to \infty} \mu_T$

In view of the energy inequality the inequalities (3.35) and (3.36) imply the a priori bounds in terms of initial data and F

$$(3.37) \quad \int_0^\infty \mu_T^2(\gamma) d\gamma \leq \nu^{-2} L^3 T^{-1} \int |u_0(x)|^2 dx + \nu^{-1} L^3 T^{-1} \int |\omega_0(x)| dx$$

$$+ \nu^{-3} L^5 T^{-1} \int_0^T \int |F(x,t)|^2 dx\, dt + \nu^{-1} L^3 T^{-1} \int_0^T \int |curl F(x,t)| dx\, dt$$

and

$$(3.38) \qquad \int_0^\infty \mu_\infty^2(\gamma) d\gamma \leq$$

$$\nu^{-1} L^3 limsup_{T \to \infty} T^{-1} \int_0^T \int (L^2 \nu^{-2} |F(x,t)|^2 + |curl F(x,t)|) dx\, dt$$

Note that the right hand sides of (3.35)-(3.38) are finite even if the solution becomes singular.

Both the left hand and right hand sides of (3.35)-(3.38) have the dimension $< L >^4 < T >^{-1}$. The left hand sides have this dimension because μ represents an area and γ is a vorticity magnitude and thus scales like $< T >^{-1}$.

We define average values of μ by the formulas

$$(3.39) \qquad < \mu >_T = \left([\omega]_T^{-1} \int_0^\infty \mu_T^2(\gamma) d\gamma \right)^{\frac{1}{2}}$$

and

$$(3.40) \qquad < \mu >_\infty = \left([\omega]_\infty^{-1} \int_0^\infty \mu_\infty^2(\gamma) d\gamma \right)^{\frac{1}{2}}$$

where

$$(3.41) \qquad [\omega]_T = sup_{0 \le t \le T} sup_x |\omega(x,t)|$$

and

$$(3.42) \qquad [\omega]_\infty = sup_{0 \le t \le \infty} sup_x |\omega(x,t)|$$

These definitions are natural; the reason for dividing by the numbers defined in (3.41), (3.42) is that, for smooth solutions, the range of γ is $[0, [\omega]]$ and not $[0, \infty)$. We emphasise that the numbers defined in (3.39) and (3.40) do not involve any ensemble averaging. They are well defined for any individual smooth, deterministic solution of the Navier-Stokes equations and represent, in some sense, an average value of the area of vorticity level sets.

Let us consider now the particular case of $F = 0$. Then (3.35) becomes

$$(3.43) \qquad < \mu >_T^2 \le \nu^{-1} L^6 \left(\sqrt{\frac{\varepsilon}{\nu}} + T^{-1} \right)$$

where the energy dissipation rate is given by

$$(3.44) \qquad \varepsilon = \nu T^{-1} L^{-3} \int_0^T \int |\omega(x,t)|^2 dx dt$$

We define the Reynolds number Re by

$$(3.45) \qquad Re = T^2 \varepsilon \nu^{-1}$$

(see [B], (7.3.3)). One can think of (3.45) as defining, for fixed Reynolds number, a time T which is large enough for turbulence to reach statistical universality but which is smaller than the viscous dissipation time scale $L^2\nu^{-1}$. We recall that the Kolmogorov dissipation length η is defined by

$$(3.46) \qquad \eta = \varepsilon^{-\frac{1}{4}} \nu^{\frac{3}{4}}$$

The inequality (3.43) becomes

$$(3.47) \qquad L^{-2} < \mu >_T \leq \frac{L}{\eta}(1 + Re^{-\frac{1}{2}})^{\frac{1}{2}}$$

or, because

$$(3.48) \qquad \eta = \varepsilon^{\frac{1}{2}} T^{\frac{3}{2}} Re^{-\frac{3}{4}}$$

$$(3.49) \qquad L^{-2} < \mu >_T \leq Re^{\frac{3}{4}} L \varepsilon^{-\frac{1}{2}} T^{-\frac{3}{2}}(1 + Re^{-\frac{1}{2}})^{\frac{1}{2}}$$

The inequality (3.47) can be re-written as

$$(3.50) \qquad < \mu >_T \leq C(\frac{L}{\eta})^3 \eta^2$$

with $C = (1 + Re^{-\frac{1}{2}})^{\frac{1}{2}}$. The right hand side of (3.50) has a simple geometric interpretation. It is proportional with the area of a union of non-overlapping spheres of radii η which fills the cube of side length L.

We proved thus

Theorem 3.3. *The average area $< \mu >_T$ of a level set of the vorticity magnitude of a smooth solution of the Navier-Stokes equation (2.1) with $F = 0$ is bounded above by a multiple of $Re^{\frac{3}{4}}$ where Re is the Reynolds number. When expressed in terms of the Kolmogorov dissipation length η this upper bound ((3.50)) represents the area of a union of non-overlapping spheres of radii η which fill a fraction $(\frac{1}{4\pi}\sqrt{1 + Re^{-\frac{1}{2}}})$ of the volume of the spatial domain.*

4. Area of isotherms

We investigate the problem of estimating a priori the average area of portions of level sets of scalars. As an application we will consider the area of portions of isotherms in Rayleigh-Benard convection.

We consider a real scalar function $T(x,t)$ defined for x in a bounded, open domain D of $\mathbb{R}^3$ with smooth boundary and t nonnegative. We assume that T satisfies the equation

$$(4.1) \qquad \left(\frac{\partial}{\partial t} + u \cdot \nabla - \kappa \Delta\right) T = 0$$

in D where κ is a positive constant and the $\mathbb{R}^3$ valued function u satisfies

$$(4.2) \qquad div\, u = 0$$

and

$$(4.3) \qquad u_{|\partial D} = 0$$

In the case of Rayleigh-Benard convection u satisfies a partial differential equation which couples its evolution with that of T. We will study the area of level sets of T and bound them in terms of certain average bounds on u. At this stage we do not make more assumptions on u than (4.2), (4.3) and the minimal regularity assumptions needed to solve (4.1). We supplement (4.1) with initial data

$$(4.4) \qquad T(x,0) = T_0(x)$$

We assume that T_0 is bounded. We will not specify boundary conditions on T. Rather, we will assume a maximum principle:

$$(4.5) \qquad 0 \le T(x,t) \le M$$

with $M > 0$ a given number. The inequality (4.5) is a consequence of standard boundary conditions on T. In particular, the usual boundary conditions in convection, $T = 0$ on the top plate, $T = M$ on the bottom one and either Neumann or periodic boundary conditions on the lateral walls imply (4.5) provided, of course, that T_0 obeys (4.5). Thus (4.5) does not impose any restriction if T satisfies one of these standard boundary conditions; in many geophysical problems, however, the boundary conditions are somewhat unclear and therefore it seems worthwhile to make minimal assumptions on the boundary behavior of T. We take a smooth, compactly supported, nonnegative function ψ whose support is at distance $\geq \delta > 0$ from the boundary of D. We take also the function G defined in (3.10) except that now we can set already $\tau = \infty$. Thus

$$(4.6) \qquad G(\beta) = \int_0^\beta (\beta - \lambda)\phi^2(\lambda)d\lambda$$

with ϕ an arbitrary smooth function supported in $(0, \infty)$. We multiply (4.1) by $\psi(x)G'(T(x,t))$. We obtain

$$(4.7) \qquad \left(\frac{\partial}{\partial t} + u \cdot \nabla - \kappa\Delta\right)g_\psi + \kappa\psi G''|\nabla T|^2 = R$$

where

$$(4.8) \qquad g_\psi(x,t) = \psi(x)G(T(x,t))$$

the derivative G'' is computed at $T(x,t)$ and R is given by

$$(4.9) \qquad R = G(T)(u \cdot \nabla\psi - \kappa\Delta\psi) - 2\kappa G'(T)\nabla T \cdot \nabla\psi$$

Integrating over D we obtain

$$(4.10) \qquad \frac{d}{dt}\int_D g_\psi(x,t)dx + \kappa \int_D \psi(x)G''(T(x,t))|\nabla T(x,t)|^2 dx = \mathcal{R}(t)$$

with

$$(4.11) \qquad \mathcal{R}(t) = \int_D G(T(x,t))(u \cdot \nabla\psi + \kappa\Delta\psi)(x,t)dx$$

The relations (3.11)-(3.15) regarding G are valid. It follows that

$$(4.12) \qquad \mathcal{R}(t) \leq \|\phi\|^2 M \int_D |u \cdot \nabla\psi + \kappa\Delta\psi|dx$$

Integrating (4.10) in time we deduce

(4.13)
$$\int_0^t \int_D \psi(x)\phi^2(T(x,s))|\nabla T(x,s)|^2 dx ds \leq \kappa^{-1}\|\phi\|^2 E(t)$$

with

(4.14)
$$\|\phi\|^2 E(t) = \int_D (g_\psi(x,0) - g_\psi(x,t))dx + \int_0^t \mathcal{R}(s)ds.$$

It follows that

(4.15)
$$t^{-1}\int_0^t \int_D \psi(x)\phi(T(x,s))|\nabla T(x,s)|dx ds \leq \|\phi\|\left(\kappa^{-1}t^{-1}E(t)\int_D \psi(x)dx\right)^{\frac{1}{2}}$$

Using Theorem 3.1 it follows that the left hand side of (4.15) equals

(4.16)
$$\int_0^M \phi(\beta)\left(t^{-1}\int_0^t \mu_\psi(\beta,s)ds\right)d\beta$$

where

(4.17)
$$\mu_\psi(\beta,s) = \int_{\{x;T(x,s)=\beta\}} \psi(x)dH^{(2)}(x)$$

Let us denote the time average of μ_ψ by $\mu_{\psi,t}$:

(4.18)
$$\mu_{\psi,t}(\beta) = t^{-1}\int_0^t \mu_\psi(\beta,s)ds$$

Then, because ϕ is arbitrary in (4.15) it follows that

(4.19)
$$\int_0^M \mu_{\psi,t}^2(\beta)d\beta \leq \kappa^{-1}t^{-1}(\sup_{\|\phi\|\leq 1}E(t))\int_D \psi(x)dx$$

Now it is time to choose ψ. There are several types of choices, depending on the region in D we want to investigate. We will present two choices: the first one is suited for the situation in which the average value of $|u(x_0, t)|$ is measured at a fixed x_0 in D while the second one regards behavior in the whole of D minus a thin layer around the boundary.

Let us take a nonnegative smooth function ψ_0 which is identically equal to one in $\{y; |y| \le 1\}$ and vanishes identically for $|y| \ge 2$. Let $x_0 \in D$ be fixed and set

(4.20)
$$\psi(x) = \psi_0(\frac{x - x_0}{r})$$

where $r > 0$ is a small number. Clearly, we have

(4.21)
$$|\psi(x)| \le C_1$$

(4.22)
$$|\nabla\psi(x)| \le C_2 r^{-1}$$

and

(4.23)
$$|\Delta\psi(x)| \le C_3 r^{-2}$$

for some fixed C_1, C_2, C_3 and all x. It follows from (4.19) and the definition (4.14) of $E(t)$ that

(4.24)
$$\tau^{-1} \int_0^M \mu_{\psi,t}^2(\beta)d\beta \le C^2\kappa^{-1}\left(r^2 t^{-1} \int_0^t \int_{B(x_0,2r)} |u(x, s)| ds dt + \kappa r^4 + \sigma r^5\right)$$

where we denoted by $B(x_0, 2r)$ the ball of radius $2r$ centered at x_0, C is a constant, τ is the maximum oscillation of T

(4.25)
$$\tau = \sup_{x,y \in B(x_0,2r), 0 \le s \le t} |T(x, s) - T(y, s)|,$$

and σ is given by

(4.25A)
$$\sigma = rt^{-1}(2 + \tau^{-1}\sup_{0 \le s \le t}(T(x_0, 0) - T(x_0, s))_+)$$

Now, because $\psi = 1$ on the ball $B(x_0, r)$ it follows that

(4.26)
$$\tau^{-1} \int_0^M \mu_{r,t}^2(\beta)d\beta \le C^2\kappa^{-1}\left(r^5[|u|_{2r,t}(x_0) + \sigma] + \kappa r^4\right)$$

where

$$(4.27) \qquad \mu_{r,t}(\beta) = t^{-1} \int_0^t ds \int_{\{x;|x-x_0|\leq r, T(x,s)=\beta\}} dH^{(2)}(x)$$

and

$$(4.28) \qquad |u|_{2r,t}(x_0) = \frac{2}{\pi} t^{-1} r^{-3} \int_0^t \int_{B(x_0,2r)} |u(x,s)| \, dx \, ds$$

Let us assume, for the sake of clarity, that $u(\cdot, s)$ is continuous at x_0 for all s under consideration. This assumption is not needed: if $u(\cdot, s)$ is locally in L^p then the Hardy-Littlewood maximal function computed at x_0 and averaged in time majorizes $|u|_{2r,t}(x_0)$. We define $< \mu >_{r,t}$ by

$$(4.29) \qquad < \mu >_{r,t} (x_0) = \left(\tau^{-1} \int_0^M \mu_{r,t}^2(\beta) d\beta \right)^{\frac{1}{2}}$$

We think of $< \mu >_{r,t} (x_0)$ as representing an average value of the area of the piece of a surface $T = $ constant contained in $B(x_0, r)$. We proved

Theorem 4.1. *Let T be a solution of (4.1) and x_0 a fixed point in D.*
Then the average area $< \mu >_{r,t} (x_0)$ of a piece of a surface $T = $ constant contained in a ball of radius r centered at x_0 is bounded by

$$(4.30) \qquad < \mu >_{r,t} (x_0) \leq C\kappa^{-\frac{1}{2}} r^{\frac{5}{2}} \left(|u|_{2r,t}(x_0) + \sigma \right)^{\frac{1}{2}} + C r^2$$

for small enough r.

Let us introduce the Hardy-Littlewood maximal function

$$(4.31) \qquad < |u|(x) > = \sup_{0 < 2r \leq \delta} |u|_{2r,t}(x).$$

With it we associate a local length scale

$$(4.32) \qquad \lambda = \kappa < |u| >^{-1}$$

Then (4.32) implies that, for $r \geq \lambda$,

$$(4.33) \qquad < \mu >_{r,t} (x_0) \leq C r^{2.5} \kappa^{-\frac{1}{2}} (< |u| > (x_0) + \sigma)^{\frac{1}{2}}.$$

Therefore, although both sides of (4.30) have the correct dimension of area, (4.30) would suggest that the local (ensemble) average level set of T has fractal dimension larger or equal to 2.5. It is perhaps significant to note that experimental evidence ([S]) seems to indicate that 2.35 is a lower bound for this dimension.

Now we will consider the problem of estimating the average area of a level set of T contained in D_δ, the set of points in D situated at a distance of at least δ from the boundary of D. Taking the characteristic function of $D_{\frac{\delta}{2}}$ and using a standard mollifier we can find a smooth, nonnegative ψ which is identically equal to one in D_δ, has compact support in D and satisfies

$$(4.34) \qquad\qquad \psi(x) \leq C_1$$

$$(4.35) \qquad\qquad |\nabla\psi(x)| \leq C_2\delta^{-1}$$

and

$$(4.36) \qquad\qquad |\Delta\psi(x)| \leq C_3\delta^{-2}$$

We will assume, for simplicity, that D has aspect ratio equal to one and diameter L.

Using the inequalities (4.34)-(4.36) in (4.19) we obtain

$$(4.37) \qquad\qquad M^{-1}\int_0^M \mu_{\psi,t}^2(\beta)d\beta$$

$$\leq C^2\kappa^{-1}L^3\left(\delta^{-1}t^{-1}\int_0^t\int_D |u(x,s)|dxds + \kappa\delta^{-2}L^3 + t^{-1}L^3\right)$$

We define $\mu_{\delta,t}(\beta)$ by

$$(4.38) \qquad \mu_{\delta,t}(\beta) = t^{-1}\int_0^t ds \int_{\{x\in D_\delta; T(x,s)=\beta\}} dH^{(2)}(x)$$

and $< |u| >$ by

$$(4.39) \qquad\qquad < |u| >= t^{-1}L^{-3}\int_0^t\int_D |u(x,s)|dxds$$

Then we deduce from (4.37)

$$(4.40) \qquad M^{-1} \int_0^M \mu_{\delta,t}^2(\beta)d\beta \leq C^2\kappa^{-1}L^6\left(\delta^{-1} < |u| > +\kappa\delta^{-2} + t^{-1}\right)$$

Now we set

$$(4.41) \qquad < \mu >_{\delta,t} = \left(M^{-1}\int_0^M \mu_{\delta,t}^2(\beta)d\beta\right)^{\frac{1}{2}}$$

and conclude from (4.40) that

$$(4.42) \qquad < \mu >_{\delta,t} \leq CL^3\left(\kappa^{-\frac{1}{2}}\delta^{-\frac{1}{2}} < |u| >^{\frac{1}{2}} +\delta^{-1} + \kappa^{-\frac{1}{2}}t^{-\frac{1}{2}}\right)$$

We proved thus

Theorem 4.2. *The average area* $< \mu >_{\delta,t}$ *of a piece of a surface* $T =$ *constant contained in* D_δ *satisfies (4.42)*

As an application let us consider the Boussinesq equations for Rayleigh-Benard convection. The equation (4.1) for T is supplemented with

$$(4.43) \qquad \left(\frac{\partial}{\partial t} + u \cdot \nabla - \nu\Delta\right)u + \nabla p = g\alpha Te_3$$

Also (4.2)-(4.5) hold. In (4.43) g is the acceleration of gravity, α the volume thermal expansion coefficient e_3 is the vertical direction and ν is the kinematic viscosity. The Rayleigh number is

$$(4.44) \qquad Ra = \frac{\alpha g M L^3}{\kappa\nu}$$

We want to give an upper bound for $< |u| >$ in terms of the Rayleigh number. We multiply scalarly (4.43) by u and integrate in D. Using the Poincare inequality we obtain

$$(4.45) \qquad \frac{d}{dt}\|u\| + \nu L^{-2}\|u\| \leq g\alpha M L^{\frac{3}{2}}$$

where

$$(4.46) \qquad \|u\|^2 = \int_D |u(x,t)|^2 dx$$

Consequently

$$(4.47) \qquad < |u| > \leq \kappa L^{-1}Ra + t^{-1}\nu^{-1}L^{\frac{1}{2}}\|u_0\|$$

Direct application of Theorem 4.2 yields

Theorem 4.3. *The average area* $< \mu >_{\delta,t}$ *of a piece of surface* $T =$ *constant contained in* D_δ *satisfies*

(4.48) $$< \mu >_{\delta,t} \leq 2C \left(L^3 \delta^{-1} + L^{\frac{5}{2}} \delta^{-\frac{1}{2}} Ra^{\frac{1}{2}} \right)$$

for t *sufficiently large.*

ACKNOWLEDGMENTS

Partially supported by NSF grant DMS-860-2031.

REFERENCES

[B]   G. K. Batchelor, The theory of homogeneous turbulence, Cambridge University Press, Cambridge.

[C]   P. Constantin, Navier-Stokes equations and area of interfaces, to appear in Commun. in Math. Phys.

[Chi] B.Castaing, G. Gunaratne, F. Heslot, L. Kadanoff, A. Libchaber, S. Thomae, X-Z. Wu, S. Zaleski, G. Zanetti, Scaling of hard thermal turbulence in Rayleigh-Benard convection, preprint.

[C-F] P. Constantin, C.Foias, Navier-Stokes Equations, The University of Chicago Press, Chicago.

[C-F-T] P. Constantin, C. Foias, R. Temam, Attractors representing turbulent flows, Memoirs of the AMS, 53(1985), number 314.

[C-K-N] L. Caffarelli, R. Kohn, L. Nirenberg, Partial regularity of suitable weak solutions of the Navier-Stokes equations, Comm. Pure Appl. Math., 35(1982), 771-831.

[C-S] G. H. Cottet, J. Soler, Three dimensional Navier-Stokes equations for singular filament initial data, J. Diff. Eqns., 74(1988), 234-253.

[D]   G. F. D. Duff, Derivative estimates for the Navier-Stokes equations in a three dimensional region, preprint (1988).

[F]   H. Federer, Geometric Measure Theory, Springer, Berlin, Heidelberg, New York.

[F-G-T] C. Foias, C. Guillope, R. Temam, New a priori estimates for Navier-Stokes equations in dimension 3, Comm. in P.D.E., 6(1981), 329-359.

[G-M] Y. Giga, T. Miyakawa, Navier-Stokes flow in $\mathbb{R}^3$ with measures as initial vorticity and Morrey spaces, preprint (1988).

[K]   A. S. Kronrod, On functions of two variables, Usp. Math. Nauk., 5(1950), 24-134 (Russian).

[L]   J. Leray, Sur le mouvement d'un liquide visqueux emplissant l'espace, Acta Math., 63(1934), 193-248.

[Li] J. L. Lions, Sur la regularite et l'unicite des solutions turbulentes des equations de Navier-Stokes, Rend. Sem. Mat. Padova, 30(1960), 16-23.

[M] V. G. Maz'ja, Sobolev Spaces, Springer, Berlin, Heidelberg, New York.

[S] K. R. Sreenivasan, The physics of fully turbulent flows: Some recent contributions motivated by advances in dynamical systems, preprint (1989).

[Sc] V. Scheffer, A solution to the Navier-Stokes inequality with an internal singularity, Comm.in Math. Phys., 101(1985), 47-85.

[T] R. Temam, Navier-Stokes equations: theory and numerical analysis, North Holland, Amsterdam, New York.

9

Scaling and Structures in the Hard Turbulence Region of Rayleigh Bénard Convection

Leo P. Kadanoff

Abstract

Experimental and theoretical studies of Rayleigh-Bénard convection at high Rayleigh number ($10^8 < Ra < 10^{13}$) were performed by Bernard Castaing, Gemunu Gunaratne, François Heslot, Leo Kadanoff, Albert Libchaber, Stefan Thomae, Xiao-Zhong Wu, Stéphane Zaleski, and Gianluigi Zanetti (J. Fluid Mech. (1989)). The results of these studies are further examined in the light of visualization in Rayleigh-Bénard flow in water (Steve Gross, Giovanni Zocchi, and Albert Libchaber, C.R. Acad. Sci. Paris t. 307, Série 2, 447 (1988)). The previously developed theory is shown to be incomplete in leaving out many of the structures in the flow. We take special note of the coherent flow throughout the entire water tank. Despite the many omissions of geometrical structures from the scaling analysis, most of the order of magnitude estimates seem right.

A recent paper[1], describing convection flow in Helium gas at low temperatures, about 5 degrees Kelvin, included an overall theoretical description of the flow. We considered that the cell could be divided into regions (see the cartoon of Figure 1) which have different characteristic scaling behaviors. As in the classical descriptions of the system[2], the regions include a boundary layer, where the viscosity and thermal conductivity dominate and a central region, in which convection effects and buoyancy forces are important. In addition there is a mixing zone in which we imagined structures composed of hot fluid (see the cartoon of Figure 2) coming up and merging into the central region flow.

The scaling behavior can be specified in terms of the Rayleigh number

$$Ra = \frac{g\alpha\Delta L^3}{\kappa\nu} \tag{1}$$

Here, g is the acceleration of gravity, α is the volume thermal expansion coefficient, Δ is the temperature difference between the bottom and the top of the cell, κ and ν are respective the thermal diffusivity and the kinematic

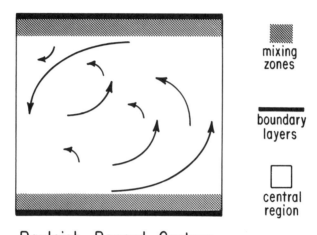

Rayleigh Benard Cartoon

FIGURE 1. A cartoon of the cell.

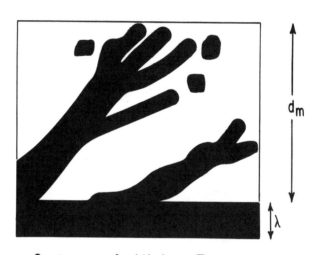

Cartoon of Mixing Zone

FIGURE 2. Cartoon of mixing zone.

viscosity, and L is a characteristic size of the cell. In the theory described in reference (1), the thickness of the boundary layer and of the mixing zone are respectively given by

$$\lambda = L\, Ra^{-2/7} \qquad\qquad (2a)$$

$$d_m = L\, Ra^{-1/7} \qquad\qquad (2b)$$

In addition, the Nusselt number is

$$Nu = Ra^{2/7} \qquad\qquad (2c)$$

a typical fluctuation in the central region is given by

$$\Delta_c = \Delta\, Ra^{-1/7} \qquad\qquad (2d)$$

while the velocity in the central region has a characteristic size

$$u_c = \frac{\kappa}{L} Ra^{3/7} \qquad\qquad (2e)$$

The Helium experiment gave direct support for relations (2c) and (2d). The observed fact that temperature difference across the boundary layer were of order $\Delta/2$ and expression (2c) implies (2a). Other experiments[3] give support for (2e). However, there was at the time that paper was written (January, 1988) no evidence for the existence of the mixing zone. In addition, the helium experiment measured a characteristic frequency of oscillations in the cell, given by

$$\omega_p = \frac{\kappa}{L^2} Ra^{1/2} \qquad\qquad (2f)$$

The theory's prediction of this frequency seems at bit *ad hoc*.

Robert Kraichnan has pointed out that 'the wonderful thing about scaling is that you can get everything right without understanding anything'[4]. I think he means that one can estimate the orders of magnitudes involved without really having a good picture of the geometrical structures. So next, we should look at the geometry of the flow.

It is very hard to see anything in the helium system. However, after the helium paper of reference 1 was written, the experimental group constructed an experiment in which they could see the temperature variations in a water cell at a similar Rayleigh number[5]. It is interesting to compare a picture they took (see Figure 3) with the result of the theory. The picture was obtained by shining light through the cell and observing the shadows cast by the variations in the water's index of refraction as in Figure 1B of reference 5.

The picture clearly shows the three different regions, that had been anticipated by the theory. The dark shadows at the very bottom and top are caused by the gross variations of temperature in the boundary layers. In the center, there is a region of relatively slow variation in temperature

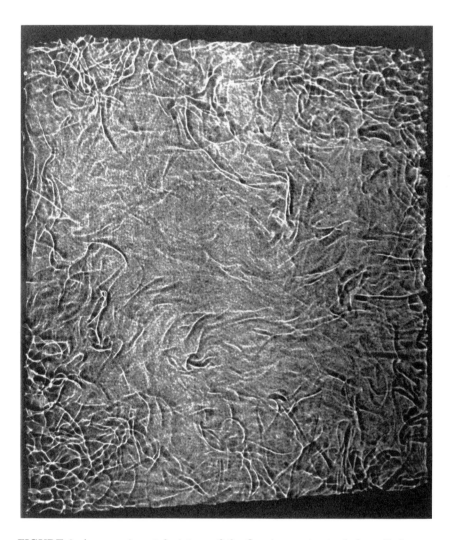

FIGURE 3. An experimental picture of the flow in a water tank from Reference 5.

corresponding to the central region of Figure 1. Furthermore, the mixing zone hypothesized by the theory is seen in the picture as a relatively active region near the top and bottom walls. So far so good.

However, notice also the heightened activity near the right and left-hand walls in comparison with the central region. By looking at the flow, one sees that there is a rapid and coherent directed upward motion of fluid on the right hand wall and a downward motion on the left. These motions are concentrated in the jet-like regions which are seen in Figure 3 as the areas of heightened activity. This coherent flow is closed by left to right motion in the bottom half of the cell and corresponding right to left motion of the top. We call those motions 'wind'. These portions of the vortical flow are apparently rather more spread out then the jets near the walls. These jets were certainly left out of the original scaling picture. But, they were seen in a follow-up study of the Helium system[6] in which average velocities of the fluid were measured. The authors of reference 6 pointed out that these jets provided a quantitative explanation of the frequency ω_p which they interpreted as simply the result of convecting hot blobs round and round the cell. A consequence of this picture and of the observed value of ω_p is that in addition to the velocity scale (2b) the overall coherent motion of the system would have to have the velocity $\omega_p L$, which would then scale roughly as $Ra^{1/2}$. This scaling behavior was not mentioned in reference 1 and in fact it is different from the $Ra^{3/7}$ behavior of Eq. (2e). I find it disconcerting to have two nearby scaling indices for two closely related quantities[7].

Next notice the boundary layers look bumpy. Reference 5 contains some pictures which show a wave-like disturbance being advected along the bottom of the container in the direction of the wind. Apparently, these waves are propagating deviations in the thickness of the boundary layer. Moreover, one sees in the mixing zone some concentrated disturbances which perhaps have a character like that in the cartoon of Figure 2. A more characteristic shape is that of a mushroom which is apparently rooted in the boundary layer and then is advected by the flow. Neither wave nor mushroom was a portion of the original scaling theory. The picture indicates that some thermal have come loose from their moorings and are moving through the central region, but these free plumes are few and far between. Many features of this picture have been seen before. Van Dyke[8] shows pictures of mushroom-like thermals while Chu and Goldstein[9] say "As the Rayleigh number is increased, the thermals show a persistent horizontal movement near the bounding surface and the majority of the thermals are dissipated without reaching the opposite surface; the central region of the layer becomes an isothermal core".

I have drawn in the waves and thermals in the cartoon of Figure 4. This particular drawing shows the thermals as rising from the crest of the wave. It is not implausible that there is a secondary three-dimensional instability that breaks up the two-dimensional wave-structure. However it

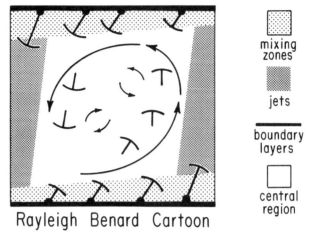

Rayleigh Benard Cartoon

mixing zones

jets

boundary layers

central region

FIGURE 4. Cartoon of cell revised after observation of the water tank.

is also plausible that thermals might arise independently of the wave[10]. Because the pictures are two-dimensional projections, it is not yet known just how the waves and plumes might be associated.

All in all, the hard-turbulence cell appears to be a very complicated object with many working parts. Nonetheless a simple scaling description close to that given in Equs. (2) seems to hold.

Acknowledgements

This description of experimental observations in this report is a report upon knowledge I have gained from many hours of talking with Albert Libchaber and with his students and colleagues. The research reported here was supported by the University of Chicago MRL.

References

1. Bernard Castaing, Gemunu Gunaratne, François Heslot, Leo Kadanoff, Albert Libchaber, Stefan Thomae, Xiao-Zhong Wu, Stéphane Zaleski, and Gianluigi Zanetti (J. Fluid Mech. (1989)).

2. Malkus W. V.R.:Discrete transitions in turbulent convection, Proc. Roy. Soc. **A 225** 185-195 (1954), The heat transport and spectrum of thermal turbulence Proc. Roy. Soc. **A 225** 196-212 (1954).; Spiegel, E.A.: On the Malkus theory of turbulence, Mécanique de la turbulence (Colloque International du CNRS à Marseille), Paris,'Ed CNRS 181-201, (1962); Kraichnan R.H.: Turbulent thermal convection at arbitrary Prandtl number, Phys. Fluids. **5**, 1374, (1962).

3. H. Tanaka and H. Miyata, Int. J. Heat and Mass Transfer **23**, 1273 (1980); A.M. Garron and R.J. Goldstein, Int. J. Heat and Mass Transfer **23**, 738 (1980).

4. private communication.

5. Steve Gross, Giovanni Zocchi, and Albert Libchaber, C.R. Acad. Sci. Paris t. 307, Série 2, 447 (1988).

6. Masaki Sano, Xiao-Zhong Wu, and Albert Libchaber 'Study of Turbulence in Helium Gas Convection'.

7. Sano, Wu, and Libchaber (reference 6) point out that there is a loop-hole in that the unknown Prandtl number dependence of the various measured quantities might perhaps produce changes in the observed critical index values.

8. M. Van Dyke, *An Album of Fluid Motion*, The Parabolic Press, Stanford, California, (1982). p. 63.

9. T.Y. Chu and R.J. Goldstein: Turbulent convection in a horizontal layer of water, J. Fluid Mech., **60**, 141-159 (1973).

10. Andrew J. Majda, private communication.

10

Probabilistic Multifractals and Negative Dimensions

Ashvin B. Chhabra and K.R. Sreenivasan

Abstract

We propose that negative dimensions can be best understood using the concept of level-independent multiplier distributions and show that, by utilising them, one can extract the positive and negative parts of the $f(\alpha)$ function with exponentially less work than by using conventional boxcounting methods. When the underlying multiplicative structure is not known, both methods of computing negative dimensions can give spurious results at finite resolution. Applications to fully developed turbulence are discussed briefly.

1 Introduction

Fractal and multifractal concepts [1, 2, 3, 4, 5] are now widely used in the study of nonlinear systems. The specific idea of decomposing a singular measure into interwoven sets of singularity strengths α, with dimension $f(\alpha)$ [6, 7], has turned out to be a useful and compact way of describing the scaling properties of such a measure.

However, the $f(\alpha)$ analysis of measures constructed from experiments (e.g. the dissipation field of turbulence) yields two surprising results. First, the analysis of long data sets yields negative numbers for the $f(\alpha)$ of sets of extremal iso-α values [8, 9]. Second, although the data show unambiguous scaling (often over several decades), the actual exponent or dimension fluctuates from sample to sample by an amount greater than the (least square) error bars on any one sample [10, 11]. This seems to contradict the implicit assumption that unambiguous scaling indicates a well-defined and convergent $f(\alpha)$ function.

The purpose of this paper is two-fold: First, we wish to provide a simple way of understanding the above observations using the concept of level-independent multiplier distributions. This has been discussed briefly in [12, 8]. Second, we show that the positive and negative parts of the $f(\alpha)$ function can be computed by suitably manipulating these multiplier distributions. We will demonstrate that, for probabilistic multifractals (those generated by random multiplicative processes), this method (hereafter referred to as the multiplier method) is significantly more accurate and requires exponentially less work compared to conventional boxcounting methods. Finally, for energy dissipation in the atmospheric surface layer, the $f(\alpha)$ deduced from the multiplier distribution is compared with previous measurements [8] by conventional boxcounting methods.

In the traditional multifractal formalism, one starts with a singular measure that may, for example, be the amount of dissipation in various regions of the dissipation field of fully developed turbulence [10, 13, 14], or the density of points in regions of phase space [7]. Cover the entire region with boxes of size ϵ and denote the integrated probability in the ith box of size ϵ as $P_i(\epsilon)$. Then a singularity strength α_i can be defined by

$$P_i(\epsilon) \sim \epsilon^{\alpha_i}. \tag{1}$$

Denote $N(\alpha)$ as the number of boxes where P_i has singularity strength between α and $\alpha + d\alpha$. Then $f(\alpha)$ can be loosely defined [7] as the fractal dimension of the set of boxes with singularity strength α [4] and written as

$$N(\alpha) \sim \epsilon^{-f(\alpha)}. \tag{2}$$

This formalism leads [6, 7] to the description of a multifractal measure in terms of interwoven sets of Hausdorff dimension $f(\alpha)$ and singularity strength α. In practice, to compute $f(\alpha)$ directly, one must take into account the ϵ-dependent prefactors in Eq. 2. The canonical way of doing so is described in Ref. [15].

An equivalent way of computing $f[\alpha(q)]$ is to use the partition function and evaluate the average value of $< \tau(q) >$ via

$$Z(q) = \sum_i P_i^q(\epsilon) = \epsilon^{<\tau(q)>}, \tag{3}$$

and then use the relation

$$f(q) = q < \alpha(q) > - < \tau(q) > . \tag{4}$$

The refinement process of a fractal measure can be mapped onto a cascading process where each interval contracts into smaller pieces and thus divides the total measure among them according to some process. The scaling properties of measures created by simple models of deterministic cascading processes have been shown to provide good approximations to the scaling properties of the *low order moments* of various intermittent fields in turbulence [13]. It is believed that for cases like turbulence, a more realistic model for such cascading process would be probabilistic, where multipliers are picked randomly from some probability distribution reflecting the underlying physics. We will elaborate on this point to show that such models can incorporate sample to sample fluctuations as well as negative dimensions, both of which have been observed in turbulence.

2 Negative dimensions in random multiplicative processes

Consider the scaling properties of measures arising from deterministic systems e.g. period doubling. Every observed α value will by definition occur at least once and thus $f(\alpha)$, which is the logarithm of the number of times an α value occurs, will always be a non-negative function. In addition, to within some small numerical error, the $f(\alpha)$ function should be identical for two different realisations (of the measure) of the same size.

On the other hand, consider measures created by finite realisations (samples) of a random multiplicative process (hereafter referred to as RMP [4]). The $f(\alpha)$ curve will fluctuate from sample to sample depending on the particular collection of multipliers picked from the probability distribution. One now has two distinct possibilities regarding the averaging procedure. First, one can average the exponents $f(\alpha)$ and α from each sample (quenched averaging). Since in any single sample, every *observed* value of α will occur at least once and hence have a positive dimension, the resultant $f(\alpha)$ curve will always be non-negative. The second procedure is to average the partition function (annealed averaging). One adopts the view that $f(\alpha)$ is the logarithm of a histogram, and defines the dimension [16, 17] by

$$f(\alpha) \sim -\frac{\log < N(\alpha) >}{\log(\epsilon)} \qquad (5)$$

where the averages are arithmetic (not Boltzmann), taken over various samples. This procedure has been called supersampling [18]. Now consider values of α that occur rarely e.g. less often than one per typical sample. The dimension assigned to these values by Eq. 5 would be negative. One physical example would be randomly oriented one-dimensional cuts through three-dimensional turbulent dissipation fields [10, 13, 14]. Events occurring in three-dimensional space with low probability would be missed in any given one-dimensional cut. However, they can be recovered by sampling over many such cuts and using Eq. 5. This enables one to gather information about higher-dimensional spaces from low-dimensional measurements. Mandelbrot [18] thus stressed that the analysis of single samples suppresses valuable information about these rare events.

On the other hand Cates and Witten [17] recognised that, in principle, the negative $f(\alpha)$ should be computable from a single sample. To this end they suggested breaking up the measure into smaller pieces, normalising each such sample and then supersampling or averaging the new (sub)partition functions (with the correspondingly reduced scaling range). This reasoning is motivated by the presumed self-similarity of the partition function. Thus, both ways of observing negative dimensions involve supersampling of either different realisations or parts of the same realisation. One would, however, like to understand the occurrence of negative dimensions from a microscopic point of view i.e., from the underlying multiplicative or refinement process, just as we do for the positive dimensions [7].

The method to be described is based on the view that scaling properties reflecting self-similarity in the measure can be described by a distribution of multipliers that are level-independent [4, 7, 18]. These multipliers define how the measure in a given piece will rearrange into smaller pieces. For stochastic or random multiplicative processes (including randomly oriented lower-dimensional cuts of deterministic processes) all the scaling properties of the measure, including $f(\alpha)$ and sample to sample fluctuations can be understood in terms of the properties of the probability distribution of level-independent multipliers. The multiplier distribution in stochastic systems is the natural analog of the scaling function [19] in deterministic systems. For the latter, the scaling function contains information about the level to level contraction ratios (multipliers) and, in addition, organises them correctly in time. For RMP, there is no natural ordering in time, but the multipliers are characterised by their value as well as by the probability with which they occur.

To understand the relationship between the probability distribution of the multipliers $P(M)$ and $f(\alpha)$, consider a binary RMP where at every level of refinement an interval breaks up into two equal pieces, but the measure is distributed in the ratio M and $1 - M$ where M is either 0.7 or $1 - 0.7 = 0.3$ [13]. If we assign the larger ratio to any one piece randomly, then we have two rules $[0.7, 0.3]$ and $[0.3, 0.7]$, which are applied with equal probability. If the process proceeds to n levels, the redistributed measure will consist of 2^n pieces and one can compute its $f(\alpha)$ curve. Clearly the left extreme of the $f(\alpha)$ curve ($\alpha = \frac{\log(0.7)}{\log(0.5)} = 0.514...$) will be given by the box containing the string of multipliers $0.7, 0.7, 0.7....n$ times. Since, in any sample, such a string will occur exactly once (the probability of such a piece is 2^{-n} and the number of pieces is 2^n), the dimension of this iso-α set will be $log(1) = 0$. Similarly the iso-α set corresponding to the string of multipliers $0.3, 0.3, 0.3...$ will also occur exactly once and have a dimension of zero. All the other strings will occur more often and thus the entire $f(\alpha)$ curve will be positive. Consider a simple generalisation [20] of the binomial measure, where now we have four sets of multipliers $[0.7, 0.3]$, $[0.3, 0.7]$, $[0.8, 0.2]$, and $[0.2, 0.8]$, which are chosen with equal probability. Then the smallest singularity strength ($\alpha = \frac{\log(0.8)}{\log(0.5)} = 0.321...$) will correspond to the box containing the multipliers $0.8, 0.8, 0.8....n$ times. However, such a box will occur with probability $(\frac{1}{4})^n$. Since there are 2^n boxes per sample, one expects such a singularity strength to be observed only once every $(2^n 4^{-n})^{-1} = 2^n$ samples. Using Eq. 5, one finds that the dimension of the singular set ($\alpha = 0.321...$) is $\frac{\log(2^{-n})}{\log(2^n)} = -1$. The $f(\alpha)$ function corresponding to this probability distribution now ranges from -1 to 1, with the number $f(\alpha)$ quantifying the relative frequency of observing a singularity strength α in a given number of samples of finite size [18].

Consider now a more general process [21, 4] where the multipliers are randomly picked from a given distribution $P(M)$. To derive a general relation between the $f(\alpha)$ function defined by Eq. 5 and the distribution of multipliers, average the partition function over K samples (supersampling) of equal size so that

$$
< \tau(q) >= \frac{\log < Z(q) >}{\log(\epsilon)} = \frac{\log[(\frac{1}{K}) \sum_{j=1}^{K} \sum_{i=1}^{a^n} P_{ij}^q(\epsilon)]}{\log(\epsilon)} \tag{6}
$$

where $\epsilon = a^{-n}$ (2^{-n} for binary cascades) and P_{ij} is the measure in the ith box of the jth sample. But, due to the self-similarity of the measure, this relation should hold at any n so *we put $n = 1$*. We can do this as we are dealing with a model with no level to level correlations. Thus we have a collection of K sets of boxes, where the measure in any one box is simply a multiplier picked randomly from $P(M)$ (subject to the constraint of conservation of the measure). Denoting the multiplier by M, and remembering that for $n = 1$ the measure in the ith box is $P_{i1} = M_i$, we can write

$$
< \tau(q) >= -D_0 - \frac{\log(< M^q >)}{\log(a)}, \tag{7}
$$

and correspondingly

$$
< \alpha(q) >=< \frac{\partial \tau(q)}{\partial q} >= -\frac{< M^q \log(M) >}{< M^q > \log(a)} \tag{8}
$$

with $f(q)$ given by Eq. 4. Note that the averages are over the distribution of the multipliers $P(M)$.

The explanation for the occurrence of negative dimensions is a little more complicated here than for the binomial measure. Now we have multipliers picked randomly from a distribution and multiplied together to create an effective value of α. In the binomial measure the strings containing an equal number of 0.7 and 0.3's would behave like a string of average multiplier value of $\sqrt{0.7 * 0.3}$. The logarithm of the number of ways that such strings can occur is the dimension of that iso-α set. Such an α value and the dimension of the corresponding iso-α set can also be calculated by using the parameter q. This is done by evaluating the partition function according to Eq. 3 at some fixed q to compute $\tau(q)$, taking the derivative of the partition function with respect to q to compute $\alpha(q)$ and, finally, by computing $f(q)$ using Eq. 4. Similarly, the various random values of multipliers from $P(M)$ also multiply to produce different α values. The number of different ways they can do so depends on the multiplier distribution, and the logarithm of this number suitably normalised is the dimension of that iso-α set. Once again $\alpha(q)$ and $f(q)$ can be evaluated by using Eqs. 7, 8 and 4.

These equations provide a recipe for relating the distribution of multipliers with the scaling properties of the measure. The problem of computing the positive and negative parts of the $f(\alpha)$ function is thus reduced to the problem of computing $P(M)$. We will now demonstrate that the $f(\alpha)$ curve computed using the multiplier method converges exponentially faster to the asymptotic $f(\alpha)$ curve than that computed from conventional boxcounting methods using Eqs. 3, 4 and 5.

3 Advantages of the multiplier method over conventional boxcounting

For convenience we will elucidate our arguments with a binary RMP. The simplest way of computing $P(M)$ for such a process is to cover a measure at the nth stage of refinement, with boxes of size $2^{-(n-1)}$, compute P_i $(i = 1, 2, 3, ...2^{(n-1)})$, then subdivide each of these boxes in two pieces and compute the ratios of the measures in the original box to any one of the two subdivided boxes. Each subdivided box will give a value for M, and using the entire measure one can compute $P(M)$. Clearly the computation of $P(M)$ is helped by considering samples at more levels of refinement (increasing n), for at the nth level, we have 2^n realisations of M with which to construct $P(M)$. One gets better statistics by averaging $P(M)$ over different levels in addition to averaging over different samples (i.e. supersampling).

Now, in the conventional boxcounting method described by Eq. 6, given a single sample at the nth level (consisting of 2^n pieces for a binary process), one would see only those α values that had a probability greater than or equal to 2^{-n}. However, this α value comes from a string of n multipliers, each of which must (on the average) be picked with a probability of at least $1/2$. (In general, for a process that subdivides a piece into a smaller pieces at each level of refinement, one would only see α values corresponding to string of n multipliers, each of which (on the average) is picked with a probability of $1/a$.) To see α values consisting of strings of multipliers of lower probability, the currently used procedure is to supersample [18, 8]. In this procedure, to observe α values which occur with a probability of, say, 4^{-n} one would need $\frac{4^n}{2^n} = 2^n$ samples; that is, as one refines the measure more and more, i.e increases n, one needs an *exponentially increasing* number of samples to see the same α value in that ensemble of samples. Thus one understands the statement [18] that any increase in the level of the cascade must be accompanied by an exponential increase in the number of samples to achieve the same supersampling effect. A mathematically transparent way of understanding this statement is to notice from Eq. 2 that the number of occurrences of an iso-α set with a negative dimension will decrease as $\epsilon \to 0$. Setting $\epsilon = a^{-n}$, one notices that the number of such α values will decrease exponentially with the level of refinement of the measure. This statement, although correct, is paradoxical. If one is interested in describing a measure, it stands to reason that refining it should lead to better information about its scaling properties. However, following the supersampling procedure one does increasingly worse as the level of refinement increases. The attempt to resolve this apparent paradox is what has led us to emphasize the multiplier method.

The multiplier method described by Eqs. 7, 8 and 4 assumes that the scaling properties of the measure arise from the repeated composition of multipliers from the same distribution. This means that, for each value of α, there exists a value of the multiplier M^* which, when composed n times, would produce the same α value. That is,

$$< \alpha > = \lim_{n \to \infty} \sum_{j=1}^{n} \frac{\log(M_j)}{\log(2^{-n})} = -\frac{\log(M^*)}{\log(2)}. \tag{9}$$

The probability of choosing the multiplier M^* is related to the dimension of the iso-α set by

$$< f(M^*) > = - \lim_{\epsilon \to 0} \frac{\log < N_\epsilon(M^*) >}{\log(\epsilon)} = 1 + \frac{\log(P(M^*))}{\log(2)}, \tag{10}$$

where $N_\epsilon(M^*)$ is the number of times a string with the average multiplier M^* occurs at resolution ϵ. Note that $P(M^*)$ is a scale-invariant multiplier distribution that is derived from $P(M)$ but different from it.

In order to compute α and $f(\alpha)$ from the multiplier distribution, one makes use of the parameter q (see Eqs. 7 and 8). As q in Eq. 7 moves from $-\infty$ to ∞, different multipliers ranging from M_{min} to M_{max} get accentuated, thus reproducing the entire $f(\alpha)$ curve. From Eq. 10 one can see that $< f(M^*) >$ is negative for a binary RMP if $P(M^*) < 1/2$. Now if one increases the number of levels n (i.e. sample size), then one can better approximate $P(M)$. In particular, with this method we will be able to detect any multiplier with a probability of more than 2^{-n}. Thus as we increase the number of levels in the cascade, the multiplier method gets better by computing $P(M)$ to a precision of 2^{-n}, in contrast to conventional boxcounting which needs exponentially larger number samples to maintain its precision. Even at a fixed level n, a single sample of boxcounting will see only α values corresponding to $P(M^*) > 1/2$, while the method of multipliers will pick up α values corresponding to $P(M) > (1/2)^n$. Thus, we expect the multiplier method to be correspondingly more accurate for computations of positive dimensions as well. It is because of these two improvements that the multiplier method requires exponentially less work and is commensurately more accurate.

To fix these ideas, consider a binary RMP, where the multipliers are chosen from a uniform distribution [18] (hereafter referred to as a uniform RMP). One can analytically compute its α and $f(\alpha)$ as

$$\alpha(q) = \frac{1}{(q+1)\log(2)} \tag{11}$$

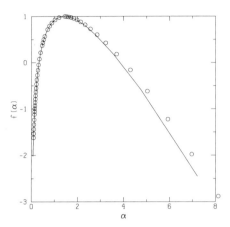

FIGURE 1. Shows that one can extract both the positive and negative parts of the $f(\alpha)$ curve from a single sample. The sample is generated from a binary cascade (of 15 levels) where multipliers were chosen randomly from a uniform distribution (uniform RMP). The solid line is the exact $f(\alpha)$ curve. The circles are from the multiplier method with $-1 < q < 15$.

and

$$f(q) = 1 + \frac{q}{(q+1)\log(2)} - \frac{\log(q+1)}{\log(2)} \tag{12}$$

i.e, the $f(\alpha)$ curve goes all the way down to $-\infty$ with α ranging from 0 to ∞ [22]. We apply the multiplier method on a single sample (of the uniform RMP) at $n = 15$. Fig. 1 shows that it is indeed possible to compute both the positive and the negative parts of the $f(\alpha)$ curve from a single sample. Fig. 2 shows the convergence of the exponent $f(q = 5)$ to its exact value as one refines the box size. Since the box size can only shrink to a value where each box would contain just one point, to improve accuracy one should increase the number of levels in the cascade. Also, to decrease local fluctuations, one can calculate the exponent by averaging multiplier distributions from several different box sizes.

The most stringent way of comparing supersampling using boxcounting and the multiplier method is to fix both the level of the cascade in each sample (n) and the number of samples in the ensemble. Let $n = 12$ and consider 32 samples i.e. $32,768$ points in all. Fig. 3 compares these two methods with the known theoretical $f(\alpha)$ function for the uniform RMP. Clearly, the multiplier method is capable of yielding far smaller dimensions with greater accuracy. If one now increased the level of refinement of the measure to, say, $n = 15$ but kept the number of samples (32) the same, for reasons mentioned earlier, we would expect the results of conventional box-counting to get worse. This is demonstrated in Fig. 4, where the two arrows demarcate the minimum dimensions found by the boxcounting method (us-

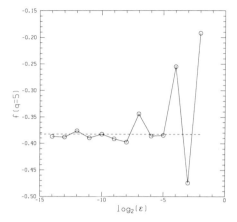

FIGURE 2. The quantity $f(q = 5)$ using the method of multipliers and varying the box size for the same multiplicative process (uniform RMP) as in Fig. 1. The dashed line is the exact value.

ing $-1 < q < 15$). (The multiplier method does not suffer from this defect, and improves as the number of levels is increased: as more statistics are gathered, exponents from higher and higher q values become more reliable as demonstrated in Fig. 2.)

Let us discuss the case where one does not understand how to partition the measure correctly and covers the measure arbitrarily with boxes of uniform size. Here one has a trade-off between the number of errors caused by improper boxing (which increases as the number of boxes) and the magnitude of the finite-size errors caused (which decreases as the box size becomes smaller). In the multiplier method with uniform partitioning, the obvious course is to go to more levels of refinement. In doing so the errors due to improper boxing decrease. One also has the advantage that the increased number of boxes results in correspondingly improved statistics for computing multiplier averages. On the other hand, since the number of boxes increases so do the number of errors. Thus there is a tradeoff and one may have to go to very fine resolution (a large number of levels) to produce accurate results. This may not always be possible, especially if one is dealing with experimental data. Fig. 5a shows the results of using the multiplier method on the uniform RMP (for a single sample at $n = 15$) with multipliers computed by comparing measures from boxes one third their size. Since the underlying structure of the cascade process is binary, one expects convergence problems and spurious results. Comparing this with Fig. 1 one notes, however, that the fit is surprisingly good. There is some error in the negative q region (right-half of the curve corresponding to large α values), but the positive q region (left-half of the curve corresponding to small α values) seems, if anything, better at finding the negative dimensions. This fortuitous result comes from improper boxing, as demonstrated in Fig. 5b,

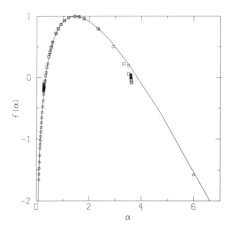

FIGURE 3. Comparison between boxcounting (squares) and the multiplier method (circles) at $n = 12$ for the uniform RMP described in Fig 1. Both methods have been supersampled (averaged) over 32 different realisations.

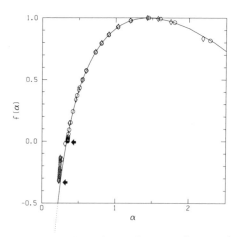

FIGURE 4. Illustrates the degradation boxcounting results with increasing refinement level. Diamonds represent data from cascades with $n = 10$, while the circles are for $n = 15$ (for the uniform RMP described in Fig. 1). The data have been averaged over 32 samples in both cases.

where we have performed the same improper boxing (with base 3) on the generalised binomial measure discussed earlier in the text. From theoretical arguments we know that the minimum dimension for this measure is -1. Note in Fig. 5b both the existence of spurious dimensions (which are less than -1) and the increased error for the negative q region. One reason for the increase in error is that the uniform RMP arises from a smooth distribution of multipliers so improper boxing computes a slightly shifted distribution with largely similar scaling properties. This is not true for the generalised binomial measure whose multiplier distribution consists of four delta functions. Improper boxing here replaces the true distribution with a poorly overlapping smooth one. The spurious negative dimensions arise from the scaling properties of the non-overlapping region which also contaminates the negative q portion (large α portion). (One however might expect that multiplier distributions arising in nature would have smooth multiplier distributions.) Thus the results of improper boxing vary depending on the distribution as well as the level under consideration. It follows that, while attempting to compute negative dimensions for experimental data with an unknown multiplicative structure, it is highly advisable to use different bases for the computation of the multiplier distribution and check the convergence of $f(\alpha)$ with the number of levels in each sample. In addition, one might also average data over several samples.

What about the effects of improper boxing in the supersampling method? Here one runs into an even more serious problem. As one increases the number of levels (to reduce errors due to improper boxing) one must supersample over an exponentially larger number of samples to observe the same negative dimensions. However, this has the consequence of increasing the number of errors exponentially (due to improper boxing), which in turn can easily be mistaken for real events. Let us note, generally, that the idea of negative dimensions is to describe rarely occurring events. Boxcounting with boxes of arbitrary size creates errors, which in turn can be mistaken for precisely the events we are seeking to quantify. The reason that these spurious values (errors) scale with box size (just like the real exponents), is that their number is proportional to that of the boxes, which in turn increases exponentially with increasing levels of refinement. Thus the spurious values also increase exponentially with some smaller exponent. The occurrence of such spurious dimensions for boxcounting is shown in Fig. 6a and Fig. 6b, where the 10,000 samples of the generalised binomial measure (at $n = 5$) have been boxcounted using several different box sizes. The canonical method for computing $f(\alpha)$ and α directly [15] has been employed, and the entropy and energy of the measure have been plotted with increasing box size. The slopes yield $f(q = 25)$ and $\alpha(q = 25)$ respectively. In both figures the lower solid line is the correct exponent (to within a percent) obtained from the least squares fit using box sizes of the form 2^{-n}. The upper solid line is the least square fit obtained from box sizes other that 2^{-n}. These data from incorrect box sizes also scale but with spurious

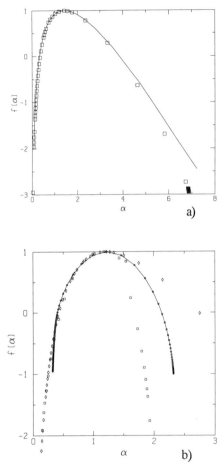

FIGURE 5. (a) Demonstrates the effect of using the wrong base for the multiplier method on a single sample of the uniform RMP at $n = 15$. The multipliers were computed by computing the ratio of the measure in boxes with those in boxes a third of the original size (base 3). The figure demonstrates that the computed $f(\alpha)$ is still reasonable. The solid line shows the analytical solution.(b) Demonstrates the effect of using the wrong base (base 3) for the multiplier method on a single sample of the generalised binomial measure at $n = 10$ (squares) and $n = 15$ (diamonds). Notice the spurious dimensions ($f < -1$) and the large error in the negative q region. The solid line is the theoretical curve.

exponents. Here improper boxing yields $f(q = 25) = -2.24$ which is clearly incorrect. The dashed line in the two figures is the least square fit to all the data (from proper and improper boxing) and is incorrect for both f and α.

4 Applications to turbulence

We will now apply the multiplier method for the determination of the negative dimensions [23] for the distribution of energy dissipation in the atmospheric surface layer several meters above the ground [10, 8].

To obtain negative dimensions in the atmosphere using the supersampling method, one needs an enormous amount of data. For example, to be able to observe a dimension of -2 in the atmospheric dissipation field one needs to distinguish multipliers that have a probability of occurrence $1/8$. Following Ref. [8] we estimate an integral length scale to consist of $\sim 10^4$ data points (sampled at about 6000 Hz). Thus assuming a binary cascade we can estimate from the appropriate Reynolds number that $n \sim 12$ and solve $(1/8)^{12} * (\text{ number of samples }) * (2^{12}) \sim 1$ which gives us an estimate of about ten million samples. Remembering that each sample is about 10^4 points we arrive at an estimate that one needs roughly 10^{11} data points which at the sampling rate of 6000 Hz. would require several years of data acquisition alone. So, we conclude that supersampling ill-suited for the purpose of measuring negative dimensions in atmospheric (or other high Reynolds number) flows.

On the other hand, the measurement of their scaling properties is rather important in order to be able to make statements about universality. In addition, the examination of the multiplier distribution itself may be quite useful in understanding the underlying fractal structure of turbulence.

Figure 7a shows the probability distribution of the multipliers for the energy dissipation at several levels of the cascade process in the inertial range. This was obtained by assuming that a random multiplicative binary cascade in one dimension models the scaling properties of the one dimensional signal. In addition it was assumed that box-averaging the energy dissipation duplicated the splitting process in reverse. To compute the multiplier distribution, the data (a component of the energy dissipation [10, 8]) were divided into bins of m points each. Each bin was then subdivided into two bins of $m/2$ points each and the ratio of the the dissipation contained in the smaller bin to that of the larger bin was computed. Since a conservative binary cascade was assumed, the multipliers from the two bins must add to unity and hence the resulting multiplier distribution is symmetrical. Extension to cascade of bases other than binary is trivial and has been done, but will not be disussed here. The multiplier distributions shown here were obtained from a record length of $409,600$ points of the atmospheric dissipation field obtained by hot-wire measurements. In spite of the scatter in the data, it appears that there is rough self-similarity in the cascade process.

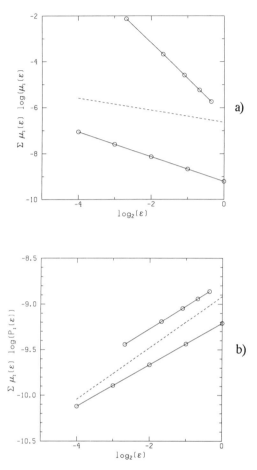

FIGURE 6. (a) Spurious scaling of negative dimensions that should not exist, arising from improper boxcounting of a generalised binomial measure at $n = 5$ averaged over 10^5 samples. Plotted is the entropy of the measure, whose slope with respect to different box sizes yields $f(q = 25)$. The lower solid line is the least square fit to data obtained from using boxes of 2^{-n} and is the correct value within a percent. The upper solid line is a least square fit obtained from box sizes other that 2^{-n} (boxes consisted of $5, 10, 15, 20, 25$ points each). The data from these box sizes also scale but with incorrect exponents. The dashed line is the least square fit to all the data (from proper and improper boxing) and also gives incorrect results. (b) The same as Fig. 6a but plots the internal energy of the measure to compute $\alpha(q = 25)$.

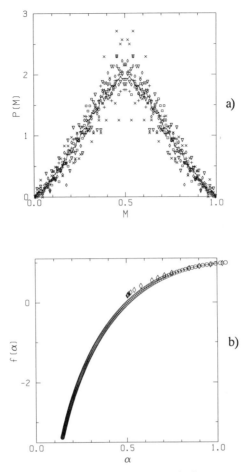

FIGURE 7. (a) The multiplier distribution $P(M)$ obtained by assuming a binary cascading process on the data (409,600 points) from the dissipation field of an atmospheric surface layer. Different symbols correspond to different levels of coarse-graining of the data. The symbol + represents multipliers determined by comparing the total dissipation in boxes of 50 points each with those containing 100 data points. The other symbols are as follows ($\Box$)100 : 200, ($\Diamond$)200 : 400, (∇)300 : 600, ($\times$)500 : 1000. The joined solid circles are the mean distribution. The symmetry about 0.5 is a consequence of assuming a conservative binary cascade. (b) Comparison between conventional boxcounting (combined with supersampling) (diamonds) and the multiplier method (joined circles) for the dissipation field of fully developed turbulence in the atmospheric surface layer using $409,600$ data points.

In addition Fig. 7a shows the mean multiplier distribution (continuous line). From this mean distribution, we obtain the $f(\alpha)$ curve as discussed in the earlier sections. For the present, we show in Fig. 7b only the singular part of the curve ($\alpha < 1$). Box-counting methods using the same length of data yields only positive dimensions (diamonds), whereas both the positive and negative dimensions (circles) are easily obtained using the multiplier method. Furthermore, the negative dimensions obtained for this atmospheric flow are in good agreement with the negative dimensions obtained by supersampling data from laboratory flows at lower Reynolds numbers [8]. (Note that supersampling becomes a feasible method at lower Reynolds numbers; even so, Meneveau & Sreenivasan [8] had to use record lengths containing ten million points to obtain reliable results.) We conclude that the multiplier method is much more economical for computing negative dimensions, and that the concept of universality in the multiplicative process is a reasonable one for turbulence.

5 Conclusions

We have discussed a simple way of computing the scaling properties of fractal measures arising from probabilistic processes. The only assumption concerns the existence of a scale-invariant probability distribution of multipliers, which in any case is always necessary for the existence of self-similarity. In cases where the underlying multiplicative process is understood, the multiplier method requires exponentially less work than the boxcounting method. Where one does not know how to partition the measure, the reader is strongly warned about the pitfalls of quantifying such events. However, if there exists other information that allows one to check against spurious scaling, the accuracy of the multiplier method will improve as the refinement gets finer. On the other hand, conventional boxcounting methods are inherently doomed in the search for such rare events.

In this paper, we have not touched upon the important case where level to level correlations exist. One then has to compute these correlations and incorporate them into the relevant equations for calculating dimensions. Alternatively one can increase the base of the process (thus effectively increasing the amount of coarse-graining per level) to decrease these correlations.

We hope that this paper will motivate examination of such scale invariant multiplier distributions whenever one observes sample to sample fluctuations in the dimension of objects or measures - as indeed one does in turbulence and growth models like DLA. The application of these ideas to turbulence has been discussed briefly under the assumption that a binary multiplicative process occurs. More details will be discussed elsewhere.

Acknowledgments

We thank C.J. Evertsz, R.V. Jensen, P.W. Jones, B.B. Mandelbrot and N. Read for useful discussions.

References

[1] B.B. Mandelbrot. *The Fractal Geometry of Nature.* W. H. Freeman & Co., New York, 1982.

[2] J. P. Eckmann and D. Ruelle. *Rev. Mod. Phys.* **57**, 3–617 (1985).

[3] G. Paladin and A. Vulpiani. *Physics Reports,* **156**, 147 (1987).

[4] B.B. Mandelbrot. *J. Fluid Mech.,* **62**, 331 (1974).

[5] H.G.E. Hentschel and I. Procaccia. *Physica,* **8D**, 435 (1983).

[6] U. Frisch and G. Parisi. In *Turbulence and Predictability of Geophysical Fluid Dynamics and Climate Dynamics.* (Eds. M. Ghil, R. Benzi, and G. Parisi). North-Holland, New York, 1985, page 84.

[7] T.C. Halsey, M.H. Jensen, L.P. Kadanoff, I. Procaccia, and B.I. Shraiman. *Phys. Rev. A,* **33**, 1141 (1986).

[8] C. Meneveau and K.R. Sreenivasan. *To appear in the Journal of Fluid Mechanics,* 1990.

[9] The measurements of negative dimensions reported in [8] were motivated by conversation with Benoit Mandelbrot.

[10] C. Meneveau and K.R. Sreenivasan. *Nucl. Phys. B (Proc. Suppl.),* **2**, 49 (1987).

[11] A.B. Chabra and K.R. Sreenivasan. *In preparation,* 1990.

[12] B.B. Mandelbrot. *STATPHYS 17: Proc. of the Rio de Janeiro meeting.* North-Holland, Amsterdam, 1989.

[13] C. Meneveau and K.R. Sreenivasan. *Phys. Rev. Lett.,* **59**, 797 (1987).

[14] R.R. Prasad, C. Meneveau, and K.R. Sreenivasan. *Phys. Rev. Lett.,* **61**, 74 (1988).

[15] A.B. Chabra and R.V. Jensen. *Phys. Rev. Lett.,* **62**, 1327 (1989).

[16] B.B. Mandelbrot. *J. Stat. Phys.,* **34**, 895 (1984).

[17] M.E. Cates and T.A. Witten. *Phys. Rev. A,* **35**, 1809 (1987).

[18] B.B. Mandelbrot. *Fractals: Proc. of the Erice meeting.* (Ed. L. Piettronero). Plenum, New York, 1989.

[19] M.J. Feigenbaum. *J. Stat. Phys.,* **25**, 669 (1978).

[20] C.J. Evertsz. *Laplacian Fractals* [PhD. Thesis]. The Cheese Press, Edam, The Netherlands, 1989. We thank C.J. Evertsz for bringing to our attention this analytically tractable example.

[21] E.A. Novikov. *P.M.M.,* **35**, 266 (1971).

[22] In this example $\alpha \to \infty$ as $q \to -1$. Thus the entire $f(\alpha)$ curve is reproduced by letting q vary from ∞ to -1.

[23] One should in principle allow for the possibility of completely deterministic processes in turbulence, in which case the need for negative dimensions disappears; so far, however, nobody has been able to determine one such. The closest attempt, which yields results in good agreement with the positive part of the $f(\alpha)$ curve, is given in [13].

11

On Turbulence in Compressible Fluids

M.Y. Hussaini, G. Erlebacher, and S. Sarkar

There is no prospect of a comprehensive theory of turbulence, especially compressible turbulence. At best there would be hypotheses based on intuition and empiricism, scaling laws based on similarity methods, analogy and dimensional analysis, and physical and mathematical models which partially explain some turbulence phenomena. An example of a hypothesis is that due to Morkovin (1964) who enunciated that the compressibility effects on turbulence structure are negligible if the root-mean-square density fluctuations are small relative to the absolute density. This was based on the empirical data which was then available, and the implicit assumption was that the pressure fluctuations and total temperature fluctuations were both small. In his excellent review article on compressible turbulent shear flows, Bradshaw (1977) discusses the compressibility effects on turbulence when these assumptions are not satisfied as, for instance, in the case of boundary layers and wakes for Mach numbers beyond 5.

The practical utility of scaling laws cannot be overemphasized. According to Saffman (1977), "practically everything that is useful in turbulence theory is a scaling law". The most well-known of such scaling laws is the 5/3 law due to Kolmogorov (1941) and Obukhov (1941), and it has been extended to include Mach number effects (Zakharov and Sagdeev 1970, Kadomtsev and Petviashvili 1973, Moiseev, Sagdeev, Tur and Yanovskii 1977, and L'vov and Mikhailov 1978) Although the Kolmogorov-Obukhov law has been observed in reality, there is no hope that any of the modifications thereof proposed to account for compressibility effects will ever be validated by experimental observation. Some of the other useful scaling laws are the law of the wall, similarity laws of boundary layers, wakes and jets (Bradshaw 1977, Schlichting 1979, Barnwell 1989). Many of these laws are extensions of the incompressible counterparts, where the compressibility effects are essentially incorporated through mean density. These extensions may not be applicable if features peculiar to compressible flows such as the "hypersonic conditions" involving large pressure and total temperature fluctuations or shock waves are present.

Homogeneous turbulence is a mathematical model which is strictly inadequate for describing any real turbulent flow. However, it appears to retain the important basic mechanisms that are active in a real turbulent flow except for transfers due to the spatial dependence of the moments. For this reason and also because of its relative simplicity, it has been extensively studied both theoretically and experimentally. However, the experimental studies of homogeneous turbulence have necessarily been confined to low speed flows. Moyal (1952) appears to be the first one to look at the spectra of isotropic turbulence in compressible fluids. He decomposed the velocity field in the spectral space into a longitudinal component (random noise) and a transverse (eddy turbulence) component. This is in fact equivalent to a Helmholtz decomposition (gradient of a scalar and curl of a vector) in physical space. The longitudinal component in physical space is variously known as acoustic component or compression turbulence; the transverse component is also termed incompressible component, solenoidal component or shear turbulence. A broad conclusion of his work was that the interaction between these two components is exclusively due to the nonlinear terms, and such interactions are the strongest at high levels of turbulence and at high values of Reynolds number. Analogous to incompressible turbulent flows, the dynamic equations for the two-point double and triple correlations have been derived from the equations of motion (Krzywoblocki 1951), but these equations are hopelessly complicated and of little use in understanding compressibility effects. An evolution equation for the two-point density correlation was obtained by Chandrasekhar (1951) who showed that any change in the density distribution would propagate with a finite velocity proportional to sound speed. He further pointed out that compressibility would act as a source of energy dissipation. Recently, Marion (1988) extended the Direct Interaction theory (DIA) of Leslie (1973) and Eddy Damped Quasi Normal Markovian (EQDNM) theory to include compressibility effects. He computed from EQDNM the inertial range characteristics of the compressible components of the flow in the case of isotropic turbulence and obtained results consistent with those of L'vov and Mikhailov (1978).

In order to interpret and analyse the hot-wire data from turbulence measurements in supersonic flows, Kovasznay (1953) proposed a decomposition of compressible turbulence into three modes— the vorticity mode, the entropy mode and the acoustic mode. He showed how to determine their levels and correlations from mass flow and stagnation temperature fluctuations measured by a hot-wire anemometer. Subsequently, Chu and Kovasznay (1958) have outlined a consistent successive approximation procedure in terms of an amplitude parameter, and have provided explicit formulae for second-order interactions among these three modes. They provide no explicit solutions since their main purpose was to provide a consistent framework to assess the nonlinear interactions in the experimental data. Tatsumi and Tokunaga (1973) and Tokunaga and Tatsumi (1974) study the interactions of weakly nonlinear disturbances such as compression waves, expansion waves and contact discontinuities using a reductive perturbation method due to Taniuti and Wei (1968). The key result is that the interaction between waves of different families of characteristics leads to alterations in their amplitudes, phase velocities and propagation directions, whereas the interaction between waves of the same family of characteristics causes merger or coalescence. They further inferred that the statistical properties of two-dimensional shock turbulence are similar to one-dimensional shock turbulence which in turn are identical to those of Burgers turbulence. A lucid discussion of the invariants of compressible isotropic turbulence, the quadratic effects of interaction among the vorticity, acoustic and entropy modes, and the linear theory for the final decay of energy may be found in Monin and Yaglom (1967).

The computational approach to turbulence is based on the Navier-Stokes equations. In recent times, it has assumed a status equal in importance to theory and experiment. The first attempt at the direct numerical simulation of compressible homogeneous turbulence is due to Feiereisen, Reynolds and Ferziger (1981). They assumed the divergence of the initial flow field and its time-derivative to be both zero. It was therefore not surprising that their results for homogeneous isotropic compressible turbulence did not show any significant departure from the corresponding incompressible data. Recently, Passot and Pouquet (1987) have carried out numerical simulations of two-dimensional homogeneous compressible turbulent flows. They show that the behavior of the flow beyond an initial turbulent Mach number 0.3 differs sharply from the lower Mach number cases which are characterized by the absence of shocks/shocklets.

In what follows, we present a synopsis of the currently available results of the ongoing research work on compressible turbulence at NASA Langley Research Center. Specifically, we draw attention to the symbiotic relationship between mathematical theory, direct numerical simulation (DNS) and turbulence modeling. These three components are described in turn in the following sections. The theory describes low-Mach number turbulence as an initial-value problem and provides estimates of compressible flow variables on the acoustic time scale. DNS results are presented next for two-dimensional compressible turbulence. Both shocked and shock-free results are presented. They verify and are verified by the theory. Next, we discuss some recent Large-Eddy Simulation (LES) work applied to three-dimensional decaying isotropic compressible turbulence. Comparisons against DNS are made. The theoretical formulation leads naturally to a new Reynolds-Stress turbulence model which accounts for the energy dissipation due to compressibility effects. This model has been tested for a supersonic mixing layer and shown to predict correctly the decrease of mixing rate with increasing convective Mach number. Finally, some conclusions are drawn.

1 Theoretical Analysis

In this section we present the theoretical results of Erlebacher, Hussaini, Kreiss, and Sarkar (1990). The objectives of the theoretical analysis are first to classify the various types of equilibrium turbulent regimes (distinguished by the presence or absence of shocks and also by the fraction of the total kinetic energy solely due to compressibility effects), and second, to predict the range of initial conditions that leads to each one. The effect of viscosity is felt either on a viscous time scale (much greater than the acoustic time scale), or during the formation of shocks. In the latter case, viscosity serves to prevent the formation of singularities. Although there is a distinct possibility that shocks will form as a result of certain types of initial conditions, viscosity does not help initiate the processes (wave steepening) which eventually lead to shocks. Rather, the viscosity diffuses the sharp gradients to generate shocks of finite thickness. The above considerations lead to the neglect of all viscous effects in the theoretical formulation and hence we consider just the compressible Euler equations. This assumption is of course verified a posteriori by the direct numerical simulations based on compressible Navier-Stokes equations.

In their analysis, the turbulent Mach number is assumed to be very much less than unity, so that the sound velocity is much greater than the flow velocity. Under these circumstances, the acoustic time scale is much smaller than the convective time scale, which in turn is much less than the viscous time scale (if the Reynolds number is sufficiently large). The different regimes emerge on the acoustic time scale, during which time any inconsistency in the acoustic component of the flow is washed away. In other words, the initial transients only extend over a time period of $O(M_R)$ where M_R is the initial turbulent Mach number of the fluctuating motion.

Erlebacher et al. (1990) depart from the standard asymptotic analysis which requires formal expansions of all dependent variables in terms of a small (or large) parameter. Rather, they reduce the original problem to several simpler sets by decomposing the dependent variables into one set which satisfies a known problem, and a new unknown variable vector which satisfies new evolution equations. This introduces no approximations, and does not require that the unknown vector be of smaller magnitude than the first one. This process can be repeated as often as desired. At each step, the order of the neglected terms is assessed, and the procedure stops when the desired order of accuracy is achieved. This approach was used in Kreiss, Lorenz and Naughton (1990) in their description of compressible flows under the Boussinesq approximation. Rigorous convergence proofs are given therein. At each order, the new dependent variables are rescaled so that they are initially bounded by unity. The objective is then to obtain a set of equations in which all terms are $O(1)$ in the time frame under study here, the acoustic time scale. This determines the required change of independent/dependent variables. If required, the whole process is then repeated. The advantage of this approach is that the order of the neglected terms is precisely known at each step, and a standard formal expansion procedure is not required.

As mentioned earlier, the starting point of the analysis are the Euler equations for an ideal gas. The pressure P is decomposed into a mean and a small fluctuating component p and the flow is assumed to be homentropic. The density is then expanded in powers of p and substituted into the Euler equations.

It is convenient to pursue the analysis by decomposing the velocity vector into two components which reflect in some sense the solenoidal and irrotational properties of the flow. One approach is to apply a Helmholtz decomposition to the velocity vector field (Moyal 1952):

$$\mathbf{u} = \nabla\Phi + \nabla \times \boldsymbol{\psi} \tag{1}$$

where

$$\begin{aligned} \nabla\Phi &= \mathbf{u}_H^I \\ \nabla \times \boldsymbol{\psi} &= \mathbf{u}_H^C. \end{aligned}$$

The evolution equations for $\mathbf{u}_H^I$ and $\mathbf{u}_H^C$ can be replaced by a pair of coupled time-dependent equations for $\nabla \cdot \mathbf{u}_H^C$ and $\nabla \times \mathbf{u}_H^I$, i.e. dilatation and vorticity. These equations include coupling terms which describe vorticity generation due to acoustic effects, and conversely, sound generation due to vorticity sources.

Erlebacher et al. (1990) split the velocity vector according to

$$\mathbf{u} = \mathbf{u}^I + \mathbf{u}^C, \tag{2}$$

where the solenoidal component $\mathbf{u}^I(\mathbf{x}, t)$ satisfies the time-dependent incompressible Navier-Stokes equations which in turn defines the incompressible pressure component p^I. If the time evolution of the incompressible component of velocity is known, it remains to calculate the compressible velocity vector $\mathbf{u}^C$. Note that the terminology "compressible velocity" is chosen for convenience to emphasize that it becomes zero in the limit of incompressibility. Note that $\mathbf{u}^C$ is not of a fully acoustic nature. In fact, $\mathbf{u}^C$ acquires a small degree of vorticity. The advantage of this split, as will become clear, is that the solution to the full Navier-Stokes equations is simply a superposition of a time-dependent incompressible solenoidal flow and a compressible component which has a small degree of vorticity. This vorticity is, by nature of the velocity split, a consequence of compressibility effects.

The perturbation pressure is decomposed according to

$$p = \gamma M_R^2 p^I + \delta p^C. \tag{3}$$

This particular decomposition removes p^I from the evolution equation for $\mathbf{u}^C$. The parameter δ is defined to insure that $p^C(\mathbf{x}, 0)$ has unit norm. The full initial velocity $\mathbf{u}_0$ is also normalized to unity. Therefore, both $\mathbf{u}_0^C$ and $\mathbf{u}_0^I$ are $O(1)$. Incompressible variables vary on a time scale which is slower than the acoustic time scale, the time scale of interest, so that their order of magnitude is invariant in time. In contrast, certain combinations of initial conditions can lead to initial transient behavior which can change the order of magnitudes of the compressible variables on a time scale of $O(M_R)$.

From the Euler equations, it is clear that there are two different time scales: the convection scale of order $O(1)$, and the acoustic scale of order $O(M_R)$. Since the initial compressible velocity field $\mathbf{u}_0^C(\mathbf{x})$ is vorticity-free, $\mathbf{u}^C$ varies on the fast acoustic time scale. Substitution of Eqs. (2)–(3) into the Euler equations and subtracting off the incompressible Euler equations leads to hyperbolic evolution equations for the compressible components of the flow variables.

The strongly asymmetric nature of the evolution equations indicates that one component of the solution vector in general grows rapidly, thus possibly invalidating the original scalings (Erlebacher et al. 1990). Only when the equations are symmetrized is the solution vector bounded by the order of magnitude of the initial solution vector.

Thus, after symmetrizing the equations, and some further analysis, it can be shown that there are two possible regimes. These are characterized by the ratio of the initial compressible pressure norm to the reference turbulent Mach number. The two cases are considered in turn.

$\delta \leq \gamma M_R$

This regime is characterized by low levels of initial acoustic pressure fluctuation. To lowest order in M_R, the Euler equations reduce to the linearized wave equations

$$\frac{\partial \mathbf{u}^C}{\partial t'} + \frac{\delta}{\gamma M_R} \nabla p^C = 0,$$

$$(4)$$

$$\frac{\partial p^C}{\partial t'} + \frac{\gamma M_R}{\delta} \nabla \cdot \mathbf{u}^C = 0$$

which describe linear waves decoupled from the underlying solenoidal velocity field. Kreiss, Lorenz and Naughton (1990) have shown that the wave equation still describes the compressible component of the flow on the convective time scale. Here, $t' = t/M_R$ is $O(1)$ on the acoustic time scale.

The order of magnitude of vorticity production can be assessed from Eqs. (4). Taking the curl of the momentum equations gives

$$\frac{\partial \boldsymbol{\omega}^C}{\partial t'} + M_R \left[\mathbf{u}^I \cdot \nabla \boldsymbol{\omega}^C + \mathbf{u}^C \cdot \nabla \boldsymbol{\omega}^I + \mathbf{u}^C \cdot \nabla \boldsymbol{\omega}^C \right] = \qquad (5)$$

$$+ M_R \left[\boldsymbol{\omega}^I \cdot \nabla \mathbf{u}^C + \boldsymbol{\omega}^C \cdot \nabla \mathbf{u}^I - \boldsymbol{\omega}^I \nabla \cdot \mathbf{u}^C + \boldsymbol{\omega}^C \cdot \nabla \mathbf{u}^C \right]. \qquad (6)$$

where $\boldsymbol{\omega}^c = \nabla \times \mathbf{u}^C$ and $\boldsymbol{\omega}^I = \nabla \times \mathbf{u}^I$. On an acoustic time scale, it is clear that the extra vorticity $\boldsymbol{\omega}^C$ generated from dilatational effects is

$O(M_R)$ since both $\mathbf{u}^C$ and $\mathbf{u}^I$ are $O(1)$. At the lowest order, incompressible-compressible interactions generate $\boldsymbol{\omega}^C = O(M_R\sqrt{\chi(1-\chi)})$, while the compressible–compressible interactions generate $\boldsymbol{\omega}^C = O(M_R\chi)$ (χ is the ratio of the compressible kinetic energy, $||\mathbf{u}^C||^2$, to the total kinetic energy $||\mathbf{u}||^2$). At low compressibility levels ($\chi << 1$), the solenoidal-compressible mode interactions are the dominant source of vorticity production. More information can be found in Sarkar et al. (1990).

$\delta > \gamma M_R$

In this regime, pressure fluctuations are much greater than the turbulent Mach number. The evolution equations become to lowest order in M_R,

$$\frac{\partial \mathbf{u}^C}{\partial t'} + M_R \frac{\delta}{\gamma} \mathbf{u}^C \cdot \nabla \mathbf{u}^C + \frac{\delta}{\gamma M_R} \nabla p^C = 0,$$

(7)

$$\frac{\partial p^C}{\partial t'} + M_R \mathbf{u}^C \cdot \nabla p^C + \frac{\gamma M_R}{\delta} \nabla \cdot \mathbf{u}^C = 0.$$

It is shown in Erlebacher et al. (1990) that

$$\mathbf{u}^C(\mathbf{x}, t') = O(\frac{\delta}{\gamma M_R}).$$

(8)

When $\delta = O(1)$, the convective terms balance the time derivative terms, in which case wave steepening may occur on a $t' = O(1)$ time scale, and there is a propensity for shocks to form. Although $\delta << 1$, $\delta = O(1)$ should be interpreted as $M_R << \delta << 1$, which is possible when the fluctuating Mach number is very low. If M_R is 0.1 and above, this condition becomes increasingly hard to satisfy. Note that when shocks do occur, the turbulent Mach number, M_t is no longer small, but has increased by a factor $1/M_R$ from its initial value. Estimates of vorticity proceed similarly to the previous case. In this case, the vorticity generated is $O(M_R\chi)$ on the acoustic time scale $t = O(M_R)$.

RESULTS

From the results of the previous section, one can estimate the long time behavior ($t' \to \infty$) of all the flow variables. Of particular relevance to the issue of turbulence modeling is the compressible ratio χ and the ratio of total dilatational to total solenoidal dissipation rate.

Earlier, the variable

$$\chi = \frac{||\mathbf{u}^C||^2}{||\mathbf{u}^C||^2 + ||\mathbf{u}^I||^2}$$

(9)

which is the fraction of the total kinetic energy solely due to compressible effects was introduced. Two different expressions for $\chi(t' = \infty)$ are then obtained for δ respectively $O(1)$ and $O(M_R)$.

When $\delta = O(1)$, Eq. (8)implies

$$\chi_\infty = \frac{1}{1 + \frac{\gamma M_R^2}{\delta^2}||u^I||^2}, \tag{10}$$

and since $||u^I|| = O(1)$, $\chi_\infty = O(1)$. Consequently, the lower the initial level of compressibility at a fixed M_R (i.e. the lower χ_0), the stronger the initial transient (which extends over the time period $O(M_R)$). As this imbalance intensifies, the propensity for shocks to form also increases.

When $\delta = O(M_R)$, the solution to the linear system of equations (4)leads to the result

$$\chi_\infty = \frac{\chi_0}{2} \frac{1 + F_0}{F_0 + 0.5\chi_0(1 - F_0)}. \tag{11}$$

as explained more fully in Sarkar, Erlebacher, Hussaini, and Kreiss (1989).
The function

$$F(t) = \frac{\gamma^2 M_t^2 \chi}{\delta^2 ||p^C||^2} \tag{12}$$

was introduced by Sarkar et al. (1989) where it was shown to play a fundamental role in the description of compressible turbulence. It is now easy to see that a necessary condition for the appearance of shocks is $\delta = O(1)$ (see previous section). If the latter condition is not satisfied, the level of compressibility will remain small for all time, which is inconsistent with the appearance of an isotropic distribution of shocks.

In Fig. 1, the ratios of $||\mathbf{u}^C||^2$ and $||p^C||^2$ (normalized with respect to their initial values) are plotted versus $F_0 = F(0)$. For low F_0, the $||\mathbf{u}^C||$ has strong transients and $||p^C||$ is almost unaffected. On the other hand, when F_0 is greater than one, the situation is reversed: the compressible velocity remains at its initial order of magnitude while large compressible pressure waves appear.

If $\delta = O(1)$, $F_0 \sim M_R^2\chi_0 << 1$. Therefore, $F_0 << 1$ is a necessary, but not sufficient condition for shocks to form. This is easily seen by setting $\delta = O(M_R)$, which implies that $F_0 = \chi_0$. In this case, χ_0 might be very small, but the pressure fluctuations are not strong enough. In fact, under these conditions, Eq. (11)predicts that $\chi_\infty \approx 0.5$.

It is interesting to establish the conditions under which p^C is a good approximation to the total perturbation pressure p. From previous results,

$$\frac{||p^I||}{||p^C||} = O(\frac{\gamma M_R^2||\mathbf{u}^I||^2}{\delta p^c}) \tag{13}$$

$$= O(\sqrt{\frac{F_0}{\chi}} M_R(1 - \chi))$$

So that under equilibrium conditions, p^I is non-negligible only when

$$\frac{M_R(1 - \chi)}{\sqrt{\chi}} >> 1. \tag{14}$$

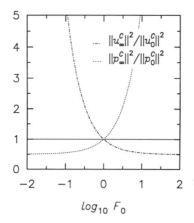

FIGURE 1. Ratios of $||\mathbf{u}^C||^2$ and $||p^C||^2$ to their initial values as a function of $\log(F_0)$.

In other words, the incompressible pressure field is dominant for the higher values of M_R, and for lower χ. Otherwise, p^C is a good approximation to p.

The expression (3) for the pressure immediately suggests two different limiting cases. If $M_R^2 >> \delta p^C$, the flow will remain quasi-incompressible, and one obtains

$$\frac{p-1}{\gamma M_R^2} \to 1. \tag{15}$$

On the other hand, if $M_R^2 << \delta p^C$, the flow is dominated by compressibility effects and the pressure P satisfies

$$p - 1 \to \delta p^C. \tag{16}$$

In this paper, we relate the value of p^C after several acoustic periods, to the remaining flow parameters, and prove that

$$p - 1 \to \gamma M_t \sqrt{\chi}. \tag{17}$$

2 Direct Numerical Simulation

The DNS of two and three dimensional compressible turbulence are based on the compressible Navier-Stokes equations. The conductivity and viscosity fluctuations are neglected for simplicity. As the flow is assumed to be isotropic, a Fourier representation of all variables in both coordinate directions is appropriate. Spectral methods, by nature of their high accuracy and low dispersive and dissipative errors are ideally suited for the direct simulation of isotropic flows. The present solution technique employs a Fourier collocation method (Canuto, Hussaini, Quarteroni and Zang 1988). A sufficient number of collocation points are used to resolve all significant scales. The nonlinear convective terms are written in a form which ensures numerical stability. This issue is discussed in Erlebacher et al. (1987).

We are interested in the low subsonic regime where the turbulent Mach number $M_t \ll 1$. The sound speed is in this case much greater than the flow velocity and this imposes a severe restriction on the time step of any explicit time marching numerical scheme. To remove this constraint, a splitting method is adopted. The first step integrates explicitly the Navier-Stokes equations from which the fast time scales have been subtracted. The second step integrates the "acoustic equations" which vary on the fast time scale. These equations are linear in the pressure, momentum and density, and are amenable to an analytical solution in Fourier space (Erlebacher, Hussaini, Speziale and Zang 1987). This splitting is employed at each stage of a low-storage third-order Runge-Kutta method. The advantage of this splitting is that the principal terms responsible for the acoustic waves have been isolated, and the explicit time scale is only determined by the flow speed. This is clearly most efficient at low Mach numbers. Moreover, since the second fractional step is integrated analytically, it does not generate dispersion or dissipation errors. The only errors incurred are time splitting errors.

If one is truly interested in the detailed time-evolution of the acoustic components, the time-step must be small enough to resolve the temporal evolution of these waves. Note however, that because the acoustics are treated exactly in the second step, simulations with Δt on the convective time scale still produce accurate results, but the data is no longer available on the acoustic time scale.

The initial conditions for the direct numerical simulations presented here are similar to those of Passot and Pouquet (1987). They are sufficiently general to produce the various turbulent regimes expected to occur when $M_t \ll 1$. For both analysis and data reduction purposes, it is convenient to impose a Helmholtz decomposition on the velocity. Thus

$$\mathbf{u} = \mathbf{u}^I + \mathbf{u}^C \tag{18}$$

where $\mathbf{u}^I, \mathbf{u}^C$ are respectively the solenoidal and irrotational components of velocity. At $t = 0$, one must separately specify the reference turbulent Mach number M_{t0}, the Reynolds number Re, the Prandtl number Pr, the rms levels of $\mathbf{u}_0^C$, $\mathbf{u}_0^I$, ρ_0, T_0, and the autocorrelation spectrum for ρ, T, $\mathbf{u}^C$ and $\mathbf{u}^I$. The zero subscript refers to the initial state. The energy spectrum of the density, temperature and velocity fluctuations are all given by

$$E(k) = k^4 e^{-\frac{2k^2}{k_0^2}} \tag{19}$$

where k_0 corresponds to the spectrum peak. Strictly speaking the initially irrotational component of velocity should have an autocorrelation spectrum proportional to k^2 for low k to guarantee analyticity. However, Erlebacher et al. (1990) did not feel that this would qualitatively influence the numerical results.

RESULTS

In this section, we show results of 2-D direct numerical simulations on grids of 64×64 and compare some of the results against theoretical predictions. This validation is accomplished by comparing the value of χ after several acoustic periods as predicted by the computation, against that given by the theory. χ is a measure of the degree of compressibility in the flow. It does not give any indication of the nonlinear interactions that occur between the irrotational and the solenoidal components of the flow. Another quantity of interest is the ratio of the compressible potential energy to the compressible kinetic energy F, defined by Eq. (12)(see Sarkar et al. 1989).

Figure 2 shows time histories of χ for four different numerical simulations, along with the final equilibrium level predicted by theory (drawn as solid horizontal lines). These simulations are referred to as 112, 221, 222, and 331 in Erlebacher et al. (1990). The three digits respectively refer to the initial rms levels of χ, M_t and p. The four triplets correspond respectively to

$$(\chi, M_t, p) = (\{0.1, 0.03, 0.07\}, \{0.5, 0.08, 0.014\},$$

$$\{0.5, 0.08, 0.07\}, \{0.9, 0.14, 0.014\}) \qquad (20)$$

Figures 2 (a)–(b) contrast the short time scale and long time scale properties of χ. Note that the (convective/viscous) time scale decay rate of χ is a function of the input parameters. These rates cannot yet be predicted from theory. As shown by Sarkar et al. (1989), $F(t)$ oscillates around unity after some initial transient behaviour. This is illustrated in Figs. 3 (a)–(b) which again contrast acoustic and convective/viscous time scales. Even on the longer time scales, $F(t)$ is close to unity. Sharp estimates of the rms of these oscillations as a function of the input parameters are not yet available. Fig. 4 shows the extrema of $\nabla \cdot \mathbf{u}$ over the entire numerical grid. The energy spectrum peaks at $k_0 = 10$, which explains the large values of divergence, and its subsequent rate of dissipation. Note that the minimum and maximum are approximately symmetric about the origin indicating an absence of shocks. On the other hand, a simulation with $k_0 = 1.85$ and a high ratio of pressure fluctuations to fluctuating Mach number can generate shocks. This is shown in Fig. 5 where the compression regions translate to regions of strong negative divergence. A two-dimensional plot of pressure is shown in Fig. 6. Note the formation of shocks. The figure corresponds to $t = 4$. At later times, two parallel shocks will move at the sound velocity in opposing directions. The large structure of the shocks is directly related to our choice of the initial energy spectrum which peaks at $k_0 = 1.85$. Fig. 6 corresponds to $\delta = 0.14$, and $M_R = 0.03$ which corresponds to $\delta/M_R = 4.6$.

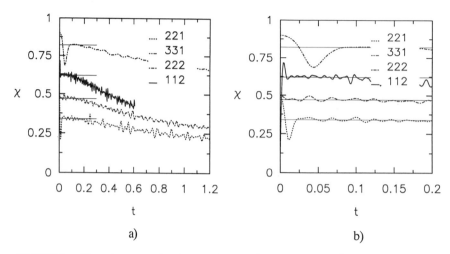

FIGURE 2. Time history of χ for DNS cases (112,222,331,221). Horizontal lines are the equilibrium values predicted by the theory. (a) acoustic time scale, (b) viscous time scale.

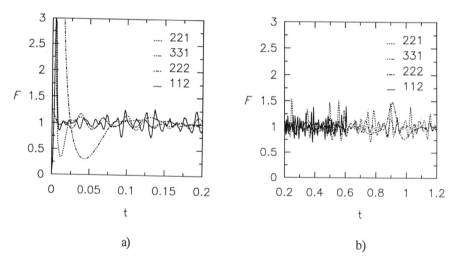

FIGURE 3. Time history of F for DNS cases (112,222,331,221). (a) acoustic time scale, (b) viscous time scale.

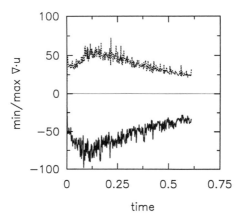

FIGURE 4. Time history of minimum and maximum $\nabla \cdot \mathbf{u}$ over the entire grid. DNS case (112), $k_0 = 10$.

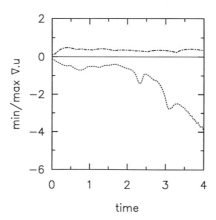

FIGURE 5. Time history of minimum and maximum $\nabla \cdot \mathbf{u}$ over the entire 256^2 grid. Two-dimensional DNS results with shocks. Parameters are $M_0 = 0.028$, $\delta = 0.14$, $k_0 = 1.85$, $Re = 150$.

FIGURE 6. Contour plot of pressure. The parameters are the same as for Fig. 5.

3 Large-Eddy Simulation

This section is a recapitulation of the results presented at the 12th International Conference on Numerical Methods in Fluid Dynamics (Erlebacher et al. 1990). Direct numerical simulations of high Reynolds number turbulent flows of technological interest will not be possible in the foreseeable future. The calculation of such flows will be done primarily by Reynolds stress models or by large-eddy simulations. Among these two approaches, large-eddy simulations appear to have better predictive capabilities. In LES, the large eddies (which are directly influenced by the flow geometry) are computed directly and only the small scales (which are more universal in character) are modeled, whereas in Reynolds stress models, all scales must be modeled. Most applications of LES have been to incompressible flows (Ferziger 1984). There have been few applications of LES to compressible turbulent flows. Recently, Yoshizawa (1986) and Speziale, Erlebacher, Zang and Hussaini (1988) have derived subgrid scale models for compressible turbulent flows. This section discusses the large-eddy simulations of compressible isotropic turbulence based on the subgrid scale models of Speziale et al. (1988).

The large-eddy simulations are based on the following Favre-filtered continuity, momentum, and energy equations (Speziale et al. 1988):

$$\frac{\partial \overline{\rho}}{\partial t} + \nabla \cdot (\overline{\rho} \tilde{\mathbf{u}}) = 0 \tag{21}$$

$$\frac{\partial}{\partial t}(\overline{\rho} \tilde{\mathbf{u}}) + \nabla \cdot (\overline{\rho} \tilde{\mathbf{u}} \tilde{\mathbf{u}}) = -\nabla \overline{p} + \nabla \cdot \overline{\boldsymbol{\sigma}} + \nabla \cdot \mathbf{g} \tau$$

$$\frac{\partial}{\partial t}(C_v \overline{\rho} \tilde{T}) + \nabla \cdot (C_v \overline{\rho} \tilde{\mathbf{u}} \tilde{T}) = -\overline{p \nabla \cdot \mathbf{u}} + \overline{\mathbf{g} \sigma \cdot \nabla \mathbf{u}} + \nabla \cdot \overline{\kappa \nabla T} - \nabla \cdot \mathbf{Q} \tag{22}$$

where ρ is the density, $\mathbf{u}$ is the velocity vector, T is the absolute temperature, $p = \rho R T$ is the thermodynamic pressure, R is the ideal gas constant,

C_v is the specific heat at constant volume, κ is the thermal conductivity,

$$\mathbf{g}\sigma = -\frac{2}{3}\mu\nabla\cdot\mathbf{u}\,\mathbf{I} + \mu(\nabla\mathbf{u} + \nabla\mathbf{u}^T)\tag{23}$$

is the viscous stress tensor, $\mathbf{g}\tau$ is the subgrid scale stress tensor, $\mathbf{Q}$ is the subgrid scale heat flux, and $\mathbf{I}$ is the identity tensor. In Eqs. (21)–(22), an overbar represents a spatial filter whereas a tilde represents a mass weighted or Favre filter, i.e.

$$\tilde{\mathcal{F}} = \frac{\overline{\rho\mathcal{F}}}{\overline{\rho}}\tag{24}$$

where $\mathcal{F}$ is any flow variable. These equations are obtained by applying a Gaussian filter to the full Navier-Stokes equations. The subgrid scale stress tensor and subgrid scale heat flux are modeled as follows (Speziale et al. 1988):

$$\mathbf{g}\tau = -\overline{\rho}(\widetilde{\tilde{\mathbf{u}}\tilde{\mathbf{u}}} - \tilde{\mathbf{u}}\tilde{\mathbf{u}}) + 2C_R\overline{\rho}\Delta_f^2 II_{\tilde{S}}^{1/2}(\mathbf{S} - \frac{1}{3}(\mathbf{S}:\mathbf{I})\,\mathbf{I}) - \frac{2}{3}C_I\overline{\rho}\Delta_f^2 II_{\tilde{S}}\mathbf{I}\tag{25}$$

$$\mathbf{Q} = C_v\overline{\rho}(\widetilde{\tilde{\mathbf{u}}\tilde{T}} - \tilde{\mathbf{u}}\tilde{T} - \frac{C_R}{Pr_T}\Delta_f^2 II_{\tilde{S}}^{1/2}\nabla\tilde{T})\tag{26}$$

where $II_{\tilde{S}} \equiv \tilde{\mathbf{S}}:\tilde{\mathbf{S}}$, Δ_f is the filter width, and C_R and C_I are constants which assume the values of 0.012 and 0.0066, respectively. The turbulent Prandtl number Pr_T is taken to be 0.7. In the incompressible limit, the subgrid scale stress model Eq. (25) reduces to the linear combination model of Bardina et al (1983). Of course, in the limit as the filter width $\Delta_f \to 0$: $\tilde{\mathcal{F}}$, $\overline{\mathcal{F}} \to \mathcal{F}$ (where $\mathcal{F}$ is any flow variable) and hence the unfiltered continuity, momentum and energy equations are recovered. The solution technique is essentially the numerical method discussed in Erlebacher et al. (1990) modified to account for the eddy viscosity term.

RESULTS

The computed results from the coarse-grid LES are validated by comparisons with the results of a fine grid direct numerical simulation (DNS) of isotropic turbulence for the same initial conditions. The model constants are evaluated by comparing stresses obtained from DNS results against the model stresses calculated based on the DNS.

The results of the simulations to be shown correspond to the initial conditions: $Re_\lambda = 26$, $< M_t^2 >^{1/2} \equiv < q^2/\gamma RT >^{1/2} = 0.1$, $\rho_{rms}=0$, T_{rms} $< T >= 0.0626$ and $Pr = 0.7$ (where $\gamma \equiv C_p/C_v = 1.4$, $q^2 = \mathbf{u}\cdot\mathbf{u}$, and $< \cdot >$ denotes a spatial average). A direct numerical simulation was performed on a 96^3 mesh for these initial conditions. The results were then filtered and injected onto a coarse 32^3 mesh for comparison with the LES. Large-eddy simulations were performed on a 32^3 mesh with filter widths

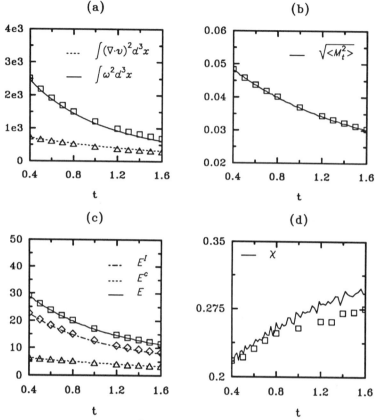

FIGURE 7. LES results on 32^3 grid with $\Delta_f = 2$. Parameters are stated in the text.

$\Delta_f = 0$ and $\Delta_f = 2$ (here the former value of $\Delta_f = 0$ corresponds to a coarse-grid direct simulation). In Figs. 7 (a)-(d), the integrated average rms divergence of velocity and vorticity, the average turbulence Mach number, the compressible and incompressible average turbulent kinetic energy ($E = < q^2 >$), and $\chi = E^c / E$ obtained from the LES and DNS are compared. It is clear that these results — which correspond to a filter width $\Delta_f = 2$ and are given for a few eddy turnover times — are in excellent agreement. We now demonstrate that the excellent results obtained from the LES are due to the adequate performance of the subgrid-scale models. In Figs. 8 (a)-(d), the same turbulence statistics are shown for a filter width $\Delta_f = 0$ — namely, the case where the LES is actually a 32^3 coarse grid DNS. It is clear from Fig. 8 that the quantitative accuracy of the results degrades considerably. Hence, it follows that the success of the LES shown in Fig. 7 is largely due to the good performance of the subgrid-scale models in draining from the large scales the proper amount of energy which would have otherwise cascaded to the unresolved scales.

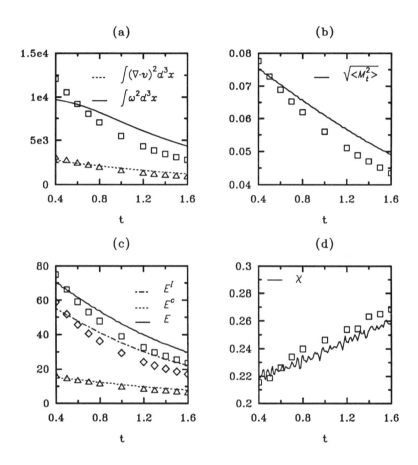

FIGURE 8. LES results on 32^3 grid with $\Delta_f = 0$. Parameters are stated in the text.

4 Second-order modeling of compressible turbulence

A common approach to quantitative computation of high Reynolds number compressible turbulent flows employs approximations to the compressible Navier-Stokes equations based on certain averaging processes. Each field variable is decomposed into a mean component and a fluctuating component. Usually Favre averages (density-weighted averages) are used for the mean velocity and temperature, while conventional Reynolds averages are used for the mean pressure and density so as to minimize the complexity of the temporal derivative and convective terms in the averaged equations. Thus, the velocity $\mathbf{u}$, the density ρ, the temperature T and the pressure p are expressed as the sum of mean and fluctuating parts by

$$\mathbf{u} = \tilde{\mathbf{u}} + \mathbf{u}'' \qquad \rho = \bar{\rho} + \rho'$$
$$T = \tilde{T} + T'' \qquad p = \bar{p} + p'$$

The overbar denotes the conventional Reynolds average and the superscript $'$ denotes fluctuations with respect to the Reynolds average, while the overtilde denotes the Favre average and the superscript $''$ denotes fluctuations with respect to the Favre average. The Favre average $\tilde{\phi}$ of a field variable ϕ is a density-weighted Reynolds average

$$\tilde{\phi} = \overline{\rho\phi}/\bar{\rho}.$$

Before modelling compressible turbulence, it is important to determine when compressible effects become significant. Measurements in the turbulent boundary layer have lead to the hypothesis (Morkovin 1964, Bradshaw 1977) that direct compressible effects on the turbulence may be ignored when the ratio of the root-mean-square (r.m.s.) density fluctuations to the mean density is small. Turbulence modelers have used Morkovin's hypothesis to justify the retention of incompressible model formulations with minor modifications which allows the density scaling in the modeled terms to be determined by the local mean density. However, such variable-density generalizations of standard two-equation (eg. $k - \epsilon$) and second-order closures (e.g., Kline, Cantwell, and Lilley 1982) have been rather unsatisfactory in predicting the reduced growth rate of high-speed shear levels of velocity fluctuations induced by a shock-boundary layer interaction.

In high-speed flows, appreciable levels of density and temperature fluctuations can occur, and the dilatation $d = \nabla \cdot \mathbf{u}$ of the velocity is no longer negligible. These additional variables lead to new correlations which have to be modeled in order to close the averaged equations. For example, the pressure-dilatation $\overline{p'd'}$, which appears explicitly in the mean temperature equation and in the equation for the Reynolds stress tensor $\widetilde{\mathbf{u}''\mathbf{u}''}$, is well known as a correlation requiring modeling in high-speed flows. Recently

Sarkar et al. (1989) and Zeman (1990) have identified a new dilatational term, the compressible dissipation ϵ_c, which is the contribution of the dilatational velocity component to the turbulent dissipation rate, through the following decomposition,

$$\epsilon = \epsilon_s + \epsilon_c$$

where the incompressible dissipation rate $\epsilon_s = \overline{\mu}\,\overline{\boldsymbol{\omega}' \cdot \boldsymbol{\omega}'}$ and the compressible dissipation rate $\epsilon_c = \frac{4}{3}\overline{\mu}\overline{d'^2}$. Here $\boldsymbol{\omega}'$ is the fluctuating vorticity, while d' is the fluctuating dilatation.

Apart from the dilatational correlations, important correlations involving the fluctuating thermodynamic field that require modeling are the turbulent heat flux $\widetilde{T''\mathbf{u}''}$, the turbulent mass flux $\overline{\rho'\mathbf{u}'}$, and the pressure-velocity correlation $\overline{p'\mathbf{u}'}$. If the velocity statistics are of primary interest, as is usually the case, algebraic methods are used for $\widetilde{T''\mathbf{u}''}$ and $\overline{\rho'\mathbf{u}'}$. An additional transport equation is provided for the turbulent dissipation rate ϵ, both because it is a dominant term in the Reynolds stress budget and because it provides length scale information. The exact equations for the Reynolds stress tensor, and the mean variables are obtained by manipulation of the compressible Navier-Stokes equations, for example, see Cebeci and Smith (1974).

Sarkar et al. (1989) have recently used the low-Mach number asymptotic analysis of section 1 in order to identify the terms that become truly important with increasing Mach number, and then they have devised models for these terms. Adopting the approach of Erlebacher et al. (1990) they decomposed the governing equations into an incompressible and a compressible problem and found a number of implications for the modeling of the compressible component. Analysis of the compressible problem revealed that quantities, such as the dilatational component of the velocity field and the thermodynamic pressure evolve on a fast time scale t_C which is $O(1/M)$ times the characteristic incompressible time scale $t_T = k/\epsilon$. The idea of *acoustic equilibrium* was introduced by them to describe the quasi-stationary equilibrium that these compressible variables, which evolve on a fast time scale, attain. They were able to obtain analytical solutions for the acoustic equilibrium values of pressure-dilatation $\overline{p'd'}$, dilatational variance $\overline{d'^2}$, the pressure variance $\overline{p'^2}$ and the compressible kinetic energy $\overline{\mathbf{u}^{C'} \cdot \mathbf{u}^{C'}}$. Interestingly enough, the acoustic equilibrium value of the pressure-dilatation (which can be positive or negative) was found to be zero while that of the dilatational variance (which is positive definite) was found to be a positive quantity. Three-dimensional DNS of isotropic turbulence also showed that the spatial average of the pressure-dilatation, after an initial transient, oscillates around zero. When the pressure-dilatation is averaged over the acoustic temporal oscillations, it has a positive value which is much smaller than the compressible dissipation. In the case of decaying isotropic turbulence, the density variance $\overline{\rho'^2}$ decreases in time, and an inspection of the following exact equation for $\overline{\rho'^2}$

$$\frac{\partial \overline{\rho'^2}}{\partial t} = -2\overline{\rho' d'} - \overline{\rho' \rho' d'}$$

shows that the dominant term on the right-hand side, namely, the density-dilatation $\overline{\rho' d'}$, after averaging out the temporal acoustic oscillations, has to be positive. Since, the thermodynamic fluctuations in low-Mach number isotropic turbulence are approximately isentropic, the pressure and density have a large positive correlation, and therefore it follows from the positiveness of the time-averaged density-dilatation that the time averaged pressure-dilatation has also to be positive in the case of decaying isotropic turbulence. Since the pressure-dilatation is highly oscillatory and the positiveness of the average over its acoustic oscillations is flow-dependent, a simple algebraic model for the pressure-dilatation may not be generally valid. A possible alternative is to take the advantage of the smallness of the pressure-dilatation with respect to the compressible dissipation and model the combination $\epsilon_c - \overline{p' d'}$ which appears in the kinetic energy equation by a positive definite algebraic model.

A simple algebraic model for the compressible dissipation ϵ_c was derived by Sarkar et al. (1989) in the following fashion. The parameter F (Eq. (12)) was shown through analysis by Sarkar et al. (1989) to approach unity in compressible turbulence. Since F can be rewritten as the ratio of the compressible kinetic energy to the compressible potential energy, the physical interpretation of $F \rightarrow 1$ is that there is a tendency towards equipartition between potential and kinetic energies in the compressible mode. The result $F \cong 1$ was validated in the case of homogeneous turbulence through 3-D direct simulations with different initial conditions for F. Since the pressure fluctuations scale as M_t^2, it follows from $F = \gamma^2 M_t^2 \chi / p_c^2 \simeq 1$ that $\chi = O(M_t^2)$. If the ratio of the Taylor microscales of the compressible and solenoidal components is $O(1)$, then the ratio of the compressible dissipation ϵ_c / ϵ_s becomes $O(\chi)$. Therefore ϵ_c can be modeled as

$$\epsilon_c = \alpha_1 \epsilon_s M_t^2. \tag{27}$$

The effect of compressiblity on the enstrophy $\overline{\boldsymbol{\omega}^{C'} \cdot \boldsymbol{\omega}^{C'}}$ was shown, again by the asymptotic analysis, to be a factor of M_t^2 smaller than its effect on the dilatational variance, and therefore in moderate Mach number, shock-free turbulence, the solenoidal dissipation $\epsilon_s = \overline{\nu \boldsymbol{\omega}' \cdot \boldsymbol{\omega}'}$ is relatively insensitive to M_t in comparison with the compressible dissipation $\epsilon_c = \frac{4}{3}\overline{\nu d'^2}$. Therefore, for such situations, one may carry the standard ϵ equation for the solenoidal dissipation ϵ_s, and use Eq. (27) as a model for ϵ_c.

RESULTS

The high-speed shear layer is a flow which shows strong compressibility effects. The growth rate of the high-speed, compressible shear layer is significantly smaller relative to its low-speed counterpart. Due to the significant deviation of the high-speed shear layer's behavior from that of the low-speed one, the compressible shear layer has assumed a central role in the study of compressible turbulence. However, the prediction of this phenomenon of reduced mixing has been a major, unresolved issue in theoretically based turbulence modeling. Variable density extensions of incompressible turbulence models have failed to predict the significant decrease in the spreading rate accompanying an increase in the convective Mach number. This has led to attempts by Oh (1974), Vandromme (1983), and Dussauge and Quine (1988), among others, to make phenomenological modifications to incompressible turbulence models, in order to obtain successful predictions of the compressible mixing layer. The model (Eq. (27)) for the compressible dissipation was applied by Sarkar and Lakshmanan (1990) to the case of the compressible shear layer. Fig. 9, which depicts the functional dependence of the normalized growth rate $C_\delta/(C_\delta)_0$ on the convective Mach number M_c, compares experiments with computations obtained with a second-order Reynolds stress closure. The Langley experimental curve in Fig. 9 is a consensus representation of various experimental investigations prior to 1972, while the symbols denote data obtained from later experiments. The poor performance of a variable-density generalization of a standard incompressible Reynolds stress closure indicated the need for *explicitly* incorporating the mechanisms associated with turbulence compressibility. It is clear that the inclusion of the new model of the compressible dissipation ϵ_c into the Reynolds stress closure leads to agreement with the following critical trends in the experimental data: first, the significant decrease in the spreading rate when the Mach number is increased; and second, the relative insensitivity of the spreading rate to further increases in the convective Mach number. This model for the compressible dissipation, has also been applied by Wilcox (1989) to some supersonic and hypersonic flows within the framework of a $k - \omega$ turbulence closure. Wilcox's study concludes that the addition of this model of the compressible dissipation leads to the experimentally observed reduction in the growth rate of the compressible shear layer, leads to values of skin friction in adiabatic boundary layers that are slightly lower than the measured values, and results in an improved prediction of the separation bubble size in a shock-boundary layer interaction problem.

5 Conclusions

This paper discusses an asymptotic theory (with turbulent Mach number as a small parameter) for the initial-value formulation of the equations of

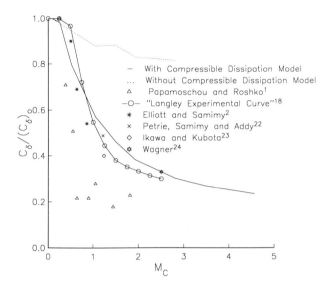

FIGURE 9. Normalized growth rate of a mixing layer as a function of the convective Mach number. Symbols represent experimental data, dotted line results without the compressible model, and the solid line, results with the compressible model.

motion for compressible isotropic turbulent flow. A key feature of the theory is the decomposition of the velocity and pressure into a solenoidal and a compressible component. The former satisfies the unsteady incompressible Navier-Stokes equations and the latter satisfies the remainder of the full compressible Navier-Stokes equations. A key result is the classification of initial conditions which lead to equilibrium states which are distinguished by the presence or absence of shocks and also by the fraction of the compressible kinetic energy. The theory is verified by the direct simulation of compressible turbulence. As expected, shocks form when the rms levels of the thermodynamic variables are sufficiently high with respect to the rms levels of velocity. When large scales dominate, the shocks appear in pairs which propagate in opposite directions to satisfy periodicity conditions. The shock/eddy interactions enhance turbulence. They also produce sound. The simulations are performed at too low a Reynolds number to sustain a good isotropic distribution of shocks for a long period of time. Until high Reynolds number simulations become feasible, it will be hard to acquire good statistics on the properties of the shock turbulence to verify the theoretical results of Tanuiti and Wei (1968).

The results discussed in this paper show that the LES approach to compressible turbulence is quite promising. Additional refinements in the subgrid scale models should be explored in order to improve the predictions for the dilatational turbulence statistics. Future efforts should be directed toward extending these models to anisotropic and inhomogeneous turbulent flows. With these further refinements—and with additional advancements in computer capacity—compressible large-eddy simulations could become an important tool for the analysis of high-speed compressible flows.

The insight gained by the combined theoretical analysis and direct numerical simulations led to the development of Reynolds stress closure models for compressible turbulence. Specifically, a new term, the compressible dissipation, was identified and modeled. Incorporation of this model of the compressible dissipation into a Reynolds-stress closure yielded the correct prediction of the dramatic decrease in the growth rate of the supersonic shear layer which occurs when the convective Mach number increases. These results are encouraging since Reynolds stress models will remain the prevailing tool for the computation of high Reynolds number complex flows.

References

Bardina, J.; Ferziger, J.H.; and Reynolds, W.C. 1983 - Improved Turbulence Models Based on Large-Eddy Simulation of Homogeneous, Incompressible Turbulent Flows. Stanford University Technical Report, TF-19.

Barnwell, D. 1989 - A skin friction law for compressible turbulent flows. AIAA Paper No. 89-1864, to appear in AIAA J., 1991.

Bradshaw, P. 1977 - Compressible Turbulent Shear Flows. *Ann. Rev. Fluid Mech.* **9**, edited by M. Van Dyck, J.V. Wehausen and J.L. Lumley (Annual Reviews Inc., Palo Alto, 1977), 33.

Canuto, C.; Hussaini, M. Y.; Quarteroni, A.; and Zang, T. A. 1988 - *Spectral Methods in Fluid Dynamics*, Springer-Verlag, Berlin.

Cebeci, T.; and Smith, A.M.O. 1974 - *Analysis of Turbulent Boundary Layers*. New York: Academic Press.

Chandrasekhar, S. 1951 - Density fluctuations in isotropic turbulence. Proc. Roy. Soc. London, **211A**, 18-24.

Chu, B.T.; and Kovasznay, L.S.G. 1958 - Nonlinear Interactions in a Viscous Heat-Conducting Compressible Gas. J. Fluid Mech. **3**, 494.

Dussauge, J.P.; and Quine, C. 1988 - A Second-Order Closure for Supersonic Turbulent Flows: Application to the Supersonic Mixing. *Workshop on the Physics of Compressible Turbulent Mixing*, Princeton.

Erlebacher, G.; Hussaini, M.Y.; Speziale, C.G.; and Zang, T.A. 1987 - Toward the Large-Eddy Simulation of Compressible Turbulent FLows. ICASE Report No. 87-20.

Erlebacher, G.; Hussaini, M.Y.; Kreiss, H.O.; and Sarkar, S. 1990 - The Analysis and Simulation of Compressible Turbulence. Theor. and Comput. Fluid Dyn. **2**, 73-95.

Erlebacher, G.; Hussaini, M.Y.; Speziale, C.G.; and Zang, T.A. 1990 - On the large-eddy simulation of compressible turbulence. Proceedings of the 12th International Conference of Numerical Methods in Fluid Dynamics, University of Oxford.

Feiereisen, W. J.; Reynolds, W. C. and Ferziger, J. H. 1981 - Numerical Simulation of Compressible, Homogeneous, Turbulent Shear Flow. Report TF-13, Dept. Mech. Eng., Stanford University.

Ferziger, J.H. 1984 - Large Eddy Simulation: Its Role in Turbulence Research. In *Theoretical Approaches to Turbulence* edited by D.L. Dwyer, M.Y. Hussaini, and R.G. Voigt (springer, New York).

Kadomtsev, B.B.; and Petviashvili, V.I. 1973 - Acoustic Turbulence. Sov. Phys. Dokl. **18**, 115.

Kline, S.J.; Cantwell, B.J.; and Lilley, G.M. 1982 - *1980–81, AFSOR - HTTM - Stanford Conference, Vol. 1*, Stanford University Press, California, 368.

Kolmogorov, A.N. 1941a - Local structure of turbulence in an incompressible fluid at very high Reynolds numbers. Doklady AN SSSR, **30**, No. 4, 299-303.

Kovasznay, L.S.G. 1953 - Turbulence in Supersonic Flows. J. Aero. Sciences **20**, No. 10, 657-682.

Kreiss, H.O.; Lorenz; and Naughton, M., 1990 - Convergence of the Solutions of the Compressible to the Solution of the Incompressible Navier-Stokes Equations. To appear in Advances in Applied Mathematics.

Krzywoblocki, M.Z.E. 1951 - On the generalized fundamental equations of isotropic turbulence in compressible fluids and in hypersonics. Proc. 1st US Nat. Congr. Appl. Mech., Chicago, New York, 827-835.

Leslie, D.C. 1973 - *Developments in the Theory of Turbulence*, Oxford University Press, Oxford.

L'vov V.S.; and Mikhailov, A.V.. 1978a - Sound and Hydrodynamic turbulence in a Compressible Liquid. Sov. Phys. J. Exp. Theor. Phys. **47**, 756.

L'vov, V.S.; and Mikhailov, A.V. 1978b - Scattering and Interaction of Sound with Sound in a Turbulent Medium. Sov. Phys. J. Exp. Theor. Phys. **47**, 840.

Marion, J-D. 1988 - Etude Spectrale d'une Turbulence Isotrope Compressible. Ph.D. Thesis, Ecole Centrale de Lyon, France.

Moiseev, S.S.; Sagdeev, R.Z.; Tur, A.V.; and Yanovskii, V.V. 1977 - Structure of Acoustic-Vortical Turbulence. Sov. Phys. Dokl. **22**, 582.

Monin, A.S.; and Yaglom, A.M. 1967 - *Statistical Fluid Mechanics*, **2**, MIT Press, Cambridge, Massachusetts.

Morkovin, M.V. 1964 - *The Mechanics of Turbulence*, edited by A. Favre (Gordon & Breach, New York), 367.

Moyal, J.E. 1952 - The Spectra of Turbulence in a Compressible Fluid; Eddy Turbulence and Random Noise. Proc. of the Cambridge Phil. Soc., **48**, part 1, 329-344.

Obukhov, A.M. 1941 - Energy distribution in the spectrum of turbulent flow. Izvestiya AN SSSR, Ser. geogr. geofiz., No. 4-5, 453-466.

Oh Y.H. 1974 - Analysis of Two-Dimensional Free Turbulent Mixing. AIAA Paper No. 74-594.

Passot, T.; and Pouquet, A. 1987 - Numerical Simulation of Compressible Homogeneous Flows in the Turbulent Regime. J. Fluid Mech. **181** 441-466.

Saffman, P. G. 1977 - Problems and Progress in the Theory of Turbulence. *Structures and Mechanisms of Turbulence II*, Lecture Notes in Physics 76, edited by H. Fieldler, Springer-Verlag, Berlin, 273.

Sarkar, S.; Erlebacher, G.; Hussaini, M.Y.; and Kreiss, H.O. 1989 - The Analysis and Modeling of Dilatational Terms in Compressible Turbulence. ICASE Report No. 89-79. Accepted for publication in J. Fluid Mech.

Sarkar, S.; and Lakshmanan, B. 1990 - Application of a Reynolds stress turbulence model to the compressible shear layer. ICASE Report No. 90-18. Accepted for publication in *AIAA Journal.*

Schlichting, H.; 1979 *Boundary Layer Theory*, McGraw-Hill, New York.

Speziale, C.G.; Erlebacher, G.; Zang, T.A.; and Hussaini, M.Y. 1988 - The subgrid scale modeling of compressible turbulence. Phys. Fluids **31**, 940-942.

Tanuiti, T.; and Wei, C.C. 1968 Reductive Perturbation Method in Nonlinear Wave Propagation. I.J. Phys. Soc. of Japan, **24**, No. 4, 941–946.

Tatsumi, T.; and Tokunaga, H. 1974 - One-Dimensional Shock-Turbulence in a Compressible Fluid. J. Fluid Mech. **65**, 581.

Tokunaga, H.; and Tatsumi, T. 1975 - Interaction of Plane Nonlinear Waves in a Compressible Fluid and Two-Dimensional Shock-Turbulence. J. Phys. Soc. Japan **38** 1167.

Vandromme, D. 1983 - Contribution to the Modeling and Prediction of Variable Density Flows. *Ph.D. Thesis*, University of Science and Technology, Lille, France.

Wilcox, D.C. 1989 - Hypersonic Turbulence Modeling without the Epsilon Equation. *Seventh National Aero-Space Plane Technology Symposium.*

Yoshizawa, A. 1986 - Statistical Theory for Compressible Turbulent Shear Flows, with the Application to Subgrid Modeling, Phys. Fluids **29**, 2152–2164.

Zakharov, V.E.; and Sagdeev, R.Z. 1970 - Spectrum of Acoustic Turbulence. Sov. Phys. Dokl. **15**, 439.

Zeman, O. 1990 - Dilatational Dissipation: The Concept and Application in Modeling Compressible Mixing Layers. Phys. Fluids A **2**, 178.

12

Statistical Aspects of Vortex Dynamics in Turbulence

Zhen-Su She, Eric Jackson and Steven A. Orszag

Abstract

Coherent features of vortex structure and dynamics are investigated statistically in pseudo-spectral numerical simulations of moderate Reynolds number turbulence. The alignment of the vorticity vector with a principal axis of the rate of strain is investigated with an emphasis on its time development in turbulence decay from a random gaussian field. In addition, a tendency in developed turbulence for velocity vectors to lie in a plane formed by the two principal stretching directions is reported, leading to an explanation of depletion of nonlinearity in turbulence. The spatial and temporal coherence of vortex structures is further studied using recently developed techniques which combine dynamical visualization with statistical sampling analysis.

1. Introduction

The study of vortex dynamics is of great importance in understanding the structure of small scale motions in turbulence. Small scale structures are responsible for energy dissipation, and therefore distinguish turbulence from equilibrium statistical mechanical systems. The major questions here concern what the typical physical processes are which lead to the generation of these small scales and what their structural properties are. The answers to these questions may be crucial to an understanding of inertial range scaling properties and intermittency phenomena (Frisch & Orszag 1989, Lundgren 1982).

There have been extensive mathematical studies of vortex dynamics during recent years (see Caflisch 1988 for a review). The general concern of these mathematical analyses is the possible formation of singularities from smooth initial conditions in the inviscid Navier-Stokes equation. Attempts have been made by Brachet *et al.*(1983) to observe such singularities in Taylor-Green flow, however, no evidence of singularity formation was found. More recent efforts have focused on the vortex reconnection process in the presence of viscosity. Intensive numerical studies of vortex reconnection conducted during the past few years (Siggia 1985, Melander & Zabusky 1987, Ashurst & Meiron 1987, Meiron *et al.* 1989, Shelley *et al.* 1989) have been essentially restricted to the description of dynamical sequences associated with simple vortical structures. Such studies provide important information on vortex structure interactions. In developed turbulence, however, one must contend with the existence of a vast, highly random background field which may lead both to much more complex vortical structures and to more complex interactions. It is unclear, for instance, whether such simple vortical structures as counter-rotating vortex tubes can actually be typical in the developed turbulence regime so as to contribute measurably to the dynamics. It is then important to conduct numerical studies to clarify what forms of vortex structures are robust in order to assess their statistical relevance in a developed isotropic turbulent flow.

From the equation of motion of the vorticity in a Lagrangian frame,

$$\frac{D\omega_i}{Dt} = s_{ij}\omega_j + \nu\nabla^2\omega_i$$

it appears that the dynamically significant processes are stretching of the vorticity at large scales by the rate-of-strain matrix, $s_{ij} = \frac{1}{2}\left(\frac{\partial u_i}{\partial x_j} + \frac{\partial u_j}{\partial x_i}\right)$, and vortex tearing at very small scales induced by viscosity. Since s_{ij} is symmetric, the (normalized) eigenvectors of s_{ij}—$s_\alpha, s_\beta, s_\gamma$—define a local coordinate system; the eigenvalues, α, β, γ (in decreasing order) then give the relative stretching/compression rate in each of the three principal axis directions. Incompressibility requires that $\alpha + \beta + \gamma = 0$. Thus, α is necessarily positive, γ negative, and β, the middle eigenvalue, may in principle be either positive or negative.

The first attempt to characterize vortex dynamics statistically was made by Betchov (1956), who derived an inequality:

$$|\langle\alpha\beta\gamma\rangle| \le \frac{1}{3\sqrt{3}}\langle\epsilon^{3/2}\rangle,$$

where ϵ denotes the dissipation rate. This relation specifies a maximum rate of vorticity production imposed by the incompressibility condition of the flow. Furthermore, he argues that, since $\langle\alpha\beta\gamma\rangle$ is related to the longitudinal velocity derivative skewness, β in a turbulent flow is typically positive.

More recent investigations begun by Ashurst *et al.* (1987a,b) and Kerr (1987) focus on the statistical behavior of the orientation of the vorticity vector. Since the eigenvectors of s_{ij} determine, in a sense, the natural coordinate system for the dynamics of the vorticity (particularly where vortex stretching is concerned), the orientation of ω can be most conveniently studied in terms of its alignment with the eigenvector directions. Surprisingly, measurements performed by the above authors using both isotropic flow and homogeneous shear flow data from direct numerical simulations show a marked tendency for alignment of the vorticity with s_β and β has a positive mean, $\bar{\beta} > 0$. The positivity of $\bar{\beta}$ implies a net stretching of vorticity statistically, and is related to energy transfer to small scales (Betchov 1956). The observed tendency of ω to align with s_β, however, provides a strong indication of coherence in turbulence. In the next section, we report some recent results of a similar kind that we have obtained, but with an emphasis on the time development of such coherent properties in turbulence decay from a random gaussian field. In addition, we have investigated the orientation of the velocity vector in the local $s_{\alpha\beta\gamma}$ coordinate system; the results provide a possible explanation of depletion of nonlinearity in turbulence (Yakhot & Orszag 1987, Kraichnan and Panda 1988).

Such single-point, single-time measurements provide a simple measure of local structural properties and provide the desired statistical treatment. Nevertheless, they are incapable of giving insight into the dynamics. It is natural, then, to extend the study by including multi-point, multi-time measurements. However, the overall pattern of turbulent structures is unlikely to be captured by traditional two or three-point/time correlation functions. We have begun parallel investigations using a new dynamical visualization technique, presented in Section 3, and new measurements of time-space correlations in a Lagrangian frame, discussed in Section 4. These techniques, used in direct numerical simulations, give a direct view of the dynamical evolution of turbulence structures.

The measurements discussed in the present work were all conducted in a series of direct numerical simulations of isotropic homogeneous incompressible turbulence. The time integration of the Navier-Stokes equations was performed using a pseudo-spectral code with resolution 96^3. 2π-periodic boundary conditions in all three directions were assumed. For measurements in stationary turbulence, we forced the system by maintaining constant energy in the first two wavenumber shells. For a more detailed description of the computer code and simulations, see She, Jackson & Orszag (1988) and references therein.

2. Time-Development of Alignment Properties and Velocity Alignment Statistics

For a random gaussian field, the velocity $\mathbf{u}$, vorticity ω, and the eigenvectors of the rate-of-strain are independent vector fields, and do not show tendencies toward any particular alignment. In this case, the middle eigenvalue β also has a zero mean. The alignment properties observed by Ashurst *et al.* (1987a) then clearly indicate significant coherence in turbulence. In order to explore further the character of this coherence, we have numerically simulated a flow which starts from a gaussian-random initial condition with excitation restricted to large scales and is then allowed to decay. In par-

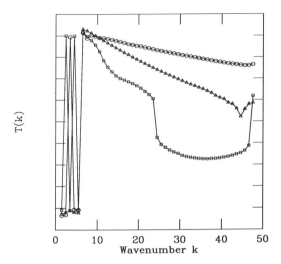

FIGURE 1. Time development of the energy transfer spectrum in turbulence decay from a random initial condition. $\square$ $t \approx 0.044$; $\triangle$ $t \approx 0.26$; $\circ$ $t \approx 0.7$.

ticular, it is interesting to investigate the time scale on which alignment is established.

Since vortex stretching is associated with the transfer of energy from large to small scales, we begin by considering the time evolution of the transfer spectrum $T(k)$. Initially, in the gaussian field, the transfer at large wavenumbers is negligibly small. The transfer to small scales is established relatively quickly, however, and the spectrum attains its characteristic turbulent shape by $t \approx 0.5 - 0.6$, well under the large-eddy turnover time (Fig. 1).

We may contrast the behavior of $T(k)$ with the development of the mean of the middle eigenvalue of the rate-of-strain, which is initially zero. The eigenvalues are normalized by $\sqrt{\epsilon/6}$ (where ϵ is the local dissipation rate) so that they lie in the ranges $-2 \leq \gamma \leq -1, -1 \leq \beta \leq 1, 1 \leq \alpha \leq 2$. The mean of β, as expected, becomes positive on approximately the same timescale as the appearance of significant transfer to small scales. This overall measure, however, is misleading. In Fig. 2 we show the mean of β conditioned on the magnitude of the rate-of-strain ϵ at several times. For high ϵ (which we may roughly consider as a measure of the small scale excitation), β develops a significant positive mean almost immediately. By $t \approx 0.2$, $\bar{\beta}$ is close to its maximum normalized value, $\bar{\beta} = 1$, over much of the range of ϵ. After the small scale excitation has been developed, the value of $\bar{\beta}$ then decreases to its typical turbulent value, $\bar{\beta} = 0.5$, at about $t \approx 0.5$. The behavior of α (Fig. 3) underscores the dynamical importance of β in establishing small-scale excitation—it tends to be close to its minimum, $\bar{\alpha} = 1$, during the

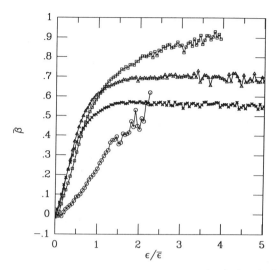

FIGURE 2. Mean of the eigenvalue β conditioned on the dissipation rate   at four different times in decaying turbulence. Here and in Figs. 3, 4, and 5 ϵ is normalized by its mean. ○ $t \approx 0.044$; □ $t \approx 0.22$; △ $t \approx 0.44$; * $t \approx 0.7$.

initial stages of transfer, then increases to its typical turbulent value as $\bar{\beta}$ decreases.

We now consider the evolution of the alignment of ω with each of the three principal axes of the rate-of-strain, $s_\alpha, s_\beta, s_\gamma$ using the variance of the cosine of the angle between ω and each eigenvector: $\Delta_{s_i,\omega} = \left\langle \left(\frac{s_i \omega}{|s_i||\omega|} \right)^2 \right\rangle$. At the initial time $\Delta_{s_i,\omega}$ is close to $1/3$ for all three eigenvectors, indicating essentially random orientation. The results plotted in Figs. 4 show that the high alignment of ω and s_β for large amplitudes of ϵ is again established on approximately the same timescale as $T(k)$, but in contrast to the evolution $\bar{\beta}$, it appears only after significant excitation has been developed at small scales.

On the basis of these results, we may characterize the meaning of the alignment of ω and the positivity of $\bar{\beta}$ somewhat more clearly. The latter property is clearly associated with the mechanism of energy transfer to small scales (indeed, it is related to the negative skewness of the longitudinal velocity derivatives). However, the alignment of ω appears to be associated with the dynamics of the small scales themselves.

Ashurst *et al.* (1987b) have also discovered that the gradient of the pressure shows a clear tendency to align with s_γ. In order to better understand the consequences of this, we have measured $\Delta_{s_i,u}$, i.e., alignment of the velocity with the principal axes. Such alignment, if it exists, would indicate correlation between large and small scales since the velocity is dominated by large-scale eddies, while the rate of strain is primarily determined by

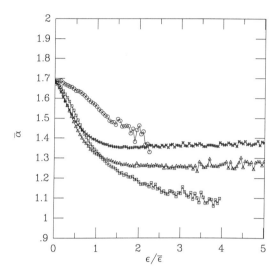

FIGURE 3. The same plot as Fig. 2, but for the eigenvalue α.

small-scale excitations. It should be noted that the velocity fluctuations in the measurements reported here are intrinsic to moderate Reynolds number turbulence.

In Figs. 5 we plot $\Delta_{s_i,v}$ conditioned on the rate of strain amplitude. There is a clear tendency for $\mathbf{u}$ to lie in the plane determined by s_α and s_β, although this tendency is not as strong as that for alignment of the vorticity. This result leads to a possible explanation of the depletion of nonlinearity in turbulence. Depletion of nonlinearity was first observed by Yakhot & Orszag (1987) and later studied by Kraichnan and Panda (1988): compared to a random gaussian field having the same energy spectrum, the variance of the total nonlinear contribution $\langle(\mathbf{u} \times \omega + \nabla P)^2\rangle$, where $P = \nabla(p + v^2/2)$, is reduced by approximately a factor of two. Kraichnan has shown that the DIA approximation is able to reproduce this feature fairly faithfully, however, understanding of the physical mechanism leading to depletion is still lacking. Our present results suggest an explanation based on physical space coherence of the flow. Since ω is remarkably aligned with s_β, the tendency for $\mathbf{u}$ to lie in the $s_\alpha - s_\beta$ plane, with a somewhat greater tendency to align with s_β, implies the existence of alignment between $\mathbf{u}$ and ω. This is the first source of depletion. In addition, $\mathbf{u} \times \omega$ will tend to align with s_γ. The observation by Ashurst et $al.$ (1987b) that the gradient of the pressure is significantly aligned with s_γ in both isotropic turbulence and homogeneous shear turbulence then implies that the pressure gradient more efficiently decimates $\mathbf{u} \times \omega$ than in a random field.

The features observed here and by Ashurst et $al.$ (1987) are a striking indication of coherence in turbulence. In addition, their association with

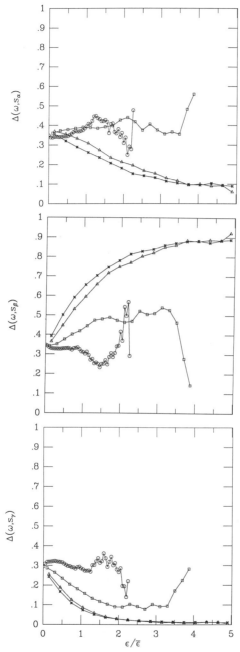

FIGURE 4. Time evolution of the alignment of ω with principal axes of the strain rate measured by $\Delta_{s_i,\omega} = \left\langle \left(\frac{s_i \cdot \omega}{|s_i||\omega|} \right)^2 \right\rangle$ and conditioned on the dissipation rate ϵ. Times corresponding to the different symbols are the same as in Fig. 2. (a) $\Delta_{s_\alpha,\omega}$; (b) $\Delta_{s_\beta,\omega}$; (c) $\Delta_{s_\gamma,\omega}$.

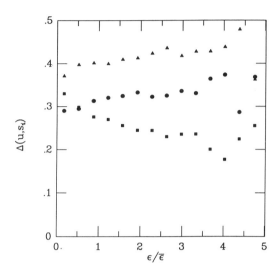

FIGURE 5. $\Delta_{s_i,\mathbf{v}}$ conditioned on ϵ, showing a tendency for velocity to align with s_α and s_β. ○: $\Delta_{s_\alpha,\mathbf{v}}$; △ : $\Delta_{s_\beta,\mathbf{v}}$; □ : $\Delta_{s_\gamma,\mathbf{v}}$.

high amplitude dissipation events clearly indicates that this coherence is closely associated with the phenomenon of intermittency in turbulence. However, as we have already indicated, these properties are a consequence of the dynamics of small scales. The single-point, single-time measurements performed here are therefore unable to shed light on their origins. In the following sections we present two new methods of observing the dynamical behavior of vorticity in turbulence.

3. Dynamical Visualization of Small-Scale Turbulent Structures

The statistical measures traditionally made in numerical simulations of turbulence are completely unable to capture three-dimensional structural information and, to date, no other measures which may reasonably be used have been available. Direct visualization of the three-dimensional turbulent field seems to be the most efficient way to capture spatial coherence in turbulence. A number of investigators (Siggia 1981, Brachet *et al.* 1987, Yamamoto *et al.* 1988, Shelley *et al.* 1989) have obtained single-time volumetric visualizations of iso-surfaces of such quantities as vorticity and dissipation. However, these too are inadequate to assess the dynamical significance of the structures captured. In order to form a better picture of what structures may be expected to play a significant *dynamical* role in turbulence, we use a new technique for visualizing the time evolution of three-dimensional continuum fields, following Russell (1989).

Full volumetric visualization of a field is computationally prohibitive to perform frequently enough to capture the time evolution of structures. Instead, we have implemented a variant of a radiative transfer method to project a picture of the field onto a moderately fine (128×128) plane. The technique, briefly, is as follows. At a given instant, we fix a plane in space from which we will view the field. A source on the opposite side of the computational box emits light along rays perpendicular to the viewing plane. The quantity to be visualized in the field is then treated as an absorbent material which reduces the intensity, $I = 1$ at the source, of the light passing through it according to the iteration function $I_{t+1} = I_t/[1. + \alpha A^2 \Delta]$, where A is the amplitude of the visualized quantity, α is a normalized opacity parameter, and Δ is the step size along the ray (other functions were tried, from exponential to linear, with consistent results—this scheme was chosen as visually optimal). After the iteration over all rays, the low and high intensity regions on the plane are reversed ($I \rightarrow 1/I$) and the data stored. Once a series of such frames at different times is generated, they are visualized as a movie on a graphics workstation by mapping the intensity to a color ramp and rapidly displaying the frames in succession. Frames were generated with sufficient frequency to capture smooth evolution of the field. We have performed these visualizations for both forced, stationary turbulence and for decay from a gaussian random large-scale initial condition with similar results. The figures discussed below are from a turbulence decay run.

The visualizations presented in Figs. 6a,b, showing typical frames for ω^2 and ϵ after several large-eddy turnover times in a decay run, shed light on an important problem which has attracted much attention recently, namely, whether small scale turbulent structures are sheet-like or tube-like. Based on the fact that the strain rate typically has two positive eigenvalues and one negative one in isotropic turbulence, it has been speculated (Betchov 1956) that small-scale structures should be sheet-like. In other words, an initially nearly spherical material element will evolve, under the action of the turbulent strain, to a "lasagne"-shaped object. The results shown in Figure 6a, however, clearly show thinner, elongated "macaroni"-like structures (She, Jackson & Orszag 1989). The strain rate plotted in Fig. 6b seems to have a more diffuse structure than in Fig. 6a, related to a faster falloff of the probability of high-amplitude events in ϵ than in ω. Use of an increased damping parameter for the dissipation reveals a core with a structure similar to that of the vorticity. What cannot be shown here is that these filament-like structures retain their coherence over *extended* times, many times longer than the local turnover times associated with their primary length scales. They move with the fluid and undulate in time, but do not break up; rather, after a certain time, they disappear entirely, maybe through rapid viscous diffusion. The long lifetime of these objects clearly

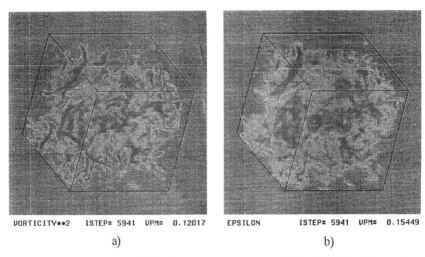

VORTICITY**2   ISTEP= 5941   UPM=   0.12017   EPSILON   ISTEP= 5941   UPM=   0.15449

a)   b)

FIGURE 6. A snapshot of (a) the vorticity intensity field and (b) the dissipation rate field of developed turbulence using a dynamical visualization technique.

implies that they may have significant dynamical consequences on the surrounding field.

4. Dynamics of Alignment Properties in Turbulence

The statistical measurements of vector alignments in a turbulent field and the dynamical 3-D visualization technique discussed above give new ways to investigate turbulence. The results obtained so far dramatically underscore the existence of much stronger structural coherence in turbulence than had previously been imagined. At the same time, however, we are immediately faced with the limitations of these techniques when we attempt to seek the origins of the behavior they capture. Visualization techniques, on the one hand, give us a gross picture of large-scale, dynamically significant structures, but cannot produce information on the details of local interactions which give rise to them. The statistics on alignment, on the contrary, give only pointwise information at a single instant and cannot explain the dynamics that lead to the alignment. In order to pursue these questions further, we are developing new measurement techniques which incorporate elements of both these approaches. Some important initial results are presented below. However, to motivate them, let us first consider what kinds of processes might lead to the observed alignment properties.

The instantaneous effect of the strain at a given point is to disproportionately increase the α-component of ω, leading to increased alignment with the α-eigenvector direction. There are only two ways in which this effect may be avoided: the diffusion of the vorticity acting particularly strongly on the α-component, or rotation of the principal axes of the rate-of-strain due to the interaction of the strain with pressure and vorticity. One possibility is to investigate numerically the contribution of the various terms involved, again using a simple single-point, single-time approach. While this may lead to interesting results, it is unlikely to capture the dynamics. Indeed, such a simple and vital question as what statistical behavior is associated with the development of alignment and what is associated with its later breakdown falls completely outside the scope of this approach.

What must be done, then, is to isolate a set of events, e.g., high alignment or high dissipation at some instant, and to study the events which surround them in time and space. We restrict ourselves in the present paper to an examination of their temporal behavior. Since small-scale events are advected by the large-scale velocity, an Eulerian framework is inappropriate. Instead, we choose a large set of particles, usually several thousand, distributed randomly in space and track them in time, recording at regular intervals all pertinent local information. With this database, we then choose classes of particles having some set of characteristics which we specify, and explore the properties associated with them at previous and subsequent times.

Since the question of how vorticity grows and whether there is any mechanism which eventually limits its growth is central both to the study of turbulence dynamics and to the mathematical theory of the Navier-Stokes equations, here we examine groups of particles according to their vorticity amplitude. Furthermore, it turns out that high vorticity regions closely parallel high amplitude (intermittent) regions of ϵ as well. Consistent with the single time measurements discussed above, these are also the regions where the tendency of ω to align with s_β is strongest.

In Figure 7a, we plot the typical behavior of the vorticity and rate-of-strain amplitudes, normalized by $\langle \omega^2 \rangle$ and $\langle \epsilon \rangle$, for a set of particles specified to have normalized vorticity greater than 0.4 at a time $t_0 = 0.8$. To see how alignments change in time, in Fig. 7b we plot Δ_{s_i,ω_0} and Δ_{ω,ω_0} using the fixed vector $\omega_0 = \omega(t_0)$ as a reference. At $t \approx t_0$, ω is almost perfectly aligned with s_β for these particles, consistent with the results of conditional measurements presented above. Thus, the reference direction is also approximately the s_β direction at t_0. It can be seen that during some period before t_0, it is s_α which aligns more with the reference direction. Since ω has an appreciable component in s_α the stronger stretching in s_α tends to both increase the vorticity amplitude and rotate it further toward the reference direction. Consistently, however, after an initial period of stretching in the s_α and s_β directions, a rapid rotation of the principal axes occurs which leaves s_β strongly aligned with ω. At no point is there any significant alignment of ω and s_γ. (It should be noted that s_α rotates to the direction previously associated with s_γ—this may be important for subsequent rotation of ω; a second candidate is the pressure which, although it does not enter the vorticity equation directly, may affect how it is stretched or compressed through the strain. We have observed that the period of peak vorticity is also associated with an increase in the magnitude of the pressure gradient.)

A striking feature of all the cases we have studied is the fact that there is a well-defined time scale associated with growth of the vorticity, rotation of the principal axes, and subsequent rapid decay of the vorticity amplitude. Since we observe no particular alignment of w with the principal axis of compression, s_γ, we may expect that viscous diffusion plays the dominant role in the decay period. This is confirmed by the results plotted in Fig. 7c where we show the time development of $|\omega|$ and the diffusion term, $\nu|\nabla^2\omega|$.

5. Conclusions and Future Work

The results which we have presented in this work represent a first step toward clarifying what kinds of vortex structures play a significant role in turbulence dynamics and what sequences of events are associated with the development of the intermittent behavior of ω and ϵ. However, we still lack a picture of what physical mechanisms give rise to the observed events and how they lead to the generation of the large-scale, long-lived structures observed using our dynamical visualization technique. In order to pursue these questions, two points seem to us to be essential. First, it seems clear that more detailed study of spatial structure must be made, but that this cannot be done in isolation from the time dimension. Second, it seems to us essential that this study embrace the inherently statistical nature of turbulence phenomena.

Acknowledgments

This work was supported by the Air Force Office of Scientific Research under Contract F49620-87-C-0036, the Office of Naval Research under Contract N00014-82-C- 0451, and DARPA under Contract N00014-86-K-0759. Computations were performed on the ETA10 at the John von Neumann Supercomputer Center.

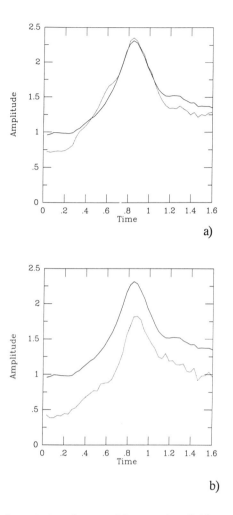

FIGURE 7. Temporal statistics of a set of Lagrangian fluid particles chosen to have large vorticity amplitude at $t = 0.8$. (a) Time variation of $\langle|\omega|\rangle$ (solid line) and ϵ (dotted line). (b) Time variation of $\Delta_{\omega,\omega_0}(\triangle), \Delta_{s_\alpha,\omega_0}$ (solid line), $\Delta_{s_\beta,\omega_0}$ (dotted line) and $\Delta_{s_\gamma,\omega_0}$ (dashed line). (c) Time variation of $\langle|\omega|\rangle$ (solid line) and diffusion, $\langle\nu|\omega \cdot \nabla^2\omega|\rangle$ (dotted line). Values of ω, ϵ and diffusion are normalized by their rms values.

References

Ashurst W. T., Chen J.-Y., and Rogers M. M., 1987b. Phys. Fluids **30** (10), 3293.

Ashurst W. T., Kerstein A. R., Kerr R. M., and Gibson C. H., 1987a. Phys. Fluids **30**, 2343.

Ashurst W. T. and Meiron D. I. 1987. Phys. Rev. Lett. **58**, 1632.

Betchov R., 1956. J. Fluid Mech. **1** (5), 497.

Brachet M. E., Meiron D. I., Orszag S. A., Nickel B. G., Morf R. H., and Frisch U., 1983. J. Fluid Mech. **130**, 411.

Caflisch R. E., 1989. In *Mathematical Aspects of Vortex Dynamics* (ed. R. E. Caflisch), SIAM, p. 1.

Frisch U. and Orszag S. A., 1989. Phys. Today **43**, 24.

Kerr R. M., 1987. Phys. Rev. Lett. **59** (7), 783.

Kraichnan R. H. and Panda R., 1988. Phys. Fluids **31**, 2395.

Lundgren T. S., 1982. Phys. Fluids **25** (12), 2193.

Meiron D. I., Shelley M. J., Ashurst W. T., and Orszag S. A., 1989. In *Mathematical Aspects of Vortex Dynamics* (ed. R. E. Caflisch), SIAM, p. 183.

Melander M. and Zabusky N., 1987. In *Proceedings, IUTAM Symposium on Fundamental Aspects of Vortex Motion*, Tokyo.

Russell G., 1989. "Volumetric Visualization of Scientific Data", Ph.D. Thesis, Princeton University.

She Z.-S., Jackson E., and Orszag S.A., 1988. J. Sci. Comp. **3**, **4**, 407-434.

She Z.-S. and Orszag S.A., 1990. Nature **344**, 226.

Shelley M. J., Meiron D. I., and Orszag S. A., 1990. J. Fluid Mech., submitted.

Siggia E. D., 1981. J. Fluid Mech. **107**, 375.

Siggia E. D., 1985. Phys. Fluids **28**, 794.

Yakhot V. and Orszag S. A., 1987. Nuc. Phys. B (Proc. Suppl.) 2.

Yamamoto K. and Hosokawa I., 1988. J. Phys. Soc. Japan **57** (5), 1532.

13

The Lagrangian Picture of Fluid Motion and its Implication for Flow Structures

J.T. Stuart

Abstract

In modern studies in fluid dynamics it is quite common to describe the velocity and pressure fields in the Eulerian way, with these quantities being measured and defined at a given point in space. Having found this Eulerian velocity field, $u_i(x_j, t)$, where i, j range over 1,2,3 and u_i is associated with the coordinate x_i, we can then study the equations

$$\frac{dx_1}{u_1} = \frac{dx_2}{u_2} = \frac{dx_3}{u_3} = dt,$$

in order to obtain the particle paths and properties associated with them. Such knowledge can be important for the understanding of flows visualized experimentally by dye or smoke.

An alternative approach is that of the Lagrangian description, in which the individual particles are *marked* and followed in a time-dependent way. A time derivative on a given marked particle gives its velocity, and this gives a connection with the Eulerian description mentioned above.

The partial differential equations for the Eulerian and Lagrangian schemes look superficially different, but are connected by the ordinary differential equations quoted above. However, there are some phenomena of relevance and importance in connection with turbulence and with transition to turbulence, in which an approach from the Lagrangian point of view gives rise to simpler and less intuitive nonlinear mathematics, and leads to illuminating insights. Thus a Lagrangian approach to such problems will be described in this paper, but with reference only to an inviscid fluid satisfying a simple pressure (p)-density (ρ) law of the form $p = f(\rho)$ or more simply still for $\rho =$ constant.

The particular class of problems to be addressed is that connected with longitudinal vorticity filaments in boundary layers or other shear flows. Of relevance here is the notion of the near collision of particles at spanwise locations where there is a convergence of the flow, and in the neighborhood of which eruptions of the flow field can take place. In an inviscid process it is shown that there is a distinct possibility of a singularity forming in a finite time, depending on the initial conditions given for the problem.

The Lagrangian picture of flows of this type will be described from a mathematical point of view but with a physical interpretation.

1 Introduction

In turbulent and transitional boundary layers and elsewhere, there are often situations which involve a flow convergence locally or even more generally. One example of this occurs in the region between longitudinal vortex placements in a boundary layer, whether this be of a transitional kind, as in Klebanoff's famous 1962 experiments, or of a turbulent nature with a sublayer. We intend here to discuss such phenomena from an inviscid point of view as initial–value problems, on the grounds that there are many processes where the time scale is so short that viscosity may not have time to act; and when it does so, effects take place initially in rather narrow regions, although over longer time scales the effects can be very significant and influential. The present author has studied these effects in the past (1,2,3) and some of the results of the present paper have been given in reference 2 in an embryonic form and from an Eulerian point of view. The so-called double-stagnation structure has been discussed in (3). Effects of viscosity on singularity development have been studied in a nice paper (4) by Childress and his co-workers. A number of references to significant past work are given in reference 3, where reference is made to work of Hoskins and Bretherton (5), Stern and Paldor (6), Calogero (7) and to Russell and Landahl (8). We wish to mention also the especial significance of work on unsteady boundary layer separation by Van Dommelen and his colleagues (9,10).

In contrast with the early work (2) of 1983 and with the developments (3) of 1987, we wish here to give a general treatment from a Lagrangian point of view. In so doing we hope to throw some light on the interesting question of the occurrence of singularities in flow fields governed by inviscid equations.

2 Basic Equations and Lagrangian Structure

We consider (Figure 1) a particle which, at time t, lies at a point P, which has coordinates x_j ($j = 1, 2, 3$), equivalent to x, y, z. The particle is *marked* by its position coordinates at time $t = 0$ (point P_0), namely $a_i(i = 1, 2, 3)$, equivalent to ϕ, ψ, χ. Thus the set a_i defines the particle uniquely. In order to follow the evolution of the field, we shall need to calculate the particle position (x_j), the pressure (p) and the density (ρ) as functions of time and of its defining particle coordinates (a_i). Thus the governing equations are

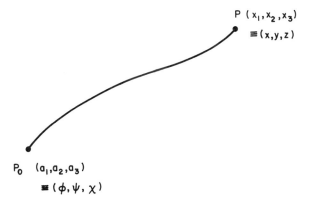

FIGURE 1. Lagrangian description.

(Lamb, (11), pp. 12-15 and 205)

$$\rho \frac{\partial^2 x_j}{\partial t^2} \frac{\partial x_j}{\partial a_i} = -\frac{\partial p}{\partial a_i}, \tag{2.1}$$

$$\rho \det\left(\frac{\partial x_j}{\partial a_i}\right) = \rho_0, \tag{2.2}$$

$$p = f(\rho), \tag{2.3}$$

$$u_i = \frac{\partial x_i}{\partial t}\Big|_{a_j \text{ fixed}}, \tag{2.4}$$

$$\omega_i = \epsilon_{ijk}\frac{\partial u_k}{\partial x_j} \tag{2.5}$$

$$\frac{\omega_i}{\rho} = \frac{\omega_{j0}}{\rho_0}\frac{\partial x_i}{\partial a_j}, \tag{2.6}$$

where ρ_0 is the density of a particle at time $t = 0$, u_i is the velocity of a particle at x_i at time t, ω_i is the corresponding vorticity and ω_{j0} is the vorticity at time $t = 0$.

The first of these formulas (2.1) is the equation of motion (Newton's second law), the second (2.2) is the condition of mass conservation (continuity), the third (2.3) is an equation of state, the fourth (2.4) is the kinematical relation for the velocity at a point, the fifth (2.5) defines the vorticity and the sixth (2.6) is Cauchy's equation for the vorticity in terms of the initial vorticity and the strain tensor $\partial x_i/\partial a_j$.

Our intention, as explained briefly in the Introduction, is to discuss a situation in which a main flow, which is along a plate parallel to its x axis (Figure 2), is perturbed by a secondary field in the $y - z$ plane (Figure 3). Thus a plane, which is defined by $z = 0$, is a plane of symmetry towards

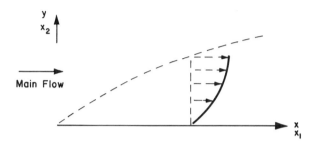

FIGURE 2. Boundary layer on a plate.

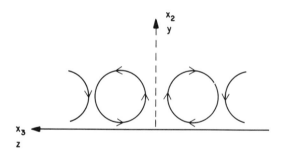

FIGURE 3. Secondary flow in cross plane.

which this secondary flow converges at lower values of y (the coordinate normal to the plate) and from which it diverges at higher values. With this concept, which is relevant for flows undergoing transition to turbulence, we can define a Lagrangian structure of the field to be discussed. It is emphasized that this assumption of plane of symmetry is essential in what follows (2.7 and onwards), and that it would be desirable to embed this flow in a more general field; this, however, remains an aspiration for the future.

Thus we write

$$x_1 = X(\phi, \psi, t), x_2 = Y(\phi, \psi, t), x_3 = \chi Z(\phi, \psi, t) \qquad (2.7)$$

from which with (2.4) we can infer the velocity field:

$$u_1 = \frac{\partial X}{\partial t}, u_2 = \frac{\partial Y}{\partial t}, u_3 = \chi \frac{\partial Z}{\partial t} = x_3 Z^{-1} \frac{\partial Z}{\partial t}. \tag{2.8}$$

The last of the formulas (2.8) shows the consistency with the concept of convergence, with a concomitant stagnation on $x_3 = 0$.

For initial conditions we shall regard the following as applying:

$$t = 0; X = \phi; Y = \psi; Z = 1;$$

$$\frac{\partial X}{\partial t} = u_0(\phi, \psi); \frac{\partial Z}{\partial t} = bw_0(\phi, \psi). \tag{2.9}$$

Here the parameter b is a rate of strain. The first three of these conditions arise from the particle current coordinates (x_i) reducing to the particle coordinates (a_i) at $t = 0$. The last two conditions are concerned with the specification of the velocity in the x and z directions at $t = 0$. The y component of velocity may be regarded as being given by the continuity condition (2.2), to which we shall turn.

Prior to doing so, however, we note that derivatives with respect to x_j and a_i are related by

$$\frac{\partial}{\partial x_j} = \frac{\partial a_i}{\partial x_j} \frac{\partial}{\partial a_i}, \tag{2.10}$$

where

$$\frac{\partial a_i}{\partial x_j} = \frac{1}{J_3} \begin{pmatrix} Y_\psi & -Y_\phi & \chi Z^{-1} J_1 \\ -X_\psi & X_\phi & \chi Z^{-1} J_2 \\ 0 & 0 & Z^{-1} J_3 \end{pmatrix}. \tag{2.11}$$

In (2.11) the Jacobians J_i are

$$J_1 = Y_\phi Z_\psi - Y_\psi Z_\phi,$$

$$J_2 = Z_\phi X_\psi - Z_\psi X_\phi, \tag{2.12}$$

$$J_3 = X_\phi Y_\psi - X_\psi Y_\phi.$$

The continuity condition, to which reference has already been made, is

$$\rho Z J_3 = \rho_0 \tag{2.13}$$

Our next object is to consider the vorticity, which can be calculated from (2.5) and then related to the initial vorticity by (2.6). Thus, using the continuity condition (2.13) above we obtain

$$-Z(X_\psi Z_{\phi t} - X_\phi Z_{\psi t}) + J_2 Z_t = Z_{\psi t o} X_\phi - Z_{\phi t 0} X_\psi, \tag{2.14}$$

$$-Z(Y_\psi Z_{\phi t} - Y_\phi Z_{\psi t}) - J_1 Z_t = Z_{\psi t 0} Y_\phi - Z_{\phi t 0} Y_\psi, \tag{2.15}$$

$$Y_\psi Y_{\phi t} - Y_\phi Y_{\psi t} + X_\psi X_{\phi t} - X_\phi X_{\psi t} = Y_{\phi t 0} - X_{\psi t 0}, \tag{2.16}$$

where knowledge of Z from Section 3 has been used to simplify (2.16). These are the first, second and third components of Cauchy's relation (2.6). A suffix zero denotes that expression at $t = 0$.

Using (2.12), the first (2.14) of these relations can be written

$$X_\phi(ZZ_{\psi t} - Z_\psi Z_t - Z_{\psi t0}) + X_\psi(-ZZ_{\phi t} + Z_\phi Z_t + Z_{\phi t0}) = 0. \quad (2.17)$$

while (2.15) is

$$Y_\phi(ZZ_{\psi t} - Z_\psi Z_t - Z_{\psi t0}) + Y_\psi(-ZZ_{\phi t} + Z_\phi Z_t + Z_{\phi t0}) = 0. \quad (2.18)$$

Since $X_\phi Y_\psi - X_\psi Y_\phi = J_3 \neq 0$ by (2.13), it follows that

$$ZZ_{\psi t} - Z_\psi Z_t = Z_{\psi t0} = bw_{0\psi}, \quad (2.19)$$

$$ZZ_{\phi t} - Z_\phi Z_t = Z_{\phi t0} = bw_{0\phi}, \quad (2.20)$$

where use has been made of the last of (2.9).

Noting that the density has disappeared from (2.19) and (2.20), as well as from (2.16), we can proceed to discuss the Z function without reference to ρ and, indeed, without reference to X and Y in the first instance.

The reader will notice a peculiarity about (2.19) or (2.20) subject to (2.9). Each equation is hyperbolic and the characteristics are $t = \text{constant}$ and $\psi = \text{constant}$ (2.19) or $\phi = \text{constant}$ (2.20). Thus initial conditions are specified on a characteristic, $t = 0$, and we cannot apply Cauchy's theorem on the properties of solutions to initial–value problems. We return to this matter in Section 3.

3 Solution for the Flow Orthogonal to a Plane of Symmetry

We consider here equations (2.19) and (2.20) which, with the conditions on Z and $\partial Z/\partial t$ of (2.9), define the function Z. Each of those equations has the form of a Wronskian expression, this being especially evident (in 2.19) if we write

$$Z_\psi = bw_{0\psi}\zeta, \quad (3.1)$$

where the ψ suffix on w_0 denotes a derivative. Then (2.19) becomes the Wronskian

$$Z\zeta_t - \zeta Z_t = 1, \quad (3.2)$$

which implies that Z and ζ each satisfy

$$Z_{tt} + q(t, \phi, \psi)Z = 0, \quad (3.3)$$

where q is a function, as yet unknown, of the variable t and the *parameters* ϕ and ψ (in the context of (3.2)).

A t differential of (2.19), namely

$$ZZ_{\psi tt} - Z_\psi Z_{tt} = 0 \tag{3.4}$$

with Z_{tt} inserted from (3.3), implies that q cannot depend on ψ. A similar argument from (2.20) with

$$Z_\phi = bw_{0\phi}\eta \tag{3.5}$$

can be used to show that q does not depend on ϕ either. Thus Z satisfies

$$Z_{tt} + q(t)Z = 0 \tag{3.6}$$

where $q(t)$ has to be determined by reference to initial (2.9) or other conditions on the problem, perhaps to the nature of the pressure field.

The nonlinear operator on the left–hand side of (2.19) is symmetrical in the t and ψ derivatives, and an alternative Wronskian relationship is suggested. Thus we define

$$Z_t = Q(\phi, \psi, t) \tag{3.7}$$

and (2.19) becomes the Wronskian

$$ZQ_\psi - Z_\psi Q = bw_0(\phi, \psi). \tag{3.8}$$

This implies that Z and Q satisfy

$$Z_{\psi\psi} - \frac{w_{0\psi\psi}}{w_{0\psi}} Z_\psi + \sigma(\phi, \psi, t)Z = 0, \tag{3.9}$$

where σ is an unknown function of the variable ψ and of the *parameters* ϕ and t (in the context of (3.8)).

A ψ differential of (2.19), namely

$$ZZ_{\psi\psi t} - Z_{\psi\psi}Z_t = bw_{0\psi\psi}, \tag{3.10}$$

with $Z_{\psi\psi}$ substituted from (3.9), implies that σ cannot depend on t. Thus we have

$$Z_{\psi\psi} - \frac{w_{0\psi\psi}}{w_{0\psi}} Z_\psi + \sigma(\phi, \psi)Z = 0, \tag{3.11}$$

where $\sigma(\phi, \psi)$ has to be determined from the initial (2.19) or other conditions. Related arguments, starting from (2.20) show that Z must also satisfy

$$Z_{\phi\phi} - \frac{w_{0\phi\phi}}{w_{0\phi}} Z_\phi + \nu(\phi, \psi)Z = 0, \tag{3.12}$$

where $\nu(\phi, \psi)$ has to be determined.

Thus we have shown that the function Z, which satisfies (2.19) and (2.20) must consequently satisfy (3.6), (3.11) and (3.12). We therefore focus attention on these three linear ordinary differential equations of the second order.

In order to make progress we first consider one initial condition (2.9), namely that $Z = 1$ at $t = 0$. But, (3.11) and (3.12) have t only as a parameter, so that any solution for Z at any time, including $Z = 1$ at $t = 0$, must satisfy (3.11), (3.12) at that time. This implies that

$$\sigma(\phi, \psi) = \nu(\phi, \psi) \equiv 0. \tag{3.13}$$

Utilizing (3.13), we see that two linearly independent solutions of (3.11) and of (3.12) are 1 and $bw_0(\phi, \psi)$, so that the solution of those two equations can be written

$$Z = \alpha(t)bw_0(\phi, \psi) + \beta(t), \tag{3.14}$$

where $\alpha(t)$ and $\beta(t)$ are *constants* of integration (albeit dependent on the *parameter* t).

We return now to (3.6), which Z must also satisfy. Thus, for that equation, $\alpha(t)$ and $\beta(t)$ are to be regarded as two linearly independent solutions, with 1 and $bw_0(\phi, \psi)$ as integration *constants* (but dependent on the *parameters* ϕ and ψ). Thus (3.14) gives the solution of (2.19) and (2.20), subject to the appropriate conditions of (2.9). There remains, however, the unknown function $q(t)$ in (3.6) and its implications for $\alpha(t)$ and $\beta(t)$. Prior to discussing its determination we can draw some inferences about (3.14).

Suppose that $\alpha(t)$ and $\beta(t)$, the solutions of (3.16), satisfy the conditions

$$\begin{aligned} \alpha(0) &= \quad \beta'(0) = 0, \\ \alpha'(0) &= \quad \beta(0) = 1; \end{aligned} \tag{3.15}$$

then (3.14) satisfies the conditions on Z and on $\partial Z/\partial t$ of (2.9). The spanwise (x_3) component of velocity is given by

$$u_3 = x_3 w(\phi, \psi, t), \tag{3.16}$$

$$w = \frac{\alpha'(t)bw_0(\phi, \psi) + \beta'(t)}{\alpha(t)bw_0(\phi, \psi) + \beta(t)}. \tag{3.17}$$

It can be seen that a singularity in the function w is possible in principle if the denominator can become zero, namely

$$\alpha(t)bw_0(\phi, \psi) + \beta(t) = 0, \tag{3.18}$$

and it would normally occur at a finite value of t. If $\alpha(t)$ and $\beta(t)$ are non-negative functions of t, the singularity will require the function $w_0(\phi, \psi)$ to have a negative range. From the definition (3.16) and Figure 3, and taking note of the condition $Y = \psi$ at $t = 0$ (2.9), we see that an appropriate form for w_0 as a function of ψ is as shown in Figure 4, where the negative region corresponds to the convergence near the plate and the positive region relates to the divergence at greater distances. We emphasize, however, that the singularity will occur, if it does, due to the region of convergence; of especial significance is the minimum in Figure 4.

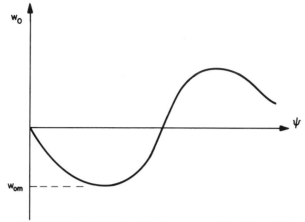

FIGURE 4. Typical profile for ω_0 as a function of ψ.

Hitherto, however, we have said relative little about X and Y as functions of ϕ and ψ (and of t). In order to be certain of the occurrence of a singularity, we need to know the expressions for X and Y and, by inversion, ϕ and ψ in terms of X and Y. Then, if this could be done, the formulas (3.14) and (3.17) would become explicit in terms of X, Y and t, and so given in our Euclidean space.

The determination of X and Y is to be pursued in Sections 4,5,6 but we can first determine $q(t)$. To this end we need to consider the equation of motion (2.1) which, after use of (2.7) followed by (3.6) to eliminate Z_{tt}, yields by integration

$$\int \frac{dp}{\rho} = \frac{1}{2}q(t)x_3^2 - \int^\psi (X_{tt}X_\psi + Y_{tt}Y_\psi)d\psi. \tag{3.19}$$

There is an equivalent form with ϕ derivatives and ϕ-integration replacing ψ operations in (3.19), the equivalence being a consequence of the third component (2.16) of Cauchy's vorticity formula.

Equation (3.19) shows that the pressure field has a stagnation–point character, given by the first term on the right side of (3.19) as a quadratic function of x_3. This is consistent with Figure 3, which shows the nature of the flow field projected on the plane of x_2 and x_3.

If there were no pressure gradient across the plane $x_3 = 0$, $q(t)$ would be identically zero, and (3.14) and (3.17) would become respectively

$$Z = 1 + btw_0(\phi, \psi), \tag{3.20}$$

$$w = \frac{bw_0(\phi, \psi)}{1 + btw_0(\phi, \psi)} \tag{3.21}$$

If, on the other hand $q(t) = -b^2$, as in a steady stagnation–point flow at the nose of a bluff cylindrical body, we would have

$$Z = \cosh bt + w_0(\phi, \psi) \sinh bt, \tag{3.22}$$

$$w = \frac{b[\sinh bt + w_0(\phi, \psi) \cosh bt]}{\cosh bt + w_0(\phi, \psi) \sinh bt}. \tag{3.23}$$

We return later to a discussion of these possibilities in Section 7.

4 Solution for the Flow Projected onto the Plane of Symmetry

The equations that we now have to study are (2.16) and (2.13) with the Jacobian given by (2.12):

$$Y_\psi Y_{\phi t} - Y_\phi Y_{\psi t} + X_\psi X_{\phi t} - X_\phi X_{\psi t} = Y_{\phi t 0} - X_{\psi t o} \tag{4.1}$$

$$\rho Z (X_\phi Y_\psi - X_\psi Y_\phi) = \rho_0 \tag{4.2}$$

Here Z is given by (3.14), or by (3.20) or (3.22) in special cases.

So far we have considered a compressible fluid with an equation of state (2.3). It is possible to make progress with the general situation, and this will be pursued elsewhere. For the present analysis, however, we restrict ourselves to the incompressible case, $\rho = \rho_0$. Thus we replace (4.2) by

$$X_\phi Y_\psi - X_\psi Y_\phi = Z^{-1} \tag{4.3}$$

We see that (4.3) involves ϕ and ψ derivatives only, with t as a parameter. The structural form of the nonlinear operator on the left–hand side is that of a Jacobian. Equation (4.1), on the other hand, is the sum of two Wronskian expressions, each with t as the relevant derivative. We wish to exploit this Wronskian property, and proceed as follows.

Let us write

$$X_\psi X_{\phi t} - X_\phi X_{\psi t} = W^{rx}(t; \phi, \psi), \tag{4.4}$$

$$Y_\psi Y_{\phi t} - Y_\phi Y_{\psi t} = W^{ry}(t; \phi, \psi), \tag{4.5}$$

where the two Wronskians, on the right-hand sides, are as yet unknown, but must satisfy

$$W^{rx} + W^{ry} = Y_{\phi t 0} - X_{\psi t 0}, \tag{4.6}$$

a relation which follows from (4.1).

Either of equations (4.4) and (4.5) is really analogous to (2.19) and (2.20), with X_ϕ and X_ψ in (4.4) and Y_ϕ and Y_ψ in (4.5) each related by a Wronskian with respect to t. Without loss of generality we can consider (4.4) only to illustrate our procedures.

It is clear that X_ϕ must satisfy

$$X_{\phi tt} - LX_{\phi t} + \sigma(t, \phi, \psi)X_\phi = 0, \tag{4.7}$$

where σ is as yet unknown, while X_ψ satisfies

$$X_{\psi tt} - LX_{\psi t} + \sigma(t, \phi, \psi)X_\psi = 0, \tag{4.8}$$

where

$$L \equiv W_t^{rx}/W^{rx}. \tag{4.9}$$

There is a consistency condition between (4.7) and (4.8), which we can obtain by differentiation with respect to ψ and ϕ respectively, followed by subtraction.

This condition is

$$L_\phi X_{\psi t} - L_\psi X_{\phi t} + \sigma_\psi X_\phi - \sigma_\phi X_\psi = 0 \tag{4.10}$$

At this stage each of the expressions in (4.10) is conceived as a function of ϕ, ψ and t. From this point, however, we shall suppose that there is a one–to–one mapping between ϕ and ψ on the one hand and $X(\phi, \psi, t)$ and $X_t(\phi, \psi, t)$ on the other hand, for a given value of t. With this assumption we shall regard L and σ as being functions of X, X_t and t, instead of ϕ, ψ, t. Then, with use of (4.4) we obtain

$$-(L_X + \sigma_{X_t})W^{rx} = 0. \tag{4.11}$$

Since $W^{rx} \not\equiv 0$, we infer that

$$L = \ell_{X_t}(X, X_t, t), \sigma = -\ell_X(X, X_t, t) \tag{4.12}$$

where $\ell(X, X_t, t)$ has yet to be determined. If we substitute (4.12) into (4.7) and (4.8), we can integrate with respect to ϕ and ψ respectively and obtain

$$X_{tt} - \ell(X, X_t, t) = 0 \tag{4.13}$$

In a similar way, from (4.5), we obtain

$$Y_{tt} - m(Y, Y_t, t) = 0, \tag{4.14}$$

where $m(Y, Y_t, t)$ has still to be determined.

At this point we see that each of the current coordinates (X, Y) satisfies its own second order differential equation. There are, however, constraints on ℓ and m, one being the continuity condition (4.3), which does not involve t derivatives. The other constraint is (4.6) between the two Wronskians, and this can be shown to imply that

$$X_{\psi t0} \left\{ 1 - \exp \left[\int_0^t \ell_{X_t}(X, X_t, t)dt \right]_{\substack{\phi, \psi \\ \text{fixed}}} \right\}$$

$$= Y_{\phi t 0} \left\{ 1 - \exp \left[\int_0^t m_{Y_t}(Y, Y_t, t) dt \right]_{\substack{\phi, \psi \\ \text{fixed}}} \right\} \tag{4.15}$$

The initial conditions on $X(t, \phi, \psi)$ for (4.13) are given by (2.9), while the conditions on $Y(t, \phi, \psi)$ are given by (2.9) and by the time derivative of the continuity condition; the latter implies

$$Y_t(0, \phi, \psi) = - \int_0^\psi [u_{0\phi} + bw_0] d\psi, \tag{4.16}$$

where the lower limit ensures that $Y_t = 0$ at the plane $x_2 = 0, \psi = 0$.

The pressure field, which is associated with the present structure, is obtainable from (3.19), (4.13) and (4.14) in the incompressible case as

$$\frac{p}{\rho} = \frac{1}{2} q(t) x_3^2 - \int^\psi [\ell(X, X_t, t) X_\psi + m(Y, Y_t, t) Y_\psi] d\psi \tag{4.17}$$

(There is an equivalent form with a ϕ integration and ϕ derivatives replacing the ψ operations in (4.17)).

Before discussing the implications of (4.13), (4.14) and (4.15) we need to note an alternative implication of (4.4), as follows. A time differential of (4.1) yields

$$Y_\psi Y_{\phi t t} - Y_\phi Y_{\psi t t} + X_\psi X_{\phi t t} - X_\phi X_{\psi t t} = 0. \tag{4.18}$$

We can regard this formula as being a relation between X_{tt}, Y_{tt}, X and Y, which are regarded here as functions of ϕ, ψ and t. Suppose, however, that there is a one–to–one mapping between ϕ and ψ on the one hand and $X(\phi, \psi, t)$ and $Y(\phi, \psi, t)$ on the other hand. (The assumption of a one-to-one mapping is given some justification by the fact of the resulting potential function Φ of (4.20) being related to the pressure by (4.25).) Then we may write

$$X_{tt} = \ell^0(X, Y, t), \quad Y_{tt} = m^0(X, Y, t), \tag{4.19}$$

where ℓ^0 and m^0 are not yet specified. Substituting (4.19) in (4.18) and using the definitions (2.12), we obtain

$$J_3(m_X^0 - \ell_Y^0) = 0 \tag{4.19a}$$

Since, by (2.13), $J_3 \neq 0$, we infer that

$$\begin{aligned} \ell^0(X, Y, t) &= \Phi_X(X, Y, t), \\ m^0(X, Y, t) &= \Phi_Y(X, Y, t), \end{aligned} \right\} \tag{4.20}$$

where Φ plays the role of a potential function. Thus

$$X_{tt} = \Phi_X, \quad Y_{tt} = \Phi_Y \tag{4.21}$$

are the equations governing X and Y; and, moreover, by comparison with (4.13) and (4.14), we note that

$$\left.\begin{array}{l} \Phi_X(X,Y,t) \equiv \ell(X, X_t, t), \\ \Phi_Y(X,Y,t) \equiv m(Y, Y_t, t). \end{array}\right\} \tag{4.22}$$

The function Φ can be interpreted in terms of the pressure field since the original momentum equation, from which (2.1) has been inferred by addition, is

$$\rho \frac{\partial^2 x_j}{\partial t^2} = -\frac{\partial p}{\partial x_j} \tag{4.23}$$

For the incompressible case with use of (2.7) and (4.17) this yields

$$\left.\begin{array}{r} \rho X_{tt} + p_X = 0, \\ \rho Y_{tt} + p_Y = 0, \\ Z_{tt} + q(t)Z = 0; \end{array}\right\} \tag{4.24}$$

comparison with (4.21) implies that

$$\Phi \equiv -p/\rho \tag{4.25}$$

Thus the result (4.21) is perfectly natural, and arises from the fact of the time differential (4.18) of the vorticity relation (4.1) being equivalent to the existence of $p \equiv -\rho\Phi$ as a potential function for the (X,Y) field.

By a utilization of (4.22) and some differentials of those formulae it is possible to show that, if the third component of vorticity is zero, which implies that each side of (4.1) is identically zero, then

$$\ell_{Xt} \equiv m_{Yt}, \tag{4.26}$$

so that the constraint (4.15) is identically satisfied. In this situation the (X,Y) velocity field is an irrotational or potential field. A Bernoulli theorem can be inferred from (4.21). An interesting and relevant reference here is to the paper of John (12), with applications to water waves.

5 A Simpler class of Rotational Flows

An intrinsic difficulty with the material described in Section 4 stems from the nonlinear constraint (4.15). One possible simplification arises if ℓ is a function of X and t only, but not of X_t. Then the left–hand side of (4.15) is zero; we require the right–hand to be zero also, and a relevant inference can be drawn from (4.22), the first of which gives

$$\Phi_X(X,Y,t) \equiv \ell(X,t), \tag{[(5.1)}$$

which implies that

$$\Phi(X,Y,t) \equiv \int \ell(X,t)dX + \int m(Y,t)dY, \qquad (5.2)$$

where the second term may be considered as an integration *constant*. From this relation (5.2) it follows that

$$\Phi_Y(X,Y,t) \equiv m(Y,t), \qquad (5.3)$$

so that m of (4.14) and (4.22) is a function of Y and t only and not of Y_t. Thus the right–hand side of (4.15) is zero and the constraint (4.15) is satisfied.

Our problem is now the following: we have (4.13) for X and (4.14) for Y, with appropriate initial conditions (2.9), together with the continuity condition (4.3). If (4.13) is solved for X with an assumed or given function $\ell(X,t)$ and subject to (2.19), then (4.3) is a linear partial differential equation for $Y(\phi, \psi, t)$, with t is a parameter, subject to the condition

$$\psi = 0, Y = 0, \qquad (5.4)$$

which ensures that a particle on the plate remains there.

Let us turn to (4.3) and return to (4.14) afterwards. If we consider the case $q(t) \equiv 0$, so that Z is given by (3.20), (4.3) becomes

$$X_\phi Y_\psi - X_\psi Y_\phi = [1 + btw_0(\phi, \psi)]^{-1}, \qquad (5.5)$$

and is subject to (5.4). The characteristics of the linear equation (5.5) for Y are given by

$$\frac{d\psi}{X_\phi} = \frac{-d\phi}{X_\psi} = [1 + btw_0(\phi, \psi)]dY, \qquad (5.6)$$

and are (i)

$$X(\phi, \psi, t) = c(t), \qquad (5.7)$$

where t is a parameter in this process and (5.7) implies by inversion

$$\phi = \Phi(c(t), \psi, t), \qquad (5.8)$$

and (ii)

$$Y = \int_0^\psi \frac{d\psi'}{\{X_\phi(t, \Phi, \psi')[1 + btw_0(\Phi, \psi')]\} \, c \equiv X(\phi, \psi, t)} \Bigg|^{\Phi(c(t), \psi', t)} \qquad (5.9)$$

Formula (5.9) gives $Y(\phi, \psi, t)$, and the form of $m(Y, t)$ must follow from Y_{tt} and (4.14), but with m independent of Y_t. This implies a constraint on $\ell(X, t)$. (The more general case of $q(t)$ is similar to (5.9), but with (3.14) replacing the second bracket in the denominator of (5.9).)

The situation may be summarized as follows

A.

$$x_3 \equiv \chi Z(\phi, \psi, t) \quad : \quad Z = \alpha(t)bw_0(\phi, \psi) + \beta(t) \atop u_3 \equiv x_3 Z^{-1} Z_t \quad : \quad Z_{tt} + q(t)Z = 0 \ \text{ with (3.15)} \Bigg\} \quad (5.10)$$

B.

$$x_1 \equiv X(\phi, \psi, t) \quad : \quad X_{tt} - \ell(X, t) = 0 \atop u_1 \equiv X_t \qquad \qquad t = 0; X = \phi \atop X_t = u_0(\phi, \psi) \Bigg\} \quad (5.11)$$

C.

$$x_2 \equiv Y(\phi, \psi, t) \quad : \quad Y \ \text{is given by (5.9).} \atop u_2 \equiv Y_t \Bigg\} \quad (5.12)$$

D. Inversion jointly of $X(\phi, \psi, t)$ from (5.11) and of $Y(\phi, \psi, t)$ from (5.9) implies $\phi(X, Y, t)$ and $\psi(X, Y, t)$.

We see, therefore that Z, X, Y can be calculated in a sequential manner, together with the corresponding components of velocity. The inversion D then enables the calculation of the fields as functions of X, Y, t, in the Eulerian sense that is to say. It is a necessary, however, that $Y_{tt} = m$ (of 4.14) should be a function of Y and t only.

There is an important implication of the assumption that ℓ is a function of X and t only, as in (5.11), and that m must be a function of Y and t only. The pressure formula (4.17) becomes

$$\frac{p}{\rho} = \frac{1}{2}q(t)x_3^2 - \int^X \ell(X, t)dX - \int m(Y, t)dY \qquad (5.13)$$

Thus our choice of $\ell(X, t)$ implies that the pressure field splits into three separate functions, one for each spatial coordinate, but each time-dependent. This is not the most general situation, however, in view of the choice of ℓ being independent of X_t.

6 Double Stagnation Case

An explicit example of the situation described in Section 5 occurs as follows (Stuart (2,3)). We suppose that the particle field has a stagnation structure in the first (x_1) as well as the third (x_3) coordinate. Thus (2.7) becomes

$$x_1 = X(\phi, \psi, t) \quad = \quad \phi\xi(\psi, t), \atop x_2 = Y(\phi, \psi, t) \quad = \quad Y(\psi, t), \atop x_3 = \chi Z(\phi, \psi, t) \quad = \quad \chi Z(\psi, t). \Bigg\} \quad (6.1)$$

The velocity field has the form

$$u_1 = \phi\xi_t(\psi, t) \quad = \quad x_1\xi_t(\psi, t)/\xi(\psi, t), \atop u_2 = Y_t(\psi, t), \atop u_3 = \chi Z_t(\psi, t) \quad = \quad x_3 Z_t(\psi, t)/Z(\psi, t); \Bigg\} \quad (6.2)$$

we can see from (6.2) that this flow has a line of symmetry at $x_1 = 0$ as well as at $x_3 = 0$, is linear in ϕ and χ and may be considered as the flow near the stagnation line $(x_1 = 0)$ on a bluff cylindrical body, with an orthogonal stagnation line $(x_3 = 0)$ associated perhaps with a longitudinal vortex structure.

In general Z is given by (3.14) but in a special case $(q(t) \equiv 0)$ it takes the simpler form (3.20). In view of the fact of Y being independent of ϕ at all times, $Y_{\phi t 0} \equiv 0$ in (4.15), so that the right–hand is zero. On the left–hand side

$$X_{\psi t 0} = \phi \xi_{\psi t 0} \tag{6.3}$$

and the expression is not zero in general. Thus we infer that ℓ is independent of $X_t = \phi \xi_t$. Then (4.13) becomes

$$\xi_{tt} - \phi^{-1} \ell(\phi \xi, t) = 0. \tag{6.4}$$

Since ξ_{tt} is independent of ϕ, this particle coordinate must cancel from the second term and we infer that

$$\ell(\phi \xi, t) = \mu(t) \phi \xi \tag{6.5}$$

and

$$\xi_{tt} - \mu(t)\xi = 0 \tag{6.6}$$

The pressure formula (5.13) takes the form

$$\frac{p}{\rho} = \frac{1}{2}q(t)x_3^2 - \frac{1}{2}\mu(t)x_1^2 - \int^Y m(Y,t)dY, \tag{6.7}$$

where m, which is necessarily independent of Y_t, is determined from (4.14) and the relationship between Y and ψ, to which we now turn.

In the present case, with Y independent of ϕ, (5.5) becomes

$$\xi(\psi,t)Y_\psi(\psi,t) = [1 + btw_0(\psi)]^{-1} \tag{6.8}$$

for the case $q(t) \equiv 0$ (the generalization is easily achieved with (3.22) on the right–hand side of (6.8).) Now $\xi(\psi,t)$ is given by (6.6) and if $\mu(t) = a^2$ (for the stagnation–point flow at the attachment line in a bluff cylindrical body) we have

$$\xi = \cosh at + u_0(\psi) \sinh at, \tag{6.9}$$

satisfying the conditions

$$\left.\begin{array}{ll} \xi = 1, & t = 0 \\ u_1 = x_1\xi_t/\xi = & x_1 a u_0(Y), \quad t = 0 \\ Y = \psi, & t = 0. \end{array}\right\} \tag{6.10}$$

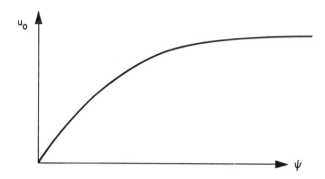

FIGURE 5. Typical boundary-layer velocity profile for u_0 as a function of ψ.

Here u_0 denotes a typical velocity profile as in Figure 5. Thus, instead of the characteristic (5.9), we have the quadrature

$$Y = \int_0^\psi \frac{d\psi'}{[1 + btw_0(\psi')][\cosh at + u_0(\psi') \sinh at]}. \qquad (6.11)$$

In this expression $w_0(\psi)$ represents the initial x_3–velocity component of a particle at $t = 0$, since

$$\left. \begin{aligned} Z &= 1, & t &= 0 \\ u_3 &= x_3 Z_t/Z = & x_3 b\, w_0(Y), & t &= 0 \\ Y &= \psi, & t &= 0 \end{aligned} \right\} \qquad (6.12)$$

The velocity field at time t for this double–stagnation example is given by

$$u_1 = x_1 \xi_t/\xi = \frac{ax_1(\tanh at + u_0(\psi))}{(1 + u_0(\psi) \tanh at)}, \qquad (6.13)$$

$$u_3 = x_3 Z_t/Z = \frac{bx_3 w_0(\psi)}{(1 + btw_0(\psi))}, \qquad (6.14)$$

while $u_2 = Y_t(\psi, t)$ is given by a time differential of (6.11) under the integral sign.

There remains of course the matter of the relationship of ψ to Y in (6.13) and (6.14), but this is given by an inversion of (6.11) for given t.

We have earlier discussed the possible occurrence of a singularity in the velocity component U_3 and found that this would happen if (3.18) were satisfied at some finite time. In the present case this means at the earliest time t_s given by

$$1 + bt_s w_{0m}(\psi) = 0, \qquad (6.15)$$

since u_3 is independent of ϕ. Here w_{0m} denotes the minimum of w_0 in Figure 4. A question raised in Section 3 was that of whether or not ψ would imply a value for Y at a given t in such a way as to permit a singularity to occur. In the present double–stagnation this can be demonstrated by asymptotics. If $w_0(\psi)$ is of the form shown in Figure 4, and which has been discussed in Section 3, a singularity is possible when $t = t_s$ is given by (6.15). If the w_0 curve is parabolic near to its minimum, then we can show that the Y position of the singularity is given by

$$Y_s = C(t_s - t)^{-1/2}, t \to t_s, \tag{6.16}$$

where C is a *constant*. (In this derivation the singular integrand in (6.11) is relevant.) If the w_0 curve were not parabolic near to its minimum, a result different from (6.16) would probably emerge. We have assumed also in the derivation of (6.16) that $u_0(\psi)$ has a monotonic velocity profile of boundary layer form as in Figure 5.

7 Discussion and Implications

The early part of the present paper, especially in Section 3, is concerned with determining the flow structure near to a line of symmetry, towards which the flow converges. Indeed it is shown that the particle coordinate (x_3) normal to and near to that line, and in the plane of the wall, can be determined independently of the other two coordinates. There is, however, a caveat to this result, which gives x_3 for the particle in terms of ϕ, ψ and t, namely that we do not know the x_1 and x_2 coordinate of the particle. It is necessary to relate ϕ and ψ, which with χ define the individual particle, to x_1 and x_2; the latter part of the paper is concerned with that problem, namely the calculation of x_1 and x_2 in terms of ϕ, ψ and t.

From formula (3.17), however, we see the distinct possibility of a singularity occurring in the x_3 component of velocity, when x_3 has a zero (see 3.18). Of importance here is the choice of the function $q(t)$, which describes the time–dependent aspect of the stagnation–point pressure field, as in (3.19). (Although we have emphasized, perhaps unduly, the case $q(t) \equiv 0$, other choices can be of great interest, and these possibilities remain to be assessed). The singularity will occur, if all conditions do allow it, at a location in x_1, x_2 to be determined for given values of ϕ and ψ by the further calculation to which reference has been made. What is clear, however, is that the particle, which at $t = 0$ is associated with the minimum on the curve of $w_0(\phi, \psi)$ in Figure 4, will be the first to achieve the condition (3.18), namely $x_3 = 0$ at $t = t_s$, at least if the special form (3.20) is appropriate. (The situation may be different if the general form (3.14) is used for Z, since then monotinicity as otherwise of α and β will be relevant, subject to a Wronskian relation of the form (3.2) between α and β). But the proviso that we need to keep in mind is that: will the coordinates x_1 and

x_2 remain defined for the particle denoted by ϕ, ψ up to and including the time t_s, or will a singularity develop in, say, the x_1 component of velocity?

The possibility of a singularity occurring in x_3 is clearly and distinctly related to the concept of flow convergence towards a line of symmetry. The local structure, which is specified by (2.7), represents this feature. Of great interest, we believe, is the matter of whether the structure (2.7) could be embedded in a field extended in the x_3 direction away from $x_3 = 0$, so that x_3 would not be linear in χ and u_3 would not be linear in x_3. For example could u_3 be periodic in x_3, as in many boundary–layer experiments, and would the singular type of structure still arise? Such questions remain to be answered. It is emphasized that this part of the discussion is valid for a compressible fluid subject to (2.3).

Sections 4,5, and 6 are concerned with developing the theory for the remainder of the flow field, but only for the incompressible case, essentially for the flow projected onto the plane of symmetry. A number of aspects of that problem are explored there, and formulae (4.13), (4.14) and (4.22) are thought to be of especial relevance. If the projected flow (u_1, u_2) is irrotational ($\omega_3 = 0$) the awkward constraint (4.15) is automatically satisfied. Otherwise it is a central difficulty, which is avoided in the special and double–stagnation situations described in Sections 5 and 6 for special forms of (4.13), namely (5.11) and (6.6) with (6.1). Crucial to those discussions is the role of the pressure, whose form (4.17 in general) is implied by the functions ℓ and m, and by the choices of them. This means, of course, that the boundary conditions in the far field are of significance for those choices. That question of choice and the concomitant role of the pressure, has not been adequately addressed here, and remains for future treatment.

Section 6 summarizes from the present point of view the solution given earlier (reference 2) for the double–stagnation case. For initial velocity profiles of the type of Figures 4 and 5, and subject to an asymptotic inversion of (6.11) it is established that a singularity can occur towards the edge of the boundary layer, or mathematically more precisely when $Y \sim (t_s - t)^{-1/2}$. The occurrence of the singularity is associated with the concept of flow convergence, as occurs between longitudinal vortex filaments in boundary layers of a transitional or turbulent nature and which occurs also in unsteady boundary layer separation.

Acknowledgements

This work was partially supported by the DARPA/URI Applied and Computational Applied Mathematics Program at Brown University, where the author was a Visiting Professor in Applied Mathematics and Engineering during 1988-89; assistance was also given by a NATO Research Grant for Travel.

References

1. Stuart, J.T., *The Production of Intense Shear Layers by Vortex Stretching and convection*, AGARD Report 514, (1965).

2. Stuart, J.T., *Instability of Laminar Flows, Non-Linear Growth of Fluctuations and Transition to Turbulence* in *Turbulence and Chaotic Phenomena in Fluids* (IUTAM Symposium Kyoto 1983), ed. T. Tatsumi, 17-26, North-Holland (1984).

3. Stuart, J.T., *Nonlinear Euler Partial Differential Equations: Singularities in Their Solution*, Proc. Symp. Honor C.C. Lin; eds. Benney, D.J., Shu, F.H., Chi Yuan; World Scientific Press, Singapore (1987).

4. Childress, S., Ierley, G.R., Spiegel, E.A., and Young, W.R., *Blow up of unsteady two-dimensional Euler and Navier-Stokes solutions having stagnation point form*, J. Fluid Mech. **203**, 1-22 (1989).

5. Hoskins, B.J. and Bretherton, F.P., *Atmospheric Frontogenesis Models: Mathematical Formulation and Solution*, J. Atmos. Sci., **29**, 11-37 (1972).

6. Stern, M.E. and Paldor, N., *Large-amplitude Long Waves in a Shear flow*, Phys. Fluids, **26**, 906-919 (1983).

7. Calogero, F., *A Solvable Nonlinear Wave Equation*, Stud. Appl. Math., **70**, 189-199 (1984).

8. Russell J.M. and Landahl, M.T., *The Evolution of a Flat Eddy near a Wall in an Inviscid Shear Flow*, Phys. Fluids, **27**, 557-570 (1984).

9. Van Dommelen, L.L. and Shen, S.F., *The Genesis of Separation*, in Num. Phys. Aspects Aero. Flows. (Proc. Symp. 1981), ed. T. Cebeci, 293-311, Springer, New York (1982).

10. Van Dommelen, L.L. and Cowley, S.J., *On the Lagrangian Description of Unsteady Boundary–Layer Separation, Part I: General theory*, J. Fluid Mech., **210**, 593–626 (1990).

11. Lamb, H., *Hydrodynamics*, 6th Edition, Cambridge University Press, 1932.

12. John, F., *Two–Dimensional Potential Flows with a Free Boundary*, Comm. Pure Applied Math. **6**, 497-503, (1953).

14

Dynamical Chaos: Problems in Turbulence

R.Z. Sagdeev and G.M. Zaslavsky

Abstract

Chaos of streamlines of steady Beltrami-type flows leads to formation of liquid stochastic web. Structural properties of the web are inherited by fluid particles. Their dynamics is intermittent. There are two origins of intermittency: *trappings* and *flights*. These effects result in anomalous diffusion. We consider random walks on multifractals of the Levy flight type and discuss differences in scaling properties of spatial and temporal averages. This reflects different fractal properties of the time evolution, and the space on which it occurs.

1 Introduction

It would be trite to talk about the difficulties accompanying any old and new attempts in the theory of turbulence. New understanding of instabilities leading to a dynamical chaos brings up a new perspective for the turbulence problem, but not only that. Many surprises happened during last decades, which are closely related to nonorthodoxal methods for turbulence analysis, came from the dynamical systems theory. Turbulence of interacting nonlinear waves and the Lorenz model can be considered as such examples, but these are very straightforward ones. Much more sophisticated is the problem of intermittency which leads to a new scaling theory [1,2]. This has brought several authors to creation of the multifractal philosophy of turbulence which has very deep roots in a dynamical chaos. Another group of problems deals with a large-scale dynamics. Fluid particles motion, passive particles motion, advection of small whirls, etc. can be treated as finite-dimensional dynamical problems for streamlines [6-9]. Chaos of streamlines is a new topic of the so-called Lagrangian turbulence. It may happen that clear understanding of the streamlines chaos will lead to resolution of some intrinsic properties of turbulence [10,11].

Transition to turbulence may pass through the set of pattern bifurcations. First steps in the analysis have been done through attempts to find out some universal scenario of turbulence [12,13]. But now we understand that the path to turbulence is much more complicated [14]. The spatio-temporal chaos occurrence is a subject of present investigations [15-18]. Streamlines dynamics in preturbulent state may provide a new view of the situation [11].

This article considers some topics, mentioned above from a dynamical viewpoint.

The first point is a discussion of possible steady patterns in continuous media and the symmetry of these patterns. The Beltrami property is the main feature of the considered flows.

The second point is a demonstration of streamlines chaos in such regular patterns. The concept of liquid stochastic web is introduced.

The third point is verification that chaos of streamlines causes their intermittency. The intermittency is strongly connected with two different singularities of dynamics - *trappings* and *flights*. From dynamical point of view, there is present a dependence of the intermittency properties on the phase portrait of the system of streamlines, i.e. on the flow pattern. From the kinetic point of view, the intermittency realizes a random walk of streamlines similar to Levy flight.

The latter point is some speculation which comes from understanding of the sources of intermittency. They have to result in different multifractal spectra of space and time structural function.s

2 Quasi-periodic flows

Quasi-periodic flows can be considered as one of the kind of stationary flows of incompressible fluid. It has the form

$$\vec{v}(\vec{z}) = \sum_{\vec{k}} \vec{v}_{\vec{k}} \exp(i\vec{k}\vec{z}) \tag{2.1}$$

where $\vec{v}_{\vec{k}}$ are the appropriate coefficients of expansion and the condition

$$div \ \vec{v} = 0 \tag{2.2}$$

is understood.

Let us give several examples of such flows.

2.1 For 2D-hydrodynamics the streamfunction $\psi = \psi(x, y)$ can be introduced:

$$v_x = -\frac{\partial \psi}{\partial y}; \ v_y = \frac{\partial \psi}{\partial x}. \tag{2.3}$$

The Euler equation

$$\frac{\partial \vec{v}}{\partial t} + (\vec{v}\vec{\nabla})\vec{v} = -\vec{\nabla}P \tag{2.4}$$

reduces to

$$\frac{\partial}{\partial t}\Delta_\perp \psi + \frac{\partial(\psi, \Delta_\perp \psi)}{\partial(x, y)} = 0 \tag{2.5}$$

where $\Delta_\perp = \partial^2/\partial x^2 + \partial^2/\partial y^2$ and $\partial(A,B)/\partial(x,y)$ is the Jacobian. It is clear from (2.5) that such function ψ is the stationary solution of (2.5), i.e. the solution of equation

$$\Delta_\perp \psi = f(\psi).$$

In particular, if $f(\psi) = -\psi$, the stream function is the solution of equation:

$$\Delta_\perp \psi + \psi = 0 \tag{2.6}$$

and may be written as the series

$$\psi = \sum_\lambda C_\lambda \psi_\lambda$$

over the set of eigenfunctions ψ_λ with eigenvalues λ. This directly leads to the form (2.1).

In [9,11] the steady-state solution of the form

$$\psi_q = \sum_{j=1}^{q} \cos(\vec{r}_\perp \vec{e}_j) \tag{2.7}$$

is considered. Here $\vec{r}_\perp = (x,y)$; the vectors $\vec{e}_j = (\cos 2\pi j/q, \sin 2\pi j/q)$ form a q-ray regular star and q is integer. The solution (2.7) results in a regular 2D-pattern of level lines of ψ. Some remarkable properties of the function ψ_q can be mentioned.

For $q = 2$ we have

$$\psi_2 = 2\cos x$$

which is the one-dimensional Kolmogorov flow [19]. For $q = 4$ we obtain a flow with square cells;

$$\psi_4 = 2(\cos x + \cos y). \tag{2.8}$$

for $q = 3$ or $q = 6$

$$\psi_6 = 2\psi_3 = \cos x + \cos(x/2 + \sqrt{3}y/2) + \cos(x/2 - \sqrt{3}y/2). \tag{2.9}$$

The flow (2.9) has hexagonal cells and coincides with the streamfunction of Rayleigh-Benard convection for hexagonal symmetry.

Thus, for $q = 2,3,4$ or 6 equation (2.7) represents all known 2D-periodic flow patterns with the structure of 2D-crystals. When the vectors $\vec{e}_j$ do not form a regular star, equation (2.7) remains a steady solution to (2.5), but with some more complicated structure of the cells.

New kinds of patterns and their symmetry appear for $q = 5,7,8,\ldots$. These structures are aperiodic and have symmetries analogous to those of quasicrystals. Numerical simulation of such patterns is considered in [11,20]. Quasicrystal-type flows can be also met in cellular authomata method of hydrodynamic simulations [21].

2.2. Beltrami flows are the particular case of steady solutions to the Euler equation (2.4). These flows satisfy the constraint

$$\vec{v} = \cos rot\vec{v} \qquad (2.10)$$

with c and arbitrary scalar function. For $c = \pm 1$ the Beltrami flows have maximal possible helicity. Some structural properties of turbulence at small and large scales are believed to be associated with flows of this type [10, 22-24]. There exist strong arguments in favor of suppression of nonlinearity in developed isotropic turbulence. A similar suppression must occur in formation of coherent structures. The field governed by (2.10) can be represented as [24]

$$\vec{v} = \sum_{\vec{k}} A(\vec{k})(\vec{n} + i[\vec{k}\vec{n}])e^{i\vec{k}\vec{z}/L} \qquad (2.11)$$

where $A^*(\vec{k}) = A(-k)$, $\vec{k}$ are the unit vectors, $\vec{n}$ is the unit vector orthogonal to $\vec{k}$, L is the characteristic scale.

The special case of six harmonics being present in (2.11) is given by [6]

$$v_x = A \sin z + C \cos y$$
$$v_y = B \sin x + A \cos z \qquad (2.12)$$
$$v_z = C \sin y + B \cos x.$$

This is the so-called Arnold-Beltrami-Childress (ABC) flow. this flow possess the cubic symmetry.

A Beltrami flow of more general symmetry is obtained in [9]. Let us define the Q-flow as

$$v_x = -\frac{\partial \phi_q}{\partial y} + \epsilon \sin z$$

$$v_y = \frac{\partial \psi_q}{\partial x} = \epsilon \cos z \qquad (2.13)$$

$$v_z = \psi_q$$

where ψ_q is defined through (2.7) and ϵ is the parameter. In particular, for $q = 4$ expressions (2.13) reduce to the ABC-flow. In general case the flow (2.13) has a stratified structure periodic in z with q-fold symmetry in (x, y) plane. For $q = 3, 4, 6$ this is the crystal symmetry while for $q = 5, 7, 8 \ldots$ the symmetry is of quasicrystal type [11].

2.3. One of generalization of (2.13) assumes replacement of ψ_q by an arbitrary function $\psi(x, y)$ governed by equation [9]

$$\Delta_\perp \psi + \psi = 0$$

In particular, for radial symmetry in (x, y) plane we obtain

$$v_z = -\frac{1}{z}\frac{\partial \phi}{\partial \varphi} + \frac{\epsilon}{r} \sin z$$

$$v_\varphi = \frac{\partial \psi}{\partial r} - \frac{\epsilon}{z} \cos z \tag{2.14}$$

$$v_z = \psi$$

where

$$\psi = \psi(r, \varphi) = \sum_n A_n J_n(z) \cos n\varphi$$

and $J_n(r)$ are the Bessel functions.

The flows (2.13) and (2.14) satisfy conditions (2.2) and (2.10) with $c = \pm 1$.

3 The chaos of streamlines

The topology of streamlines depends on the form of the velocity field, but these two properties can have strong qualitative differences. In particular, regular flows can have stochastic spatial distributions of streamlines. This property was noted in [6] for the ABC-flow while numerical simulations [8] have revealed the regions of localization of the chaos. Some further examples are given in [7] and the general analysis for Q-flows and the concept of liquid stochastic web are given in [9,11]. The idea of application of the methods of theory of dynamic systems to description of magnetic lines was used even earlier by the authors for analysis of closed magnetic traps [25].

Equations for the streamlines of the filed $\vec{v}(x, y, z)$ are given by

$$\frac{dx}{v_x} = \frac{dy}{v_y} = \frac{dz}{v_z} \tag{3.1}$$

and can be recast as

$$\frac{dx}{dz} = \frac{v_x}{v_z}; \quad \frac{dy}{dz} = \frac{v_y}{v_z}. \tag{3.2}$$

The system (3.2) determines the streamline as the parametrically given curve in three-dimensional space

$$x = x(z; x_0, y_0, z_0); y = y(z; x_0, y_0, z_0) \tag{3.3}$$

which has the *initial* condition $x = x_0, y = y_0$ for $z = z_0$.

The system (3.2) can be considered as a dynamic system (x, y) with z playing the role of time. It follows, in particular, that in two-dimensional flows, i.e. for $\vec{v} = \vec{v}(x, y)$ the chaos of streamlines is impossible. Oppositely, in three-dimensional flows presence of chaotic-streamlines regions is the generic case.

For the sake of definiteness, consider the Q-flow (2.13). Expressions (3.2) take the form

$$\frac{dx}{dz} = \frac{1}{\psi_q}\left(-\frac{\partial \psi_q}{\partial y} + \epsilon \sin z\right)$$

$$\frac{dy}{dz} = \frac{1}{\psi_q}\left(\frac{\partial \psi_q}{\partial x} - \epsilon \cos z\right) \tag{3.4}$$

where the function ψ_q is given by (2.7). Equations (3.4) have the form of generalized Hamiltonian equations [26].

The transformation

$$\frac{dz}{d\tau} = \psi_q(x(z;x_0,y_0), y(z;x_0,y_0)) \tag{3.5}$$

brings about a more common representation [9,25] with implicit time τ. Now the system (3.4) takes the form

$$\dot{x} = -\frac{\partial H}{\partial y}; \dot{y} = \frac{\partial H}{\partial x} \tag{3.6}$$

where the dot represents differentiation with respect to τ and the Hamiltonian H is given by

$$H = \psi_q(x,y) + \epsilon V(x,y,z)$$

$$V(x,y,z) = -y\sin z(\tau,x,y,) - x\cos z(\tau,x,y). \tag{3.7}$$

The non-stationary perturbation (3.7) actually solves the problem of existence of chaotic regions in the space $(x,y,\tau) \sim (x,y,z)$ [27,14] and the streamlines coincide with stochastic orbits of the Hamiltonian system (3,6), (3.7).

More direct transformation of variables is given in [28].

The representation (3.6), (3.7) allows to reach some general conclusions. The function $\psi_q(x,y)$ defined by (2.7) is an unperturbed part of Hamiltonian H. The set of trajectories determined by $\psi_q(x,y)$ in the phase space (x,y) contains separatrices. A perturbation V periodic in z results in their disruption and formation at these places of the regions of stochastic trajectories $x(z), y(z)$ [28], i.e. the regions of stochastic distribution of streamlines. Width of the regions tends to zero for $\epsilon \to 0$ but chaotic regions are present for arbitrarily small ϵ. Thus, the three-dimensionality, i.e. the dependence of V and z leads to principal modification of the problem's character in comparison with two-dimensional case.

The problem of topological pattern of chaotic regions is a more delicate problem. This problem is considered in [9]. The basis of the analysis are structural properties of stochastic webs of Hamiltonian system [29,30].

For $q = 4$ in (3.7), separatrices of unperturbed Hamiltonian ψ_4 (see (2.8)) form a square lattice. The perturbation ϵV destroys the separatrices and forms instead the region of dynamic chaos which has the form of square lattice of channels. This lattice is called the liquid stochastic web (Fig. 1a). For $q = 3$ or $q = 4$ in (3.7) (see (2.9)) the web becomes a hexagonal lattice (Figs. 1b,2). Thickness of the web is evaluated in [28,29]. For $\epsilon \ll 1$ the thickness is $\sim \epsilon$. The web exists for any arbitrarily small values of ϵ. This result has a principal importance since it implies that for $q = 3, 4$ three-dimensional structures have unremovable connected net of chaotic regions which allows an unlimited spatial transport of streamlines or passive particles.

In the cases described above, $q = 3, 4, 6$ the web has a crystal symmetry. Their symmetry group includes translations and rotations simultaneously. In all other cases, $q = 5, 7, \ldots$, the Q-flows have a quasicrystal symmetry in the (x, y) plane. The perturbation ϵV again leads to disruption of separatrices and formation of stochastic layers at their place. Once can show that for $\epsilon > \epsilon_0$ a connected infinite stochastic web also forms, which has a quasicrystal q-fold symmetry in the (x, y) plane [30]. Her $\epsilon_0 \ll 1$ is some critical value of perturbation which can be evaluated. An example of liquid web of the fifth order is shown in Fig. 3.

Proposition: A connected web exists for any small ϵ; however, the smaller is ϵ, the larger is a characteristic size of the web cell.

4 Intermittency and the Levy flights

The web width grows with the perturbation parameter ϵ and for $\epsilon \sim 1$ the width is ~ 1. However, for any ϵ there exist regions of regular motions (stability islands). They have a finite measure and this very profoundly affects the properties of the chaos of streamlines. Indeed, in this case the random walk of streamlines preserves the memory of coherent structures determined by topology of the web. The progress in understanding of the origin of chaos of streamlines allows to establish a more explicit relation between different phenomena connected with existence of coherent structures.

At the first sight, the stochastic dynamics of the system (3.7) within the web channels can be described as a random walk on the lattice of the corresponding symmetry which is produced by the web. Since a system's orbit also represents a trajectory of passive particle, this results in a spatial image of a streamline as a realization of the corresponding random process. In fact, the walk has certain specific properties.

There exist two anomalous deviations for some *average* random walk: *trappings* and *flights*. The trappings are associated with evolution of the system near a saddle singular point, i.e. with passing of the streamline near a stagnation point. The streamline can reside there for arbitrarily long. This section of the fluid particle's trajectory is very close to a regular one. This brings about the phenomenon of intermittency of the type described in [31].

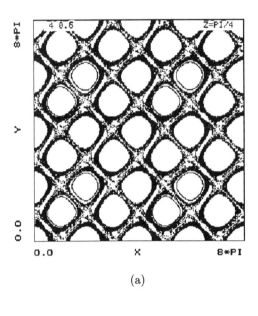

(a)

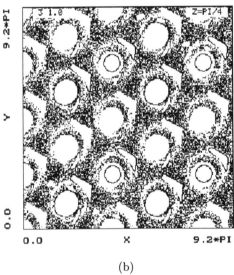

(b)

FIGURE 1. Section of a stramline by the plane $z = $ const: a) $q = 4; z = \pi/4; \epsilon = $ 0.6; b) $q = 3; z = \pi/4; \epsilon = 1$.

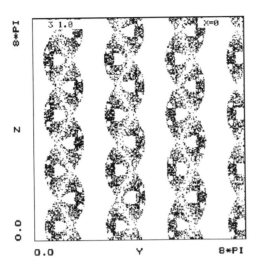

FIGURE 2. Section of a streamline by the plane $x = 0 (q = 3; \epsilon = 1)$.

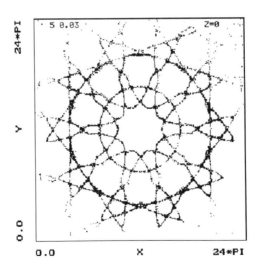

FIGURE 3. Stochastic web of streamlines with 5-fold symmetry ($q = 5; z = 0; \epsilon = 0.03$).

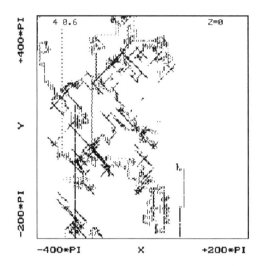

FIGURE 4. Spatial diffusion of a single streamline in the plane (x, y), for $z = 0; q = 4; \epsilon = 0.6$.

When the system (fluid particle) is near a boundary between stochastic channel and stable region, it can move along the boundary for a very long period. The resulting section of the streamline closely resembles the boundary in shape, i.e. it is practically regular. Therefore, the fluid particle passes the corresponding path at the speed of regular dynamics, without spending time for wandering in other directions. These sections of streamlines look like long tracks on the background of random walks (Fig. 4). These sections can be interpreted as Levy flights [11,32] which are typical of the so-called Levy random walk [33,34].

The flights also contribute to intermittency. Due to trappings and flights, structural properties of the phase portrait are inherited by stochastic transport of fluid particles, i.e. by the form of streamlines. It can be naturally expected that this peculiarity typical of any structure has some region of universality and, thus, can determine properties of intermittency even at very small scales in a fully developed turbulence.

From the viewpoint of the dynamic systems theory, there exists one more source of intermittency, namely the cantori. These are invariant sets of Cantor type embedded in chaotic regions in the phase space, which suppress transport of particles [35]. Flights and trappings are possible along cantori. Therefore, the role of cantori effectively reduces to the anomalous mechanisms described above.

The most characteristic property of Levy wandering manifests itself in the law of particle transport. Let r be the length which the system (particle) traverses in time t. For ordinary Brownian motion (i.e. Gaussian process),

$$< r^2 > /t = \quad \text{const.} \tag{4.1}$$

However, for Levy wandering we have

$$< r^2 > /t \sim t^\gamma, (\gamma > 0) \tag{4.2}$$

and the increment of squared length per unit time diverges for $t \to \infty$. In particular, free motion is characterized by $r \sim t$ and it follows from (4.2) that $\gamma = 1$ in this case. This clarifies the role of flights since the inequalities

$$0 < \gamma < 1 \tag{4.3}$$

imply the presence in Brownian trajectory of segments of nearly free motion.

Let $P(x,t)$ is the probability of finding the particle at position x at the moment t. The characteristic function of the process $P(x,t)$ is defined as

$$W(k,t) = \int_{-\infty}^{\infty} e^{ikx} P(x,t) dx = < e^{ikx} > \tag{4.4}$$

Thus,

$$P(x,t) = \frac{1}{2\pi} \int_{-\infty}^{\infty} e^{ikx} W(k,t) dk. \tag{4.5}$$

The Levy process is described by the following characteristic function:

$$W(k,t) = \exp(-C_\alpha t \mid k \mid^\alpha) \tag{4.6}$$

where α obeys the conditions

$$0 < \alpha \leq 2 \tag{4.7}$$

which ensure positiveness of the probability $P(x,t)$. The case $\alpha = 2$ corresponds to Gaussian distribution. This corresponds to a special distribution. For $0 < \alpha < 2$. The asymptotic form of the integral (4.5) for large x is given by [36]

$$P(x,t) \sim \frac{\alpha t \Gamma(\alpha)}{\pi \mid x \mid^{2+1}} \sin \pi \alpha/2 \tag{4.8}$$

It follows, in particular, that

$$<\mid x \mid> \sim t^{1/2}, (t \to \infty). \tag{4.9}$$

Comparison of expressions (4.2) and (4.9) shows that $\gamma- = 2/\alpha - 1 > 0$. The case $\alpha = 1$ corresponds to Cauchy distribution

$$P(x,t) = C_1 t/\pi(x^2 + C_1^2 t^2).$$

for $\alpha = 2(\gamma = 0)$ we have an ordinary Gaussian wandering law (4.1).

The result (4.9) also follows directly by integration of (4.5) with account of (4.6):

$$<\mid x \mid> = \int \mid x \mid P(x,t)dx = \frac{1}{2\pi} \int \mid x \mid .$$
$$\exp(-ikx - C_\alpha t \mid k \mid^\alpha)dxdk. \tag{4.10}$$

Transformation of variables $k \to q = kt^{1/2}, x \to y = xt^{-1/\alpha}$ leads to (4.9).

It has been shown in [34] that the exponent α correspond to the dimension of the fractal made up by the points belonging to the trajectory of wandering point, a fluid particle in our case. In other words the space of chaotic streamlines is at least fractal. In fact, this is a multifractal space. Then expressions (4.6)-(4.10) deserve an appropriate generalization [32].

5 Multifractal random walk

Consider wandering of a fluid particle on certain multifractal set S. Roughly, this process can be described as follows. The whole trajectory (i.e. the streamline) can be divided into segments, each of which belongs to a subset (fractal) S_j. The set S consists of all fractals S_j. Suppose that the individual segments are sufficiently long and neglect transitions from one fractal, S_j, to another, $S_{j'}$. By analogy with (4.6), the characteristic function of wandering on the fractal S_j is denoted as

$$W_j(k,\Delta t_j) = \exp(-C_j\Delta t_j \mid k \mid^{\alpha_j}). \tag{5.1}$$

This wandering is characterized by some fractal exponent α_j and wandering time Δt_j. Then the characteristic function of the global wandering over period t is given by

$$\begin{aligned} W(k,t) &= \Pi_{j=1}^N W_j(k,\Delta t_j) = \\ &= \exp\left(-\sum_{j=1}^N C_j\Delta t_j \mid k \mid^{\alpha_j}\right) \end{aligned} \tag{5.2}$$

where N is the total number of intervals Δt_j of the period t, i.e.

$$\sum_{j=1}^N \Delta t_j = t. \tag{5.3}$$

Denote by $\rho(\alpha)$ the fraction of wandering intervals which occur on fractals with the wandering exponent α belonging to the interval $(\alpha, \alpha + d\alpha)$. Then summation over j in (5.2) and (5.3) can be replaced by integration in α. This yields the following characteristic function of wandering on the multifractal [32].

$$W(k,t) = \exp\left\{-t \int_{\alpha_{\min}}^{\alpha_{max}} \rho(\alpha)C(\alpha) \mid k \mid^\alpha d\alpha\right\} \tag{5.4}$$

where $(\alpha_{min}, \alpha_{max})$ is the range of variation of the exponent α and the normalization

$$\int_{\alpha_{min}}^{\alpha_{max}} \rho(\alpha)d\alpha = 1 \tag{5.5}$$

is understood. Some simple results can be obtained directly from (5.4).

When $C(\alpha) > 0$ in the range $(\alpha_{min}, \alpha_{max})$, (5.4) can be represented as

$$W(k,t) = \exp\left\{-\ \text{const} \cdot t \mid k \mid^{\overline{\alpha}}\right\} \tag{5.6}$$

where $\alpha_{min} \leq \overline{\alpha} \leq \alpha_{max}$. According to (4.9) expression (5.6) leads to the following asymptotic wandering law:

$$< \mid x \mid > \sim t^{1/\overline{\alpha}} \tag{5.7}$$

If there are only a few different fractals S_j which compose the whole set S; then

$$\rho(\alpha) = \sum_j \rho_j \delta(\alpha - \alpha_j).$$

At different time intervals the wandering is described by different intermediate asymptotics

$$< \mid x \mid > \sim t^{1/\alpha_j} \tag{5.8}$$

The smaller is α_j, the later the corresponding asymptotic form develops. Thus, for $t \to \infty$ we have

$$< \mid x \mid > \sim t^{1/\alpha_{min}}. \tag{5.9}$$

This expression gives the asymptotics for random walk.

The next step in understanding of anomalous transport effects of fluid particle and intermittency is associated with determination of the distribution $\rho(\alpha)$ for specific dynamic system. In this direction, only some qualitative, but principally important results have been obtained as yet [37]. Consider again the Q-flows defined by (3.4) and consider wandering of streamlines for $q = 3$. Denote $R^2 = x^2 + y^2$. Then the wandering in the plane (x, y) is described by the asymptotic form

$$< R > \sim t^\beta \quad \text{for} \quad t \to \infty. \tag{5.10}$$

For ordinary diffusive wandering, $\beta = 1/2$. It has been shown in [37] that the exponent β strongly and non-monotonously depends on ϵ. Variations of ϵ result in bifurcations of the phase portrait and in changes of conditions for trappings and flights. Moreover, the stronger β deviates from $1/2$, the more important is the role of flights. Figs. 5 and 6 illustrate these results. Shown are Poincare sections on the plane (x, y) for streamlines at two different values of parameter ϵ. Figs. 5a and 6a show slightly more than one cell of the web. Their phase portraits differ considerably. Scales of Figs. 5b and 6b are much larger and details of individual cells cannot be seen there. Instead, flights are very evident in Fig. 6b while Fig. 5b corresponds to a common picture of Brownian motion. The values $\beta = 0.5$ in Fig. 5b and $\beta = 0.78$ in Fig. 6b confirm the mutual connection between phase portraits, flights and the anomalous transport exponent β (i.e. the Levy wandering exponent).

6 The relation between spatial and temporal structural properties

Although the problem of Lagrangian turbulence, i.e. chaos of fluid particles or streamlines, gives a rather restricted view of only some specific aspects of the general turbulence problem, the possibility to follow them rather far and deep turns out to be very important.

Having accepted the concept of dynamic chaos as the basis of turbulence, one should make the next step: the fractal properties of time history of dynamical process are, generally, different from fractal properties of that part of the phase space on which this process is realized. However, this does not exclude a relation between spatial and temporal fractal properties. Let us recast Eqs. (3-4)-(3.7) as

$$\dot{x} = -\frac{\partial H}{\partial y}; \dot{y} = \frac{\partial H}{\partial x}; \dot{z} = \psi_q \qquad (6.1)$$

where the dot again denotes differentiation with respect to time τ. The system (6.1) shows that for the velocity field of fluid particles, the intermittency in phase space (x, y) implies intermittency in the physical space. This leads to connection between structural properties of spatial and temporal dynamics. However, this relation is nonlinear due to the form of ψ_q. Consider its influence on the structural properties of the velocity field.

We introduce the spatial structure function

$$\mathcal{R}(\ell) = <| v(x + \ell, t) - v(x, t) |^p > \equiv < \delta v(\ell)^p > \propto \ell^{\zeta(p)} (\ell \to 0) \qquad (6.2)$$

where $\zeta(p)$ is the scaling parameter accounting both for fractal properties of the space and singular properties of the velocity field. The early theory of Komogorov and Obukhov predicts

$$\zeta(p) = p/3 \qquad (6.3)$$

Deviations from the law (6.3) are associated with multifractal properties of the space of realizations of the dynamic process. This problem was clarified in [4].

Introduce also the temporal function

$$T(\tau) = <| v(x, t + \tau) - v(x, t) |^p > \equiv < \delta v(\tau)^p > \propto \tau^{\eta(p)}, (\tau \to 0) \qquad (6.4)$$

where $\eta(p)$ is another scaling function which characterizes temporal spectral properties of the process.

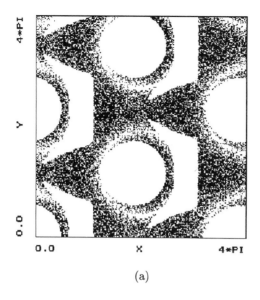

(a)

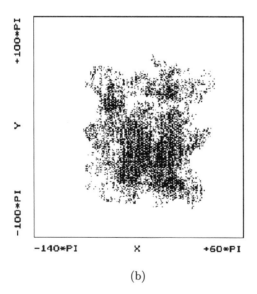

(b)

FIGURE 5. Sections of a streamline by the plane $z = 0$ (for $q = 3$; $\epsilon = 0.7$. Two different sizes are displayed.

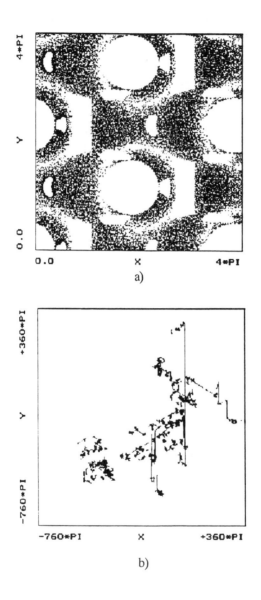

a)

b)

FIGURE 6. The same as in Fig. 5, but for $\epsilon = 1.1$.

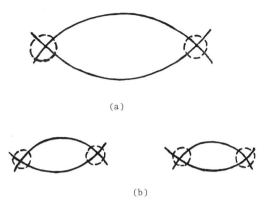

(a)

(b)

FIGURE 7. Singular dynamics regions marked by circles for a separatrix loop (a) and after its decay (b).

Normally, the relation between $\mathcal{R}(\ell)$ and $T(\tau)$ follows from the fact that dependence $\delta v(\tau)$ on τ has no singularities, that are different from (6.2), i.e.

$$\delta v(\tau) \propto \tau^{\zeta(1)} \qquad (6.5)$$

It follows also from (6.4) that

$$\eta(p) = \zeta(p) \qquad (6.6)$$

This may be considered as a form of Taylor hypothesis:

$$\tau \propto \ell. \qquad (6.7)$$

In the case of dynamical chaos, relations (6.6) and (6.7) are, generally, inapplicable. This can be illustrated in simple terms. Consider a separatrix loop which is an element of onset of chaos (Fig. 7a). Circled are the regions near saddles where trappings occur. Thus, the motion near the loop can be divided into at least two regions: the regular motion region where (6.5) holds and the singular dynamics region confined to the circles. When the separatrix loops split (Fig. 7b), this division is inherited by both resulting loops.

The cascade of singular regions near saddles should result in its particular scaling function $\eta(p)$ which does not coincide with the trivial one (6.6). This also implies that the exponent μ in relation

$$\delta v(\tau) \propto \tau^{\mu} \qquad (6.8)$$

differs form $\zeta(1)$ as in (6.5).

Thus, if the multifractal nature of the time history is accepted, there arises the necessity to introduce at least two distinct scaling function $\zeta(p)$

and $\eta(p)$. Although we do not know much about these functions and their mutual relation, verification of the qualitative aspect of this result can be accessible for an experiment or numerical simulation.

References

1. L.D.Landau and E.M. Lifshitz. Fluid Mechanics (Pergamon, L. 1959).

2. A.N. Kolmogorov. J. Fluid Mech. **13**, 82 (1962).

3. B. Mandelbrot. In *Turbulence and Navier-Stokes Equations*. Ed. R. Temam, Lecture Notes and Mathematics. Vol. 565 (Berlin, 1976), p. 121.

   B. Mandelbrot. J. Fluid Mech. **62**, 331 (1974).

4. G. Parisi and U. Frisch. In *Turbulence and Predictability in Geophysical fluid Dynamics and Climate Dynamics*, Ed. M. Ghil, R. Benzi and G. Parisi, North-Holland, Amsterdam (1985), p. 71.

5. Paladin and Vulpiani, Physics Reports, **156**, 149 (1987).

6. V. I. Arnold. Mathematical Methods in Classical Mechanics. (Springer New York, 1980).

7. H. Aref. J. Fluid Mech. **143**, 1 (1984) ibid.

   J.M. Ottino, C.W. Leong, H. Rising and P.D. Swanson. Nature, **333**, 419 (1988).

8. T. Dombre, U. Frisch, J.M. Green, M. Henon, A. Mehr and A.M. Soward. J. Fluid Mech. **167**, 353 (1986).

9. G.M. Zaslavsky, R.Z. Sagdeev, and A.A. Chernikov. Sov. Phys. JETP, **67**, 270 (1988).

10. H.K. Moffatt. J. Fluid Mech. **159**, 359 (1985).

11. V.V. Beloshapkin, A.A. Chernikov, M.Ya. Natenzon, B.A. Petrovichev, R.Z. Sagdeev and G.M. Zaslavsky, Nature **337**, 133 (1989).

12. D. Ruelle and F. Takens. Commun. Math. Phys. **20**, 167 (1971).

13. J.P. Eckmann, Rev. Mod. Phys. **53**, 655 (1981).

14. R.Z. Sagdeev, D.A. Usikov and G.M. Zaslavsky. Nonlinear Physics (Harwood Acad. Publ. N.Y. 1988).

15. Cellular structures in instabilities. Eds. J.E. Wesfreid and S. Zaleski (Springer, Berlin, 1984).

16. H. Chate and P. Manneville. Phys. Rev. Lett. **58**, 112 (1987).

17. F. Heslot, B. Castaing and A. Libchaber. Phys. Rev. A, **36**, 5870 (1987).

18. B. Nicolaenko. Los Alamos National Laboratory. 1988.

19. L.D. Meshalkin, Ya.G. Sinai, Prikl. Matem. i Mekhanika, **25**, 1140 (1961).

20. V.V. Beloshapkin, A.A. Chernikov, R.Z. Sagdeev and G.M. Zaslavsky. Phys. Lett. A, **133**, 395 (1988).

21. V. Yakhot, J.B. Bayly, and S.A. Orszag. Phys. Fluids **29**, 2025 (1986).

22. A. Tsinober and E. Levich. Phys. Lett. A, **99**, 321 (1983).

23. R.B. Pelz, V. Yakhot, S.A. Orszag, L. Shtilman, and E. Levich. Phys. Rev. Lett. **54**, 2505 (1985).

    R.H. Kraichnan and R. Panda. Phys. Fluids **31**, 2395 (1988).

24. B.J. Bayly, and V. Yakhot. Phys. Rev. A, **34**, 381 (1986).

25. N.N. Filonenko, R.Z. Sagdeev, and G.M. Zaslavsky. Nucl. Fusion **7**, 253 (1967).

26. B.A. Dubrovin, S.P. Novikov and A.T. Fomenko. Contemporary Geometry (in Russian). Nauka, Moscow, 1979, chap. 7.

27. G.M. Zaslavsky. Chaos in Dynamic Systems. Harwood Acad. Publ. N.Y., 1985.

28. A.A. Chernikov, R.Z. Sagdeev, and G.M. Zaslavsky (to be published).

29. G.M. Zaslavsky, M.Yu. Zakharov, R.Z. Sagdeev, D.A. Usikov and A.A. Chernikov. Sov. Phys. JETP **64**, 294 (1986).

30. G.M. Zaslavsky, R.Z. Sagdeev, A.A. Chernikov. Uspekhi Fiz. Nauk, **156**, 193 (1988).

31. Y. Pomeau and P. Manneville. Comm. Math. Phys. **74**, 149 (1980).

32. G.M. Zaslavsky, R.Z. Sagdeev, D.K. Tchaikovsky, and A.A. Chernikov. Zhurn. Eksp. i Teor. Fiz. (1989).

33. P. Levy, Theorie de l'addition des variables aleatoires (Gauthier-Villars, Paris, 1937).

34. E.W. Montroll and M.F. Shlesinger. In *Studies in statistical mechanics.* Eds. J. Lebowitz and E.W. Montroll, v. 11, p. 5 (North Holland, Amsterdam, 1984).

35. R.S. Mac Kay, J.D. Meiss, and I.C. Percival. Physical D, **13**, 55 (1984).

36. W. Feller. An Introduction to *Probability Theory and Its Applications.* Vol. 2 (Wiley, New York, 1966).

37. A.A. Chernikov, B.A. Petrovichev, A. Rogalskii, R.Z. Sagdeev and G.M. Zaslavsky (to be published).